Population Geography

T0276862

Population Geography: Social Justice for a Sustainable World surveys the ways in which geographic approaches may be applied to population issues, exploring how human populations are embedded in natural and social environments.

It encourages students to evaluate population issues critically, given that population topics are at the heart of many of today's most contentious subjects. Through introducing students to different lenses of analysis (ecological, economic, and social equity), the authors ask students to consider how different perspectives can lead to different conclusions on the same issue. Identifying and tackling today's population problems therefore requires an understanding of these diverging, and sometimes conflicting, perspectives. The text covers all the key background information critical to any book on population geography (population size, distribution, and composition; fertility, mortality, and migration; population and resources) but also pushes students to think critically about the materials they have covered using the perspectives of sustainability and social justice. In this way, students move beyond simple fact learning toward higher-level skills such as analysis, synthesis, and evaluation of materials.

This textbook will be a valuable resource for students of human geography, population geography, demography, and diaspora studies.

Helen D. Hazen is a teaching professor in the Department of Geography and the Environment at the University of Denver, Colorado, USA.

Heike C. Alberts is a professor of geography at the University of Wisconsin Oshkosh, USA.

Kazimierz J. Zaniewski is an emeritus professor of geography at the University of Wisconsin Oshkosh, USA.

Population Geography

Social Justice for a Sustainable World

Helen D. Hazen, Heike C. Alberts, and Kazimierz J. Zaniewski

Routledge
Taylor & Francis Group

LONDON AND NEW YORK

Designed cover image: Visitors at Shwedagon Pagoda in Yangon, Myanmar in January 2020. Photographer: Heike C. Alberts

First published 2024
by Routledge
4 Park Square, Milton Park, Abingdon, Oxon OX14 4RN

and by Routledge
605 Third Avenue, New York, NY 10158

Routledge is an imprint of the Taylor & Francis Group, an informa business

British Library Cataloguing-in-Publication Data
A catalogue record for this book is available from the British Library

ISBN: 978-0-367-69707-5 (hbk)
ISBN: 978-0-367-69796-9 (pbk)
ISBN: 978-1-003-14325-3 (ebk)

DOI: 10.4324/9781003143253

Typeset in Bembo and Helvetica
by KnowledgeWorks Global Ltd.

Contents

List of figures   x
List of tables   xiv
List of boxes   xv
About the authors   xvii

1   Introduction to population geography   1

Population geography and the pandemic 1
What is population geography? 3
Why are population issues often controversial? 10
Introducing three lenses of analysis: ecological, economic, and social equity 12
So, what's the problem … population growth or population decline? 16
Discussion questions 19
Suggested readings 20
Glossary 20
Works cited 21

2   Demographic data, visualization, and interpretation   23

Part I: Population data 23
Censuses 25
   Census errors 31
Vital statistics 34
Sample surveys 36
Challenges of collecting population data 36

Part II: Visualization of population data 39
Types of population data 39
Graphical display of demographic data 40
Cartographic display of demographic data 45
Conclusion 50
Discussion questions 50

Suggested readings 50
Glossary 51
Works cited 52

3   Population distribution and composition   55

Part I: Population distribution 55
Measures of population distribution 59
Population redistribution 59

Part II: Population composition 64
Age structure 65
  Aging populations 68
Sex and gender 70
  Sex ratios 71
  Gender-based violence 73
  Sexual orientation and gender identity 76
Race and ethnicity 78
Conclusion 83
Discussion questions 83
Suggested readings 84
Glossary 84
Works cited 85

4   Population growth and change   89

Part I: Population growth over time 89
A short history of population growth 89
Current population growth and population projections 97
Population growth in India 100

Part II: Theories of population growth 102
Malthus, Marx, and population 102
Modeling population growth and decline: the demographic transition model 104
Conclusion 111
Discussion questions 111
Suggested readings 112
Glossary 112
Works cited 113

5   Population and resources   116

Part I: Theories of population and environment 116
Large population as a problem 116
Consumption and the role of technology 119
Population, resources, and social justice 124

Part II: Applying theories of population and environment 127
Finite resources and unsustainable resource extraction 128
Food and agriculture 133

Conclusion 138
Discussion questions 139
Suggested readings 139
Glossary 139
Works cited 141

6   Fertility measures and patterns   145

Influences on fertility 145
Measures of fertility 148
High fertility and fertility decline 152
   Niger: a rapidly growing population 153
   Fertility transition in Thailand 154
Low fertility 156
   South Korea 162
Conclusion 164
Discussion questions 164
Suggested readings 165
Glossary 165
Works cited 165

7   Ethical issues in fertility   170

Contraception and abortion 170
Family planning campaigns 172
Eugenics and pro-natalism 180
   Pro-natalism in Romania 182
Fertility reduction policies 184
   China's population policies 187
Conclusion 191
Discussion questions 191
Suggested readings 192
Glossary 192
Works cited 192

8   Mortality and morbidity   197

Part I: Health indicators 197
Life expectancy 198
Mortality 201

Part II: Patterns in death and disease over time 213
Infectious disease, malnutrition, and injuries 213
The Industrial Revolution, urbanization, and the epidemiologic transition 217
Conclusion 220
Discussion questions 220
Suggested readings 221
Glossary 221
Works cited 222

9   Emerging trends in population health   225

Part I: Population health and environmental change 225
The ecology of infectious disease 225
Epidemics, pandemics, and environmental change 227

Part II: Non-infectious diseases and the social environment 235
Chronic, degenerative, and lifestyle diseases 235
Declining health and the social environment 240
Health and wellbeing 243
Conclusion 244
Discussion questions 244
Suggested readings 245
Glossary 245
Works cited 246

10   Migration patterns and theories   250

A global overview of current migration flows 253
Regional migration patterns 256
   Europe 256
   North and South America 256
   Asia and the Pacific 258
   Africa 260
   The Middle East 261
Types of migration 262
Migration theories 268
Conclusion 270
Discussion questions 271
Suggested readings 271
Glossary 272
Works cited 273

11   Social justice issues in migration   275

Structure and agency in migration 275
Slavery 276
Forced migration and ethnic cleansing 279
Refugees 283
Economic migrants 288
   Settler migrations 289
   Temporary labor migration programs 291
Illegal migrations 294
Conclusion 298
Discussion questions 298
Suggested readings 299
Glossary 299
Works cited 300

12   Sustainability and social justice in population issues   303

Part I: Cities and population 303
Ecological impacts of the city 304
Economic and social justice perspectives on the city 306
Health and the city 310

Part II: Climate Change 313
Carbon emissions, population size, and consumption patterns 313
Population impacts of global climate change 315
   Health impacts of climate change 318
Environmental migrations 319
Conclusion 322
Discussion questions 322
Suggested readings 323
Glossary 323
Works cited 323

Index   326

Figures

1.1   Global spread of coronavirus, 2020   2
1.2   Rural and urban communities in the Peruvian Andes   5
1.3   Pyramids of Giza, Egypt   7
1.4   Geographic scale using mobility and migration examples   9
1.5   Scatterplots showing the relationship between affluence and obesity
      rates at global and national scales   9
1.6   Perspectives and lenses of analysis used to consider population issues   13
1.7   United Nations' Sustainable Development Goals   15
2.1   Number of censuses conducted around the world by country since 1950   26
2.2   Changing topics covered in the US census, 1790–2020   31
2.3   Questions pertaining to race and ethnic origin from the US 2020 Census   32
2.4   Basics of a typical national vital statistics system   34
2.5   Bar graphs showing Australia's population across space and time   40
2.6   Stacked bar graphs showing the changing composition of Canada's
      foreign-born population, as population totals (A) and proportions of
      the immigrant population (B)   41
2.7   Line graphs showing birth rate and death rate trends in Chile (A) and
      age-specific fertility rates for different age groups in England and
      Wales (B)   41
2.8   Area graph showing the changing origin of European immigrants to
      the United States over time as population totals   42
2.9   Pie chart showing religious affiliation of England and Wales, 2011   42
2.10 Bar and line graph showing COVID-19 cases and COVID-related
      mortality in India, 2020–21   43
2.11 Scatterplots showing the relationship between two sets of variables
      for selected countries of the world: a weak relationship between
      urbanization and natural increase rate (A) and a stronger relationship
      between affluence and fertility rate (B)   44
2.12 Bubble graph illustrating the relationship between national affluence,
      life expectancy, and infant mortality rate   45

2.13   Choropleth maps showing population density in Poland at the
        province (A) and county (B) levels in 2020   46
2.14   Choropleth maps showing sex ratios in China by province (2018),
        illustrating four different systems for creating class boundaries   47
2.15   Dot density map showing population distribution in Switzerland in 2020 48
2.16   Proportional symbol map showing population distribution in Cuba by
        municipality in 2020   48
2.17   Flow maps showing the largest inter-provincial migration streams in
        the Czech Republic (A) and migration from one province to the rest of
        the country (B) in 2020   49
2.18   Population distribution in the Central African Republic in 2015   49
3.1   World population distribution by hemisphere, latitude, and longitude   56
3.2   World population distribution by elevation above sea level   57
3.3   Distribution of the world's population   58
3.4   Arithmetic and physiological densities by country, 2018   60
3.5   Population centroids for the United States, 1790–2020   61
3.6   Population distribution (2020) and the New Cities program in Egypt   61
3.7   Transmigration in Indonesia   62
3.8   Population distribution and change in Brazil, 1950–2020   63
3.9   Population pyramids for economically less and more developed
        regions, 2015–20   67
3.10   Population pyramids for selected US counties, 2010   67
3.11   Population pyramid for Russia, 2010   68
3.12   Population pyramid for the United Arab Emirates, 2015–20   71
3.13   Population pyramid for a locality dominated by the military   72
3.14   Global gender inequality (estimated by the Gender Gap Index)   75
3.15   Criminalization of consensual same-sex sexual activities   76
3.16   Legal protections against discrimination based on sexual orientation
        by country   78
3.17   Population distribution and composition by major ethnic groups
        in Singapore   83
4.1   Growth of global population, 10,000 BCE to present   90
4.2   Agricultural hearths and the spread of early agriculture   91
4.3   Spread of the Black Death in Europe   93
4.4   Timing of projected stabilization of population size by country   98
4.5   Proportion of global population residing in countries with different
        fertility rates   99
4.6   United Nations' world population projections by 2100   100
4.7   The demographic transition model   105
4.8   Population pyramids typical of each stage of the demographic
        transition model   108
4.9   Changing global population structure by age and sex, 1950–2100   109
4.10   Relationship between human development index and fertility, 1970
        (136 countries) and 2020 (189 countries)   110
5.1   Ecological footprint by country, 2017   120
5.2   Ecological reserve and deficit by country, 2017   121

5.3	Environmental Kuznets curve	122
5.4	The three pillars of sustainability	125
5.5	The Earth's freshwater resources	129
5.6	Freshwater withdrawals as a percentage of national water resources, 2017	130
5.7	Freshwater withdrawal per capita by country, 2017	131
5.8	Global fish catch and aquaculture, 1950–2018	132
5.9	World's undernourished population, 2020	134
5.10	Greenhouse gas emissions in the food chain	136
5.11	Ecological pyramid	137
6.1	Total fertility rates by country, 2015–20	146
6.2	Influences on fertility	148
6.3	Fertility fluctuations in the United States, 1910–2020	149
6.4	Age-specific fertility rates in the United States, 1915–2020	151
6.5	Total fertility rates in Niger and Thailand, 1950–2020	154
6.6	Fertility transition in select countries relative to replacement level, 1950–2020	156
6.7	Fertility and mortality rates in Russia, 1946–2020	160
6.8	Birth and death rates in South Korea, 1925–2020	163
7.1	Number of abortion restrictions by country, 2017	177
7.2	Government policies on fertility in relation to total fertility rates by country	178
7.3	Fertility rates and population policy in Iran	179
7.4	Fertility and mortality rates in Romania, 1960–2018	183
7.5	Fertility and mortality rates in China, 1949–2020	188
7.6	Comparison of fertility decline in several Asian countries, 1950–2020	190
8.1	Life expectancy by country, 2015–20	198
8.2	Female–male disparities in life expectancy by country, 2015–20	199
8.3	Crude death rates by country, 2015–20	203
8.4	Comparison of crude and age-adjusted mortality rates due to heart diseases, United States, 2019	204
8.5	Infant mortality rates by country, 2015–20	205
8.6	Infant mortality rates in the United States, Japan, and Western Europe, 1950–2020	207
8.7	Infant mortality rate and gross domestic product (GDP) per capita, 2019	208
8.8	Maternal mortality rates by country, 2017	211
8.9	Food insecurity	215
8.10	Epidemiologic transition model	218
8.11	Proportion of deaths from infectious causes, 2019	218
9.1	Trends in the number of Ebola deaths over time and by province in Sierra Leone, Guinea, and Liberia during the Ebola epidemic of 2014–16	230
9.2	Spread of influenza during the 1918–19 flu pandemic	232
9.3	Camel market, United Arab Emirates	234
9.4	Proportion of deaths due to chronic disease by country, 2019	236
9.5	Obesity rates by country, 2016	237
9.6	Snack vendor, Cusco, Peru	237
9.7	Black/white racial disparities in infant mortality rates in the United States	240

9.8   Life expectancy in Russia, compared with the UK and United States,
        1960–2020   241
10.1   "Many Walls Have to Be Torn Down"   253
10.2   World migrant and refugee population, 1960–2020   254
10.3   Distribution of migrants by place of origin, 2020   255
10.4   Immigrants, as a proportion of the total population, by destination, 2020   255
10.5   Foreign-born population in Europe, 2015   257
10.6   Indentured laborers by area of origin and destination   258
10.7   Destinations of Filipino migrants   260
10.8   Major sources of labor migrants to the Gulf countries   261
10.9   Refugees then and now   265
10.10   Internally displaced people, 2020   267
11.1   Degree of agency in migration process   277
11.2   The transatlantic slave trade   278
11.3   Victims of modern slavery, 2018   279
11.4   Trail of Tears   280
11.5   Major European forced population movements in the pre- and post-
        World War period   281
11.6   Expulsion of Germans from parts of Eastern Europe after World War II   282
11.7   Distribution of refugees, 2019   284
11.8   Europe's migrant crisis   286
11.9   Immigration to the United States by area of origin since 1820   290
11.10   Banner in Hamburg (Germany) in support of rescuing migrants at sea
        and against the criminalization of those who rescue people   295
11.11   Detected human trafficking in 2018   297
12.1   Agricultural and industrial innovations leading to urban growth   304
12.2   Urban population growth over time   305
12.3   Proportion of households living in slums by country, 2018   307
12.4   Squatter settlement in Cape Town, South Africa   308
12.5   Death rate from diarrheal diseases, 2019   311
12.6   Global carbon dioxide emissions, 1850–2018   314
12.7   Global emissions of carbon dioxide, 2020   315
12.8   Global climate change: predicted temperature increase and selected
        consequences by 2090 (based on RCP 2.6 Model Simulation 2006–2100)   316
12.9   Number of internally displaced environmental migrants, 2008–20   320
12.10   Coastal cities most likely to be affected by rising sea levels by 2070   321

Tables

3.1	Selected examples of gender inequality and women's rights violations	74
4.1	Global population growth	97
4.2	Examples of countries with below replacement level fertility in 2020	109
6.1	Total fertility rate by world region, 2020	150
6.2	Lowest fertility countries and territories, 2020	162
7.1	Comparison of contraceptive methods	174
7.2	United Nations world population conferences	185
8.1	Abridged life table for the United States, 2018	201
8.2	Select nutrient deficiency diseases	214
9.1	Significant vector-borne diseases	227
9.2	Infectious disease events that emerged from animals during the 21st century	233
9.3	The world's "healthiest" countries, 2020	243
12.1	World's most populous cities, 2022	307

Boxes

1.1 Three lenses of analysis for considering population issues 13
1.2 Global North and Global South—naming conventions and social justice 17
2.1 Key sources of population data 24
2.2 United Nations' recommendations for census topics that should be included in population censuses conducted around the year 2020 27
2.3 Citizenship and the US 2020 Census 30
2.4 Sex, gender, and the census 33
2.5 Ethnic identity and the 2014 Myanmar Census 37
3.1 Childhood 66
3.2 Indigenous peoples around the world 79
4.1 Plague 93
5.1 The ship-breaking industry 126
5.2 Fisheries and sustainability 132
5.3 Diet and sustainability 137
6.1 Biological influences on fecundity and infertility 147
6.2 Assisted reproductive technologies (ART) 151
6.3 Fertility decline in Russia 160
7.1 Catholic and Islamic attitudes toward contraception and abortion 173
7.2 Iran and the politicization of fertility 179
8.1 Health and inequality 202
8.2 Primary healthcare in Costa Rica 209
8.3 Maternal mortality in Afghanistan 210
8.4 Food insecurity 215
8.5 Warfare and population health 216
9.1 The "Columbian Exchange" 229
9.2 The 1918 flu pandemic 232
9.3 Health inequalities in the United States 239
10.1 Burqa bans and Islamophobia in Europe 252
10.2 The Philippines—a labor exporter 259
10.3 Lifestyle migrations 264
10.4 Internally displaced people 267

10.5	From asylum seeker to undocumented migrant	269
11.1	Modern slavery	280
11.2	Germany's experiences with refugees	282
11.3	Refugee resettlement to the United States	288
11.4	Economic opportunity in the "classic immigration countries"	289
11.5	Child soldiers	297
12.1	Climate change and social equity in Bangladesh	317

About the Authors

Helen D. Hazen is a teaching professor in the Department of Geography and the Environment at the University of Denver, Colorado. Her research interests are focused at the intersection of health and environment, exploring how health outcomes are influenced by aspects of the social and natural environment. Most recently, she has published research on home birth in Minnesota, and she is currently working on a project studying environmental influences on diarrheal disease in children in Peru. She recently co-authored an edited volume titled *Reproductive Geographies: Bodies, Places and Politics* and has worked with Routledge previously on two editions of a health geography textbook (*An Introduction to the Geography of Health*), which she co-authored. Her teaching interests include geography of health, population geography, and sustainability, with more specialist classes offered in health and environment and geographies of reproduction.

Heike C. Alberts is a professor of geography at the University of Wisconsin Oshkosh. Her research focuses on migration and internationalization issues. Her early work explored the Cuban community and enclave economy in Miami. More recently her work deals mostly with highly skilled migrations, examining the issue from the perspectives of migration (patterns and decision making), internationalization (contributions to the US education system), and pedagogy. In addition to journal articles and book chapters on these topics, she has co-edited one book and co-authored two, most recently a textbook on world regional geography. She teaches a wide variety of undergraduate geography courses including a class on population geography.

Kazimierz J. Zaniewski is an emeritus professor of geography at the University of Wisconsin Oshkosh. His major research interests include population geography and thematic cartography. He is the co-author of four books, including *The Atlas of Ethnic Diversity in Wisconsin* (University of Wisconsin Press), *The Geography of Wisconsin* (University of Wisconsin Press), and *World Regional Geography* (Cognella Academic Publishers), and several journal articles on topics such as demographic trends in Europe and Wisconsin and presidential elections in the United States. He has taught a variety of undergraduate classes, among them world regional geography, population geography, population and environment, and cartography.

Introduction to population geography

After reading this chapter, a student should be able to:

1 understand a geographic approach to population topics;

2 discuss some of the key controversies associated with population issues;

3 apply three lenses of analysis (economic, ecological, and social equity) to the topic of population growth and decline.

In November 2022, the world reached 8 billion people to surprisingly little fanfare. Adding an additional billion people in a little over a decade has now become commonplace. Anyone in their 50s or older has witnessed the global population double from 4 billion people in 1974 to 5 billion in 1987, 6 in 1999, 7 in 2011, and now 8 billion people. Though this is a remarkable human achievement in many ways, it also indicates a huge sustainability challenge, as well as raising numerous social justice issues. Population geography aims to explore just these sorts of issues by applying a suite of geographic methods to topics related to human population.

Population geography and the pandemic

Another recent event—the global COVID-19 pandemic—illustrates many of these ways in which population geographers explore the world. In 2020, the world was turned upside down when a new coronavirus spread globally. The rapidity with which it spread clearly illustrates what a **globalized** population we have become, with almost all countries except for a few remote islands reporting cases of COVID-19 within a few months of the disease's identification in December 2019 in China. The mobility of the global population spread the disease rapidly, with the best-connected parts of the world experiencing cases before those with fewer connections (figure 1.1). Understanding human population distributions and mobility were critical to addressing the COVID-19 pandemic.

DOI: 10.4324/9781003143253-1

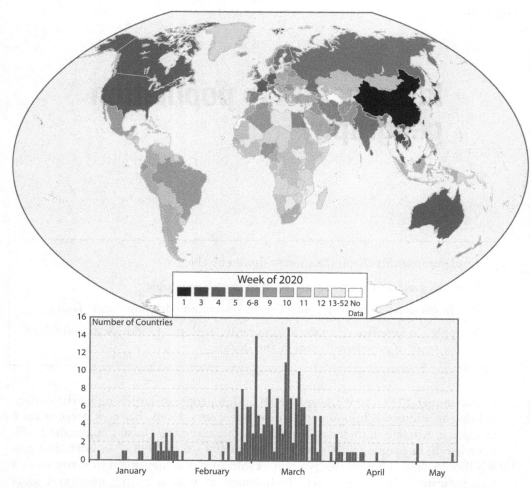

FIGURE 1.1 Global spread of coronavirus, 2020
The map shows the date of the first recorded COVID-19 case by country (China recorded its first case in late 2019 and so had cases prior to January 2020). The bar chart below the map shows the number of new countries reporting cases over time.
Source: WHO (n.d.).

Although disease is often considered a great equalizer, we know that this is far from the reality. COVID-19 was quickly revealed to affect older people in far more devastating ways than younger people, and aging populations may partially explain high death rates in the early months of the pandemic in countries such as Italy and the UK. It was not just age but also race and socio-economic status that were shown to influence COVID-19 outcomes, however. In many affluent countries, poorer populations and minority communities were initially both more likely to contract the infection (through risk factors such as larger average household size and a higher likelihood of having jobs that could not be performed remotely) and more likely to die from the disease (associated with issues such as higher rates of pre-existing conditions and poorer access to healthcare). In the UK, early in the pandemic, people of black African descent had a 3.5 times higher mortality rate from COVID-19 than the white British population; those of Pakistani descent had a 2.7 times higher mortality rate (Otu et al. 2020). Although the lower average socio-economic status of many minority groups partially explains these worse outcomes, the fact that more than 90 percent of British physicians who died in the first 6 months of the pandemic were from minority populations suggests that the impact of race and minority status was significant even in affluent populations (Adam 2020). Exploring the pandemic in ways that acknowledge diverging outcomes for different population groups has been critical to addressing this crisis.

Although the impacts of the pandemic are still evolving, we know that it has had a significant impact on population statistics. By the end of 2022, more than 6.5 million people had lost their lives to COVID-19 worldwide (WHO, n.d.). In just the year 2020, average life expectancy in the United States dropped by over a year—the largest decline since the influenza pandemic of 1918. For US black and Latino populations, the drop in life expectancy may have been three to four times that of whites (Andrasfay and Goldman 2021). It was not just mortality patterns that were affected. There was also considerable speculation over how the pandemic might affect birth rates, with some commentators predicting a baby boom associated with more time at home and disruptions to **family planning**, whereas others predicted a baby bust as couples delayed childbearing in uncertain times. In the end, patterns were place-specific, with rich countries generally reporting declining birth rates during lockdown, whereas some poorer countries experienced a baby boom associated with disrupted access to contraception. According to the United Nations' Population Fund, the pandemic interrupted access to family planning services for nearly 12 million women in more than 100 countries, generating an estimated 1.4 million unintended pregnancies (UNFPA 2021). By contrast, in Europe, countries that were hard hit by the pandemic appear to have seen significant decreases in fertility—one survey found that about 50 percent of Germans, French, and Spaniards planning to have babies in 2020 had decided to postpone pregnancy, with some abandoning the idea of parenthood altogether (Luppi, Arpino, and Rosina 2020). In the United States, a decline in births seems to have been disproportionately focused in poorer populations, leading some commentators to argue that the pandemic exacerbated pre-existing inequalities where poorer women have always questioned their ability to raise a child in a country with limited social support systems (Dockterman 2020). These sorts of place-specific population patterns are a particular focus for population geographers.

What is population geography?

Geography as a discipline explores the physical Earth and the ways in which people interact with it and with one another, and so human populations have been a long-standing and central focus—it has even been claimed that "population is the pivotal element around which all the others are oriented" (Trewartha 1953, 86). Population geography is distinct largely through its topical focus and can perhaps best be described as the application of geographic approaches and methods to topics related to human populations. This includes issues related to population size, distribution, and structure; population health; population mobility and migration; and population and resources.

The relevance of population geography is of particular significance today, given how rapidly human population patterns are changing. On one front, the question of how to support a growing population without reaching a resource crisis is critical to current discussions around **sustainability**, requiring that we consider how to foster economic development at the lowest ecological cost. On another front, population issues are central to concerns around **social justice**—how can we more fairly distribute resources, as well as ensure that human rights are not sacrificed in our efforts to meet demographic, economic, and environmental goals? **Population geography** provides a uniting approach to analyzing these varied concerns. In this book, we explore a broad spectrum of population issues, using the twin frameworks of sustainability and social justice as guiding principles in our analysis.

Though attempting to achieve the lofty goals of *sustainability* and *social justice* is admirable, many of the cases that we highlight in this book report circumstances where these aspirations are clearly *not* being met. To understand why society is falling short of achieving these goals, we suggest that considering population issues through different lenses of analysis can help make sense of some of the apparent contradictions in population concerns. For example, at the global scale, growing populations and rising consumption threaten ecological systems, and yet in some countries we are beginning to see fertility decreasing to the point where governments fear economic stagnation as the proportion of workers in the country declines. Meanwhile, migration—a potential solution to the twin challenges of population growth in some places and decline in others—has become politically charged, leading to vociferous debates over immigration. Understanding why some people are fearful of population growth, whereas others fear population decline and yet others refuse to countenance people moving from one country to another to address uneven rates of population growth, requires us to consider the different lenses through which each group is viewing the issues at hand. An ecologist may see population *growth* as the key global crisis as deforestation and landscape conversion continue apace, and yet through an economic lens we could argue that population *decline* is the greater concern if fewer workers lead to economic slowdowns. Once we begin to understand why different groups think the way they do, we can begin to work toward consensus among diverging viewpoints.

Broadly speaking, then, population geography is the application of geographic approaches to any topic associated with human populations. This incorporates many methods from the related field of **demography**, which uses statistical analysis to study the size and composition of populations. Population geography extends beyond demography, however, by emphasizing how human populations are situated in particular contexts—considering how people interact with their physical and social environments, as well as emphasizing place-based variations. For instance, urban populations often show distinctly different health and fertility patterns compared with rural communities. In this context, the **physical environment** is revealed to be an important influence on many population-scale patterns, with urban dwellers likely to be subject to higher concentrations of certain pollutants, more crowded living conditions, and less access to green spaces, among many other potential influences (figure 1.2). It is not just the physical environment that influences human health and wellbeing, however—the **social environment** is often equally important. In this context, rural dwellers may be constrained by a more conservative social environment but benefit from tighter-knit communities, for

FIGURE 1.2 Rural and urban communities in the Peruvian Andes
Population geography emphasizes the significance of context to population patterns and processes. How might the physical environments of a rural dweller and an urban dweller differ between these two Andean communities? What impacts might this have on people's lifestyles? How might social differences among individuals in each of these contexts (e.g., rich vs poor, Spanish vs indigenous heritage) mediate the impacts of these broader environmental factors?

example. The social environment consists of economic, political, and cultural frameworks that influence individuals through things like affluence, education, diet, daily activity patterns, and childrearing customs, as well as less tangible factors such as social status and peoples' ability to make decisions for themselves. All of these aspects influence fertility, mortality, resource use, and migration patterns— the key focuses of population geography.

Our interactions with our social environment are often shaped by aspects such as gender, socio-economic status, or race. For instance, we may find that women experience different outcomes compared with men (e.g., greater curbing of their earning potential if they choose to have children) or that an ethnic minority population may be subject to different pressures and expectations than the majority population (e.g., stronger expectations that children care for elderly family members). The notion of **intersectionality** argues that a layering of different aspects of **identity** may also be critical—for instance, poor white individuals may have very different outcomes from their affluent white counterparts, and some groups such as poor women of color may find themselves in particularly challenging positions because of an intersection of marginalized identities. Returning to the impacts of the coronavirus pandemic, we find that many women reported an increased work burden during the pandemic as they struggled to balance employment and family commitments. On a global scale, it was projected that COVID-19 would increase the poverty gap between men and women, given that women typically bear the brunt of caring responsibilities at home, earn less, and have less secure jobs than men (UNDP 2020). Beyond these generalizations, however, we find wide variations in women's experiences. For many women with school-age children the burden of supervising remote learning was significant, setting them apart from women without children. Socio-economic status was also significant. Affluence shielded many women from the worst impacts of the pandemic by factors such as having jobs that could be performed remotely and having access to savings. Low socio-economic status, by contrast, has been a major risk factor for contracting COVID-19 because people on a low income are more likely to have to work in person, less likely to be able to take time off work, and more likely to live in overcrowded conditions. Women from communities of color may experience yet further challenges, such as higher rates of pre-existing conditions, as well as stress associated with minority status—both of which can put them at higher risk of developing more serious disease. In short, though sharing many similarities associated with gender, the pandemic-related experiences of a low-income single mother of color are likely to have been very different from those of an affluent white woman whose children have left home. As such, we must usually explore a wide variety of potentially significant social variables, as well as the intersection of these variables, if we want to be able to identify the most pertinent ones.

This discussion begins to reveal to us what is so unique about a geographic approach to population issues: geography provides a distinct way to analyze issues, including population topics, because of its emphasis on the significance of *environment* on human activities and outcomes. This includes both the physical environment and the social environment and encourages us to ask "where" questions. Where do people have the smallest families? Where do people live the longest? Which regions have significant flows of migrants? With enquiring minds, we do not stop there, however—the next important question is "why?" Why do women

FIGURE 1.3 Pyramids of Giza, Egypt

have so few children in Russia? Why do people live so long in Japan? Why do so many people choose to emigrate from Mexico, even as others are immigrating to Mexico from neighboring countries?

As we think about how to answer these sorts of questions, two geographic principles come to the fore—**space** and **place**. Spatial analyses focus on the importance of issues such as location, distance, area, and connectivity. For instance, how does the location and relative connectivity of different countries in Central America influence migration patterns in the region? Place refers to locations with specific cultural meanings. In some cases, these meanings are understood by many people— the Egyptian pyramids are recognizable around the world as a place with historical significance, for instance (figure 1.3). In other cases, place-based meanings are deeply personal—your home is a place rich with meanings and emotional attachments for you but has little significance to most other people, for example. In other words, the meanings of places are constructed by the social significance that we as individuals and communities give them. Returning to our exploration of migration patterns in Central America, a *spatial* approach might suggest that the fact that Mexico and Guatemala share a border is significant in generating cross-border flows. However, this would not reveal why most migrants currently move from Guatemala to Mexico rather than in the opposite direction. To get at this question, it is important to investigate what it is about the place of Mexico that is attracting migrants from Guatemala or, conversely, what aspects of place in Guatemala are pushing people to leave. This will probably require interviewing migrants to explore their experiences, associations, and feelings related to particular places.

Geographers often use **case studies** to explore these sorts of place-based specifics— an approach we use in this book. A case study is used to consider how general theories can be applied to specific circumstances. They are particularly important to geographers because they reinforce the importance of place by showing how the same phenomenon can play out differently in different places. For instance, if we think again about migration patterns in Central America, we could explore how the theory that migrations are stimulated by both factors that "push" migrants out of one country and other factors that "pull" them to another country plays out in the specific setting of Central America. What are some of the place-specific push factors in Central America: political instability, economic hardship, climate change? How do these vary from one country to another or even among different regions within the same country? What sorts of pull factors are operating: economic opportunity, higher quality of life, political stability?

The complexity of many population topics means that fully answering population questions often involves a combination of **quantitative** and **qualitative analyses**. Quantitative analyses are driven by numerical data and often use big data sets to draw statistical conclusions, usually focusing on trying to make the analysis as objective as possible. From quantitative data we could, for example, state how many Guatemalans cross the border into Mexico. Although this may be important information for driving policy or providing services to vulnerable communities, further critical questions about *why* people move and how they choose where to settle can probably only be answered using qualitative methods. Qualitative methods use techniques such as in-depth interviewing and long-term observation of communities to provide more nuanced analyses. Although qualitative studies are more subjective, this is considered an appropriate price to pay for the far greater depth of understanding that can be achieved, which can help us understand intangible factors such as motivations, goals, pressures, and fears. Such analyses have additional policy implications. For instance, if we find that many Guatemalans are migrating to Mexico through lack of economic opportunities in their home communities rather than a desire to live in Mexico, we may want to focus on devising economic strategies that help Guatemalans stay in their home villages.

One final theme of relevance is the importance of geographic **scale**. Scale in this context means the geographic extent of the analysis we are undertaking, such as global, national, or neighborhood scale (figure 1.4). Global scales are important if we are looking at population growth, for example, whereas national scales may be crucial if we are focusing on policymaking. As we think about cultural aspects of population issues, smaller scales of analysis may become important, with regional-, village-, or even household-scale patterns becoming evident in things like customs and attitudes. Returning to the example of migration, we might consider movements occurring at the global scale (such as an engineer migrating overseas to take up a new position), regional movements (such as a landless farmer moving to a city in search of employment), and local migrations (such as a bride moving to a new household in a neighboring village). Although all of these movements are considered migrations, the motivations and challenges the migrants experience differ considerably and require independent analysis.

It is critical to acknowledge that patterns apparent at one scale may not hold at other scales. For instance, at the global scale we know that affluent countries typically have higher rates of obesity than poorer countries. Yet if we look at patterns of obesity *within* affluent countries, it is often the poor who experience obesity at higher rates than the rich (figure 1.5). In poorer countries, by contrast, the global-scale pattern holds, with richer individuals more likely to be obese than the poor.

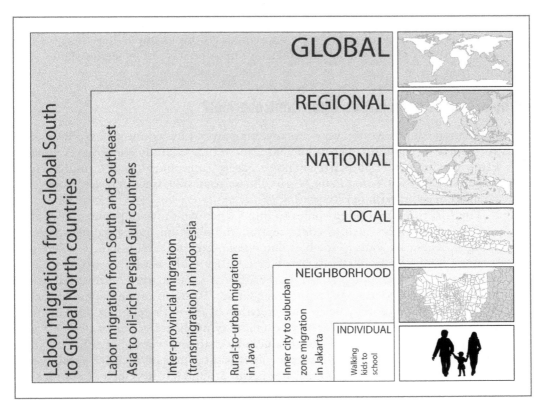

FIGURE 1.4 Geographic scale using mobility and migration examples

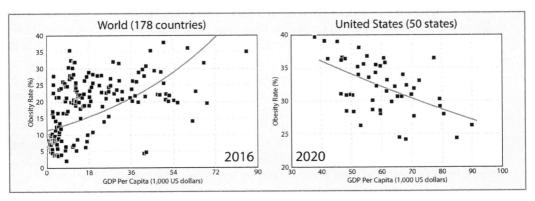

FIGURE 1.5 Scatterplots showing the relationship between affluence and obesity rates at global and national scales
The graph on the left shows a weak positive correlation between obesity and national affluence, with richer countries generally having higher obesity rates than poorer countries, although the relationship is far from perfect. The graph on the right shows this relationship at the sub-national scale in the United States, where poorer states generally have higher obesity rates than richer states—a negative correlation between affluence and obesity.
Data sources: US Bureau of Economic Analysis (2021), CDC (2020), Ritchie and Roser (2017), and CIA (n.d.).

Complicating the situation even further, we may find both undernourished and overweight individuals in the same household. As this example illustrates, patterns across scales can be very complex and so we must be wary of generalizations.

Why are population issues often controversial?

Population issues are often very controversial, given their connections to issues such as resource distribution, national defense, and intimate family concerns. Population issues also frequently pit personal choice against community goals, causing further disputes. It is worth recognizing some of these contested issues in order to approach population issues with sensitivity.

The first thing that often comes to mind when we think of population controversies is how topics such as contraception and abortion have become politically charged. These are some of the best known of numerous circumstances where religious or traditional conservative values conflict with secular (non-religious) or progressive ideas around the family. Fertility is particularly controversial, given how closely it is tied to sexual politics and taboos. Restrictions on the use of contraception, for instance, may be an interpretation of a specific religious doctrine (e.g., all children are gifts of God) or may be a more symbolic reflection of a conservative society trying to enforce traditional values in a changing world, but either way they have proved a major form of resistance to efforts to increase access to family planning. In the secular world, too, debates persist around fertility, such as whether it is appropriate for governments to make policies that attempt to influence intimate behaviors or legislate on sexual identities.

It is not just fertility policy that is religiously and culturally charged. Discussions around end of life also generate strong feelings. Efforts to allow people to choose to end their own life if they believe they have very poor quality of life have generated heated discussions around who has the right to end life, with euthanasia staunchly rejected by many under the understanding that only God has the right to end life, as well as those who believe that it is impossible to develop effective safeguards to prevent abuse of the system. Even migration policy cleaves down conservative/progressive lines, with extremely progressive viewpoints promoting the idea of **global citizenship** and freedom of movement, in contrast to more conservative ideals of the sovereign state having the right—even the responsibility—to control immigration to protect cultural values and the security of citizens.

Many social conservatives fear that changing cultural norms will have a destabilizing impact on society and point to the need to defend established customs such as traditional family structure. In response, progressive commentators have noted that ill-conceived religious or traditional practices can be very damaging, even if they have been practiced for many generations. Along these lines, critical scholars point out that we should not assume that women living in traditional villages are happy with practices that we would not consider acceptable, such as being married off at a young age and having children as a teenager, just because they are part of a particular culture (Goldberg 2009). Nor, of course, should we immediately assume that women are exploited by these customs. Instead, we must be respectful of cultural difference, while acknowledging that certain cultural practices do not promote the wellbeing of all concerned. In a social justice framework, we could add that it is usually disempowered groups such as adolescent girls who bear the brunt of the negative consequences of many cultural practices, calling into question the degree to which some groups can be said to have a genuine choice in whether they participate.

A second reason why population issues are often so controversial revolves around the significance of population size to **geopolitical** concerns. Throughout much of human history, governments have considered population growth to be a sign of military security and economic strength and have viewed population decline with dismay (Coleman and Rowthorn 2013). In this view, large populations are seen as desirable for ensuring that sufficient soldiers are available to defend borders and that a large enough body of workers exists to support the domestic economy. Russia provides a particularly stark illustration of these sorts of attitudes, with a longstanding obsession with increasing population size to try to establish itself as a global superpower (Perelli-Harris and Isupova 2013). Ironically, the breakup of the Soviet Union in 1991 meant that Russia also provides a particularly poignant example of a country *losing* power associated with the loss of a significant portion of its population (Coleman and Rowthorn 2013). Declining fertility rates associated with the political and economic upheavals only made matters worse. In response, a wide variety of **pro-natal campaigns** have been instituted by the Kremlin over the years, encouraging couples to have large families. After World War II, for instance, Russian women who bore at least ten children received the "Order of Mother Heroine" to recognize their services to the state. At about the same time, special honors were also bestowed on women who bore many children in France and Nazi Germany (Hoffmann 2000). Today, few governments have a pro-natal agenda for explicitly military reasons, but remnants of these ideas persist in concerns over the impact of falling birth rates on national security, as illustrated by Iran in recent years, and a few sparsely populated countries are still concerned about having sufficient people to secure their borders. Some commentators also continue to express concerns that slow-growing or declining populations may lose political or economic power to demographically more powerful countries. Yglesias (2020), for instance, argues for the need to encourage population growth in the United States to counter China's demographic might, even as the Chinese government is becoming increasingly concerned over the potential economic impacts of its own low birth rate.

A further traditional point of controversy has been the impact of changing population *composition*, usually expressed as concerns from a dominant population about the "threat" of the growing size of a sub-population, through either immigration or higher birth rates. Historically, low birth rates have sometimes even been referred to as "race suicide" (Coleman and Rowthorn 2013). In Israel, in the late 20th century, for instance, concerns were voiced that low birth rates in Jewish populations might lead to their eventual marginalization—presumably relative to faster-growing non-Jewish populations (King 2002). More recently, in many countries of Europe, right-wing parties have been stoking fears that Muslim immigrant populations will "overwhelm" white populations owing to traditions of larger family size. In response, pro-natal policies have sometimes been applied to try to encourage births within dominant groups. In Serbia, for instance, ethnic tensions between Serbs, Albanians, and other minority populations combined with low birth rates has led the Serbian government to repeatedly promote fertility among Serbian women (Dahlman 2019). This discriminatory history of pro-natal policies has left many Western governments wary of pro-natalist policies because of concerns that they have extremist right-wing connotations (King 2002). Nonetheless, many governments are now facing the reality of declining populations as fertility dips below replacement level in many countries. Some governments are once again embracing policies that try to encourage births but using a new suite of "family-friendly" policies that attempt to increase births without **coercion**. These include aspects such as paid maternity/paternity leave, child support payments, and free or reduced cost childcare places that try to relieve some of the burden of childrearing on parents.

The idea of demographic dominance relates not only to political power but also to concerns over resource distribution. Throughout human history, the demographic expansion of one population has often allowed it the power to claim access to resources, potentially taking assets and territory from smaller or weaker populations. Many population issues are thus also resource issues. Though this connection may today be expressed in more subtle ways than armed invasions, there are undoubtedly consequences for declining populations in terms of resource loss. Part of the role of population counts such as censuses is to guide governments in how to allocate resources, including funding for schools, roads, and healthcare, as well as to guide political representation. In addition, the tax base of a region may decline if population decreases, hindering local government's opportunity to provide services. A localized decline in population may therefore lead to the impoverishment of political representation and services.

One final point on the topic of controversy is the idea that population issues frequently require a compromise between individual rights and community goals. A person's choice to have 12 children or no children can be seen as their prerogative, but that decision undoubtedly has repercussions for the broader community, whether that be the expectation that local taxpayers will pay to educate those children or the boost that those children could provide to the local tax base when they become workers themselves. The idea that individuals have the right to determine the number of children they have is enshrined as a human right: "[…] couples have a basic human right to decide freely and responsibly on the number and spacing of their children and a right to adequate education and information in this respect" (Resolution XVIII of the Human Rights Aspects of Family Planning 1968). The degree to which it is appropriate for governments to attempt to influence their citizens' fertility choices therefore remains highly controversial, with a fine line dividing policies considered appropriate versus coercive.

Introducing three lenses of analysis: ecological, economic, and social equity

Understanding and resolving some of the controversies surrounding population matters requires critical thinking skills. One approach in this respect is to consider issues through different lenses of analysis that can help us to understand why two individuals might have contradictory, but equally valid, viewpoints on a topic, and so here we introduce the idea of ecological, economic, and social equity lenses (box 1.1). We consider these three lenses as contributing to the two major perspectives of the book: sustainability and social justice (figure 1.6).

BOX 1.1 THREE LENSES OF ANALYSIS FOR CONSIDERING POPULATION ISSUES

Ecological lens

- considers people as just one part of an interconnected ecological system
- encourages us to think about issues such as finite resources, carrying capacity, and consumption patterns
- considers "overpopulation" to be a measurable state

Economic lens

- considers the role of people in broader economic systems
- asks us to reflect on how wealth is created and distributed, with consumption seen as a key aspect of wealth creation
- focuses on economic growth as a way to improve and maintain living standards
- sees population growth in terms of more workers and more consumers

Social equity lens

- focuses on philosophical rights such as human rights and the right to existence
- provides a critical approach that highlights uneven power relations and other inequalities
- considers social equity issues related to demographic categories such as race, gender, and socio-economic status as central to understanding global patterns and processes
- questions "overpopulation" as a concept (who is the surplus population in a situation of "overpopulation"?)

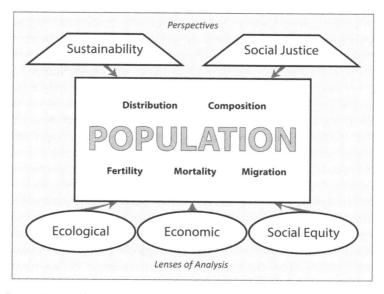

FIGURE 1.6 Perspectives and lenses of analysis used to consider population issues

An ecological lens has its basis in the science of **ecology** and sees humans as just one species among many and subject to the same ecological constraints as other animals. We know that a rapid increase in a rabbit population might lead to over-grazing of a pasture and the subsequent collapse of the rabbit population once food is in short supply. Similarly, an ecological lens sees growing human populations as putting increasing strain on **renewable resources** such as timber and food, poten-tially exhausting **finite resources** such as minerals. The ultimate extension of this viewpoint is that human populations, and the ecological systems that support them, will eventually crash if we overstep ecological bounds. In this context, **overpopu-lation** is a genuine concern, representing the population size at which ecological systems begin to degrade. To avoid these negative outcomes, we must focus on staying within the **carrying capacity** of the land—the number of people that can be sustained without damaging or diminishing the resources of that ecosystem.

For many neo-liberal economists, such views are unnecessarily doom-laden. From an economic perspective, more people can be viewed as more workers and more consumers—both good for a thriving economy. In this view, consumption is to be encouraged because it means more profits will be generated and living stand-ards can increase. Notably, the evolving field of **environmental economics** is beginning to challenge these ideas by attributing a value to environmental inputs and services within economic systems. From a conventional economic perspective, however, the environment has traditionally been seen as providing resources and services to the economy for free, with damage to the environment largely excluded from economic analyses. In conventional economic approaches, aging and shrink-ing populations pose a threat to economic stability as the ratio of workers to non-working populations declines. Meanwhile, aging populations bring additional costs associated with pensions and healthcare, further compromising economies already weakened by declining numbers of workers.

Sustainable development is often invoked in an attempt to reconcile the apparent contradictions between ecological and economic perspectives. The term was coined in the 1980s to try to resolve tensions between the increasing environ-mental aspirations of many richer countries, which wanted to impose environmen-tal regulations that were seen by many poorer countries as a threat to their efforts to bring their populations out of poverty through economic development.

Here poverty itself is considered a threat to the environment because it can lead to the destruction of natural resources by people desperately trying to meet basic needs. Although sustainable development is a fine goal, how exactly we are to achieve it remains controversial. Critics point out that economic development involves the consumption of resources, which is in direct opposition to the sustain-ability goal of conservation of environmental assets. Many hope that technology might provide a solution to this apparent impasse, by providing novel ways to maintain high living standards but with fewer impacts on the environment (e.g., through switching energy generation from fossil fuels to renewable sources). Other sustainability approaches emphasize the importance of greater efficiency of resource use (e.g., efforts to reduce waste). Despite its shortcomings, the notion of sustainable development remains powerful rhetoric in global policymaking and was brought to the fore in 2015 when the United Nations set a series of Sustainable Development Goals. These goals call on all member states to "end poverty, protect the planet and ensure that all people enjoy peace and prosperity by 2030" (UNDP 2021). The

FIGURE 1.7 United Nations' Sustainable Development Goals

Icons reproduced with kind permission of the United Nations (n.d.). https://www.un.org/sustainabledevelopment/
The content of this publication has not been approved by the United Nations and does not reflect the views of the United Nations or its officials or Member States.

goals range from ending hunger to taking climate action, improving working conditions, ending gender inequality, and developing strong institutions, indicating the vast array of aspirations that now fall under the broad umbrella of sustainable development (figure 1.7).

The Sustainable Development Goals illustrate well how social justice issues such as gender equality and strong institutions have now become central tenets of sustainable development. In simple terms, we want to achieve environmental goals but not at the cost of abandoning people to lives of poverty. Today, the broad aspirations of sustainable development have been rebranded yet again using the term **sustainability**, which explicitly recognizes the importance of three separate goals for the persistence of stable, human communities—environmental protection, economic viability, and social equity—consistent with the three lenses of analysis we use in this book.

Using a social equity lens involves a focus on human rights and uneven power relations, with issues such as race, class, and gender seen as pivotal. With respect to populations issues, the emphasis moves, for instance, from population size to also considering patterns of consumption, recognizing that an affluent North American or European is consuming resources at a much higher rate than an Indian or Ghanaian subsistence farmer. From this **critical perspective**, many scholars reject the term overpopulation as unethical, arguing that it implies that some human beings are "surplus" and that it is typically poor people and communities of color who are targeted by population campaigns as being this "excess" population.

In summary, identifying and tackling today's population problems requires generating consensus among multiple diverging, and even conflicting, perspectives. For the remainder of this chapter, we will expand on these ideas by looking at the issue of population size and growth. This will both act as an introduction to some key themes in population geography and provide a model for our approach of looking at issues through different lenses of analysis.

So, what's the problem … population growth or population decline?

Isolated voices have warned of the perils of rapid population growth for hundreds of years—including Thomas Malthus, an English clergyman who famously warned that population growth could outstrip food supply as early as the late 1700s. Despite grinding poverty and appalling living conditions for many, economic expansion and resources from the settling of new lands in this era of exploration and colonization were enabling rapid population growth in Europe. Malthus believed that the human population might be on the brink of outstripping food supply, although improved agricultural techniques and expansion of farmlands have notably kept food supplies growing faster than population, averting the food crisis that Malthus predicted to date.

The idea of population growth as an impending crisis resurfaced in the 1960s when the rapid population growth patterns seen in Europe in the 1800s and 1900s began to be repeated in the Global South (box 1.2). Improvements in agriculture, sanitation, and medicine, which had propelled growth in European populations, began to be widely adopted by poorer countries. Because many of these technologies were imported rather than having to be developed from scratch, the impact on decreasing death rates was far faster in the Global South, leading to very rapid population growth and renewed concerns over looming food shortages. In addition, an evolving environmental consciousness in the Western world generated concerns over the ecological impacts of growing populations. The situation was soon being labeled a "population explosion" with potentially devastating consequences. From a social justice perspective, many critical scholars have noted that the fact that it was communities of color that were increasing rapidly at this time was probably a significant contributor to fears around this "population explosion," with racist rhetoric emerging from white communities concerned about becoming marginalized by growing black and brown populations.

Anxiety about population growth has waxed and waned since the 1960s, but a vocal chorus has continued to warn that the Earth's resources are finite and that rapidly growing populations could potentially lead to resource shortages, ecological collapse, and even political conflict. With a global population that surpassed 8 billion in 2022, many believe we have already reached an unsustainable population size, as evidenced by the coalescing crises of global climate change, mass extinctions, habitat loss, and pollution. The term "overshoot" is sometimes used to refer to this idea that we have already exceeded the Earth's carrying capacity. Although we have temporarily been able to stretch the Earth's ecological bounds, the inevitable result of overshoot is that ecological systems will begin to degrade and could eventually collapse. From an ecological perspective, the state of the planet can seem bleak.

Although there are clearly many reasons why continued population growth is cause for concern, in the last 10 to 20 years equally vehement fears have been voiced over fertility decline. News headlines report a "'Jaw-Dropping' Global Crash in Children Being Born" (Gallagher 2020) and "The End of Babies" (Sussman 2019)

BOX 1.2 GLOBAL NORTH AND GLOBAL SOUTH—NAMING CONVENTIONS AND SOCIAL JUSTICE

The names we use for different population groups have become an important social justice issue, in recognition that how we label other people can have profound effects. You might have seen terms such as the "Third World," "developing countries," "the West," and the "Global South" and wondered about the specific meanings of these diverse labels. There are good reasons for grouping countries into categories that can be used as shorthand to reflect characteristics such as economic development and political history, but it is important that we understand why some of these terms are being retired because of negative connotations associated with them.

A stark divide has long been drawn between "the West and the rest"—the West being the countries of Western Europe, as well as those heavily settled by them (especially the United States, Canada, Australia, and New Zealand). The countries of the West share certain cultural similarities and have benefited from a longstanding dominant position in the global economy, and the term is still widely used and meaningful. During the Cold War these Western nations, as well as some newly affluent countries of East Asia, were branded the **capitalist** "First World," to emphasize their distinction from the "Second World" (the **communist** countries of the Soviet Union, China, and their allies) and the "Third World" (those countries politically unaligned to the First or Second Worlds). With the collapse of the Soviet Union in the early 1990s, the political relevance of these terms diminished, but the terms "First World" and "Third World" persist as shorthand for rich and poor countries.

In the latter part of the 20th century, new terms became popular that emphasized more subtle differences in levels of economic development, including: "developed," "under-developed," "less-developed," and "least-developed" countries. Soon, however, critical scholars were encouraging us to move away from terminology that implied that one group was more "developed" than another, arguing that it imposed value judgments on different societies and had unpleasant connotations of colonial dominance (Silver 2015). Instead, some commentators promoted the idea of "high-income," "middle-income," and "low-income" countries as reflective of economic fact, rather than value judgments. Many of these terms, including ones that refer to relative "development," are still widely used by major organizations like the World Bank and the Population Reference Bureau. In this book, we avoid terms with value judgments where possible but use them where other terms could lead to confusion.

Most recently, in an effort to move even further away from comparative judgments, the terms "Global North" and "Global South" have become popular. The Global North refers to the affluent countries of Europe, North America, and East Asia, as well as Australia and New Zealand (even though they are located south of the Equator), whereas the Global South is the poorer nations of the world, including all of Latin America and Africa and the poor and middle-income countries of Asia. Even here, scholars have pointed out serious limitations, however, with levels of affluence and economic development across the world being far more complicated than a simple binary system can accommodate (RGS, n.d.).

as demographic decline has diffused out of academic circles and into the popular press. Though attempts to lower fertility in the mid- to late 20th century were consistent with the goals and aspirations of many women, attempts to *increase* birth rates have been met with more limited success, with many women today seeing small family size as consistent with personal fulfillment, rewarding careers, and fewer domestic responsibilities. Although it may seem contradictory to be concerned about both population growth and population decline at the same time, it is here that we can use different lenses of analysis to understand the apparent paradox. Ecologically, population growth is clearly a huge problem, but economically the impacts of population decline are concerning.

For the global economy, years of population growth have contributed to an economic miracle. Living standards today are higher than ever before (with the caveat that living standards are undoubtedly far higher for some than for others). Although hunger and malnutrition persist, famine is rare today, and even relatively poor populations often have access to a range of consumer goods that would have been unthinkable even for the rich until relatively recently (TVs, cell phones, motorbikes, etc.). This huge economic growth has been premised on rising populations providing increasing numbers of workers and consumers. The wealth produced by all of these new workers not only enriches the owners and shareholders of the companies they work for but also provides taxes to support government services such as education and healthcare. In the post-war period, many countries set up sophisticated **welfare states** that use taxes to provide social support systems for vulnerable populations, including the young, the elderly, and those unemployed or with disabilities, via systems like pensions, disability benefits, and child support payments. Although countries differed widely in the degree to which they took on these programs, most people have grown accustomed to living with some sort of social safety net, as well as government involvement in provision of services like schools, police, and transportation. Even the United States, renowned for its reluctance to expand the reach of government, relies on taxes to pay for education, police, and roads, as well as healthcare for certain groups such as veterans and seniors.

Now, however, economists are warning that demographic change may be threatening this welfare system. Lower birth rates in many affluent countries, particularly in Europe and East Asia, mean fewer future workers. Indeed, many countries have already reached the point at which their birth rate has dipped below their death rate, leading to population decline unless migration makes up the shortfall. Birth rates in Japan have dropped so low that its population could be reduced by one-third by 2060 (BBC 2012). Growing life expectancies further increase the proportion of elderly people in the population, increasing the challenge of covering the high costs of retirement benefits and healthcare that welfare states provide. In a sizeable proportion of European and East Asian countries, the over-65s already make up 20 to 30 percent of the population, and this proportion is projected to increase significantly in the coming years (PRB 2020). At the same time, government tax receipts will go down because fewer people will be working. The net result is less money coming to the government, just as demands on government services are increasing. It is this mismatch that causes economists such concern.

One potential solution to this situation is migration. We currently still have growing populations in many countries of the Global South, even as birth rates are declining in much of East Asia and the West. Perhaps, then, we should allow economic migrants to redress the balance? Indeed, many commentators have made just this argument: that immigration will be critical to the economic fortunes of affluent countries with declining birth rates, while simultaneously assisting the economic development of poor countries with high birth rates as emigrants send money to support family back home. Unfortunately, this approach is complicated. People are not just workers but are also embedded in cultural contexts, and it has often proved highly controversial when people from one cultural background have migrated to another region. Indeed, the current rise in nationalist political movements is arguably partly a reaction to recent histories of migration that have seen the mixing of cultural groups all over the world. From a social justice perspective, we can certainly argue that we are all global citizens and that it makes sense to have people move freely around the world to wherever the jobs are, but we have to ask at what cost this might be achieved given the flashpoints that race, immigration, and nationalism have become.

So, where do we go from here? In this brief introduction to demographic issues, we have tried to outline some of the key changes that have occurred over time to provide a framework for the remaining chapters of this book. Each of the upcoming chapters will focus on one aspect of population geography, providing theoretical background and case studies that illustrate how theories play out in real geographic contexts. In the next chapter we discuss sources of population data and how we visualize and analyze these data, followed by a chapter on population distribution and composition. Together, these opening chapters provide the introductory information that you will need for the remainder of the book. We then consider three major realms of population geography in the following sections: population growth and sustainability, population health, and population movements. In the concluding chapter of the book, we take the examples of cities and climate change as topics through which to explore the intersection of issues raised throughout the book. In each chapter, we highlight the ways in which ecological, economic, and social equity approaches can inform the way we look at the world to model how to think critically about today's pressing population issues.

Discussion questions

1   Thinking about the COVID-19 pandemic, expand on the ways in which population issues have been central to the pandemic. Consider topics such as population growth and decline, equity among sub-populations (e.g., by race or socio-economic status), and mobility of populations.

2   We note that the impact of the COVID-19 pandemic on birth rates was highly place specific. Consider why people may have responded so differently to the pandemic in different countries with respect to their plans to have children.

3   We use the example of population size to illustrate how different lenses of analysis can bring people to very different conclusions about the same topic of study. Think about how different lenses might help us to understand why different groups of people might be for or against immigration in the country where you live. Can you think of other population issues whose analysis could be enriched by using multiple lenses?

Suggested readings

Bateman, N., and M. Ross. 2020. "Why Has COVID-19 Been Especially Harmful for Working Women?" The Brookings Institution, October 2020. https://www.brookings.edu/essay/why-has-covid-19-been-especially-harmful-for-working-women/

Frey, W. 2018. "The US Will Become 'Minority White' in 2045, Census Projects." The Brookings Institution, March 14, 2018. https://www.brookings.edu/blog/the-avenue/2018/03/14/the-us-will-become-minority-white-in-2045-census-projects/

Silver, M. 2015. "If You Shouldn't Call It the Third World, What Should You Call It?" National Public Radio, January 4, 2015. https://www.npr.org/sections/goatsandsoda/2015/01/04/372684438/if-you-shouldnt-call-it-the-third-world-what-should-you-call-it

Glossary

capitalism: economic system premised on trade and industry, designed toward profit, and controlled by private ownership of means of production

carrying capacity: the maximum number of individuals that can be supported by an ecosystem without causing degradation

case studies: real-world examples used to illustrate broader theories or principles

coercion: influencing someone using force or threats

communism: economic system in which property is publicly owned and operated for the benefit of the community

critical perspectives: approaches that question current power structures and strive for a more just society

demography: quantitative study of population distribution, composition, and dynamics

ecology: the study of the interactions of communities of organisms with their environments

environmental economics: a recent economic approach that emphasizes quantifying the valuable role of ecological systems in economic processes

family planning: the practice of controlling family size and spacing, especially using contraception

finite resources: natural resources that cannot be regenerated over a human timescale; for example, minerals

global citizenship: the idea that individuals have civic responsibilities that extend beyond their local communities to the entire global community

globalization: the growing interconnectedness of the global economy and associated cultural and economic shifts

identity: characteristics of an individual that contribute to their perception of who they are, including aspects such as race, gender, and nationality

intersectionality: recognizing and analyzing the significance of interactions among multiple identities, particularly with respect to discrimination or disadvantage

overpopulation: too many individuals to be supported sustainably in a particular region

physical environment: our natural and built surroundings

place: a location imbued with cultural significance

population geography: the study of the spatial distribution and change of human communities and their interactions with their environments and one another

pro-natal campaigns: policies designed to encourage births

qualitative analysis: research approaches that explore non-numeric information such as attitudes, opinions, or knowledge

quantitative analysis: research approaches that focus on counting and measuring

renewable resources: natural resources that can be regenerated over a human timescale, for example, timber

scale: the extent of a phenomenon being considered, such as global, national, or local scale; this usage is sometimes termed **geographic scale** to distinguish it from cartographic scale

social environment: aspects of an individual's surroundings that relate to interactions with other people, including economic, political, and cultural aspects

social justice: approaches that consider uneven power relations and strive toward more just social systems

space: geographic location as defined by facets such as proximity, direction, and area

sustainability: a social movement that strives for ecological balance, alongside economic and social justice goals

sustainable development: economic development designed to bring people out of poverty but at lowest ecological cost; often criticized for being overly idealistic and unattainable

welfare state: a political approach that uses taxes to provide support for individuals via mechanisms such as socialized healthcare and pensions

Works cited

Adam, K. 2020. "Why Is Coronavirus Hitting Britain's Minority Doctors So Hard?" *The Washington Post*, May 20, 2020. https://www.washingtonpost.com/world/europe/uk-doctors-coronavirus-deaths/2020/05/19/e9a0475e-93be-11ea-87a3-22d324235636_story.html

Andrasfay, T., and N. Goldman. 2021. "Reductions in 2020 US Life Expectancy Due to COVID-19 and the Disproportionate Impact on the Black and Latino Populations." *Proceedings of the National Academy of Sciences* 118 (5): e2014746118. doi:10.1073/pnas.2014746118

BBC. 2012. "Japan Population to Shrink by One-third by 2060." BBC News, January 30, 2012. https://www.bbc.com/news/world-asia-16787538

CDC [Centers for Disease Control and Prevention]. 2020. "Prevalence of Self-Reported Obesity Among U.S. Adults by State and Territory." BRFSS. https://www.cdc.gov/obesity/data/prevalence-maps.html

CIA [US Central Intelligence Agency]. n.d. "The World Factbook." Accessed March 26, 2021. https://www.cia.gov/the-world-factbook/

Coleman, D., and B. Rowthorn. 2013. "Population Decline—Facing an Inevitable Destiny?" In *Fertility Rates and Population Decline*, edited by A. Buchanan and A. Rotkirch, 82–101. London: Palgrave Macmillan.

Dahlman, C. 2019. "National Pasts and Biopolitical Futures in Serbia." In *Reproductive Geographies: Bodies, Places, and Politics*, edited by M. England, M. Fannin, and H. Hazen, 184–200. London: Routledge.

Dockterman, E. 2020. "Women Are Deciding Not to Have Babies Because of the Pandemic. That's Bad for All of Us." *Time*, October 15, 2020. https://time.com/5892749/covid-19-baby-bust/

Gallagher, J. 2020. "Fertility Rate: 'Jaw-Dropping' Global Crash in Children Being Born." BBC News, July 15, 2020. https://www.bbc.com/news/health-53409521

Goldberg, M. 2009. *The Means of Reproduction*. New York: Penguin.

Hoffmann, D. 2000. "Mothers in the Motherland: Stalinist Pronatalism in Its Pan-European Context." *Journal of Social History* 34 (1): 35–54.

King, L. 2002. "Demographic Trends, Pronatalism, and Nationalist Ideologies in the Late Twentieth Century." *Ethnic and Racial Studies* 25 (3): 367–89.

Luppi, F., B. Arpino, and A. Rosina. 2020. "The Impact of COVID-19 on Fertility Plans in Italy, Germany, France, Spain, and the United Kingdom." *Demographic Research* 43 (47): 1399–412. doi:10.4054/DemRes.2020.43.47

Otu, A., B. Ahinkorah, E. Ameyaw, A.-A. Seidu, and S. Yaya. 2020. "One Country, Two Crises: What Covid-19 Reveals about Health Inequalities among BAME Communities in the United Kingdom and the Sustainability of its Health System?" *International Journal for Equity in Health* 19: 189.

Perelli-Harris, B., and O. Isupova. 2013. "Crisis and Control: Russia's Dramatic Fertility Decline and Efforts to Increase It." In *Fertility Rates and Population Decline*, edited by A. Buchanan and A. Rotkirch, 141–56. London: Palgrave Macmillan.

PRB [Population Reference Bureau]. 2020. "World Population Datasheet 2020." PRB, July 2020. https://www.prb.org/wp-content/uploads/2020/07/letter-booklet-2020-world-population.pdf

Resolution XVIII: Human Rights Aspects of Family Planning, Final Act of the International Conference on Human Rights. 1968. U.N. Doc. A/CONF. 32/41, p. 15.

RGS [Royal Geographical Society]. n.d. "A 60 Second Guide to … the Global North/South Divide." Accessed January 28, 2022. https://www.rgs.org/CMSPages/GetFile.aspx?nodeguid=9c1ce781-9117-4741-af0a-a6a8b75f32b4&lang=en-GB

Ritchie, H., and M. Roser. 2017. "Obesity." OurWorldinData. https://ourworldindata.org/obesity

Silver, M. 2015. "If You Shouldn't Call It the Third World, What Should You Call It?" National Public Radio, January 4, 2015. https://www.npr.org/sections/goatsandsoda/2015/01/04/372684438/if-you-shouldnt-call-it-the-third-world-what-should-you-call-it

Sussman, A. 2019. "The End of Babies." *New York Times*, November 16, 2019. https://www.nytimes.com/interactive/2019/11/16/opinion/sunday/capitalism-children.html

Trewartha, G. 1953. "A Case for Population Geography." *Annals of the Association of American Geographers* 43 (2): 71–97.

UNDP [United Nations Development Programme]. 2020. "COVID-19 Will Widen Poverty Gap between Women and Men, New UN Women and UNDP Data Shows." UNDP, September 2, 2020. https://www.undp.org/content/undp/en/home/news-centre/news/2020/_COVID-19_will_widen_poverty_gap_between_women_and_men_.html

———. 2021. "Sustainable Development Goals." https://www.undp.org/content/undp/en/home/sustainable-development-goals.html

UNFPA [United Nations Population Fund]. 2021. "Impact of COVID-19 on Family Planning: What We Know One Year into the Pandemic." UNFPA Technical Note, March 11, 2021. https://www.unfpa.org/sites/default/files/resource-pdf/COVID_Impact_FP_V5.pdf

United Nations. n.d. "Sustainable Development Goals." Accessed March 6, 2022. https://www.un.org/sustainabledevelopment/news/communications-material/

US Bureau of Economic Analysis. 2021. "GDP by State, 1st Quarter 2021." BEA, June 25, 2021. https://www.bea.gov/sites/default/files/2021-06/qgdpstate0621.pdf

WHO [World Health Organization]. n.d. "WHO Coronavirus (COVID-19) Dashboard." Accessed November 11, 2022. https://covid19.who.int

Yglesias, M. 2020. *One Billion Americans: The Case for Thinking Bigger.* New York: Penguin Random House.

2

Demographic data, visualization, and interpretation

After reading this chapter, a student should be able to:

1 understand the strengths and weaknesses of key sources of population data: censuses, vital registration systems, and sample surveys;

2 interpret information from major types of graphs and maps;

3 discuss the benefits and shortcomings of different types of graphs and maps.

Quantitative and statistical analyses provide important contributions to population geography through using numeric data to analyze population patterns and processes. Institutions and government agencies use population data for allocating political representation and funding, monitoring demographic trends, and planning purposes such as building new roads, schools, or hospitals. Access to accurate population data is also important for organizations that respond to emergencies, assist international migrants, or consider environmental degradation and its relationship to population growth. In this chapter we introduce major types of population data, as well as some ways to visualize and interpret numerical data. We also discuss problems associated with data collection and use.

PART I: POPULATION DATA

Huge amounts of population data are collected every year, much of which is available from the major governmental and non-governmental organizations that perform this data collection (box 2.1). There are three main sources of population data: censuses, vital registration systems, and sample surveys. Modern **censuses** are complete enumerations of all people (or all citizens) and their important characteristics within

DOI: 10.4324/9781003143253-2

BOX 2.1 KEY SOURCES OF POPULATION DATA

International statistics

- United Nations. Population Division (https://www.un.org/development/desa/pd)
- United Nations. Statistics Division. Partners: National Statistical Offices (https://unstats.un.org/home/nso_sites)
- United Nations High Commissioner for Refugees (UNHCR) (https://www.unhcr.org)
- World Health Organization (WHO) (https://www.who.int)
- Migration Data Portal (https://migrationdataportal.org)
- City Population: Population Statistics in Maps and Charts for Cities, Agglomerations and Administrative Divisions of All Countries of the World (https://www.citypopulation.de)
- World Bank Open Data (https://data.worldbank.org)
- Eurostat (https://ec.europa.eu/eurostat)
- Our World in Data (https://ourworldindata.org)
- Gapminder (https://www.gapminder.org)
- World Values Survey (https://www.worldvaluessurvey.org/wvs.jsp)
- Population Reference Bureau. Data Sheets (https://www.prb.org/collections/data-sheets)

Selected national statistics

- US Census Bureau (https://www.census.gov)
- US Centers for Disease Control and Prevention (CDC) (https://www.cdc.gov)
- US Department of Homeland Security (https://www.dhs.gov/immigration-statistics)
- National Historical Geographic Information System (NHGIS) (https://www.nhgis.org)
- Statistics Canada/Statistique Canada (https://www.statcan.gc.ca)
- United Kingdom Office for National Statistics (https://www.ons.gov.uk)
- Northern Ireland Statistics and Research Agency (https://www.nisra.gov.uk)
- National Records of Scotland (https://www.nrscotland.gov.uk)
- Australian Bureau of Statistics (https://www.abs.gov.au)
- Stats NZ [New Zealand] Tatauranga Aotearoa (https://www.stats.govt.nz)
- Stats SA [South Africa] (http://www.statssa.gov.za)

Selected population organization websites

- Population Reference Bureau (https://www.prb.org)
- Population Council (https://www.popcouncil.org)
- European Research Center on Migration & Ethnic Relations (https://www.ercomer.eu)
- Australian National University. School of Demography (https://demography.cass.anu.edu.au)
- Max Planck Institute for Demographic Research (https://www.demogr.mpg.de/en)
- International Union for the Scientific Study of Population (https://iussp.org)

a country, with the goal of understanding the basic structure and trends of society. Censuses are expensive to undertake and thus only conducted periodically, usually every 10 years, resulting in significant pauses between data sets. Nonetheless, for poorer countries that lack regular record keeping, periodic censuses may provide one of the most reliable sources of population data. **Vital registration systems** are government-run systems of continuous, compulsory recording of vital events such as births, deaths, marriages, and divorces. If a civil registration system is well maintained, its data will be more current than census data, but it will not provide as much detail as can be included on the census form. Finally, **sample surveys** collect information on specific topics (e.g., fertility, HIV/AIDS, or migration). Sample surveys have several advantages, including lower costs, shorter time spans to collect and process the data, and the opportunity to collect very detailed information on specific topics. Sample surveys can also incorporate qualitative questions to get at more subjective or controversial topics. For instance, knowledge, attitudes, beliefs, and practices (KABP) surveys might explore topics such as domestic roles in the household and sexual behaviors. Their main disadvantage is that sample surveys are administered to just a sub-set of the population and so provide only an estimate of population-scale trends.

Censuses

Censuses are the oldest and best-known method of collecting basic data about a country's population. Censuses were used in ancient Babylonia, Egypt, China, and Rome, among other places. The goal of most pre-modern censuses was the identification and control of individuals for the purposes of taxation, labor force estimation, and military potential, and often only adult males were enumerated. The first modern census was probably conducted in Iceland, followed by several other Scandinavian countries in the 18th century. The 1703 Icelandic census recorded the name, age, residence, and social position of 50,366 inhabitants of the island and was remarkably accurate (UNESCO 2013; Thorvaldsen 2019).

Today, the United Nations (UN) defines a census as "the total process of planning, collecting, compiling, evaluating, disseminating, and analyzing demographic, economic and social data at the smaller geographic level pertaining, at a specified time, to all persons in a country or a well-delimited part of a country" (UNDESA 2017). Since World War II, most countries in the world have conducted at least one population census, and most affluent nations have conducted censuses on a regular basis, usually every 10 years (figure 2.1). By contrast, some poor or less politically stable countries have not had a census for several decades owing to the high cost and organization required. In other cases, censuses have been purposefully avoided for political reasons. Lebanon, for instance, has not been enumerated since 1932, related to political concerns around the changing composition of the Lebanese population. Originally majority Christian, Lebanon saw a rise in its Muslim population through **natural increase** and an influx of refugees in the mid-20th century, to the point where Christians were in the minority by the early 1970s. A disastrous civil war (1975–90) developed as Christian populations attempted to hold on to political power. Although the war led to a redistribution of power, deep religious and political divisions pose a barrier to conducting a census in the foreseeable future (Barshad 2019).

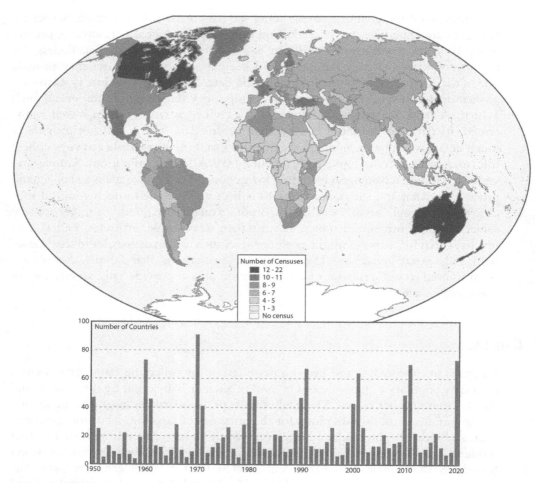

FIGURE 2.1 Number of censuses conducted around the world by country since 1950
The map shows the number of censuses taken in each country since 1950 and the graph the number of countries conducting a census by year—note the peaks every 10 years representing the large numbers of countries that conduct their census at the turn of each decade.
Data source: UNSD (2021).

A good population census should have several characteristics. It should be overseen by a reliable governmental or non-governmental agency that administers the census regularly and processes and disseminates the information collected. Each census should cover a specified territory (usually a country) and should enumerate every individual in that area at the same time, typically on a particular day. Very few censuses meet all of these criteria: some countries may need more than 1 day to enumerate their populations (e.g., India), some may opt for collecting certain information from population samples rather than the whole population (e.g., the United States), some may not conduct censuses at regular intervals (e.g., Germany), and some may not make all of the results public (e.g., the former Soviet Union).

Where most of the population is literate, respondents are typically encouraged to complete the census form themselves, traditionally on paper but increasingly online.

Census officials called "census enumerators" follow up with households that have not responded. This follow-up is very important because non-respondents are likely to come disproportionately from particular sub-populations (e.g., recent immigrants, people without a fixed address, or poorer communities), potentially leading to a systematic undercount of vulnerable groups if they are not actively encouraged to participate.

According to the United Nations, an ideal census would cover the topics shown in box 2.2. Data about where people live and household size are particularly

---

BOX 2.2 UNITED NATIONS' RECOMMENDATIONS FOR CENSUS TOPICS THAT SHOULD BE INCLUDED IN POPULATION CENSUSES CONDUCTED AROUND THE YEAR 2020

Geographic and internal migration characteristics

- Place of usual residence
- Place where present at time of census
- Place of birth
- Duration of residence
- Place of previous residence
- Place of residence at a specified date in the past
- Total population
- Locality
- Urban and rural

International migration characteristics

- Country of birth
- Country of citizenship
- Acquisition of citizenship
- Year or period of arrival

Household and family characteristics

- Relationship to the reference person of household
- Household and family composition
- Household and family status

Demographic and social characteristics

- Sex
- Age
- Marital status
- Ethnocultural characteristics

- Religion
- Language
- Ethnicity
- Indigeneity
- Disability status

Fertility and mortality

- Children ever born alive
- Children living
- Date of birth of last child born alive
- Births in the past 12 months
- Deaths among children born in the past 12 months
- Age, date, or duration of first marriage
- Age of mother at birth of first child born alive
- Household deaths in the past 12 months
- Maternal or paternal orphanhood

Educational characteristics

- Literacy
- School attendance
- Educational attainment
- Field of education and training, and educational qualifications

Economic characteristics

- Labor force status
- Status in employment
- Occupation
- Industry
- Place of work
- Institutional sector of employment
- Working time
- Participation in own-use production of goods
- Income

Agriculture

- Own-account agricultural production
- Characteristics of all agricultural jobs during the last year

Source: UNDESA (2017).

important because they provide a basis for calculating population totals that guide political representation and the distribution of resources such as funding for schools. Different enumeration methods may result in different counts for the same geographic area, however, and so it is important to think carefully about which approach to use. In the de facto method, people are counted at the location where they were at the time of the census count. For example, if someone lives in a city but was on vacation at a small coastal community on census day, that person would be considered a resident of the coastal community for census purposes. This can result in undercounts for areas of seasonal outmigration and overcounts for tourist areas. Considering the growing mobility of modern societies, the de jure method has become a more common way of counting people in recent decades. This method records individuals at their place of permanent or usual residence.

Beyond simple population counts, most censuses ask for further demographic information about household members, including characteristics such as age, sex, race, and economic and educational factors. The specific questions asked and how they are phrased varies from country to country. If certain issues are of great importance for a country, like race relations in the United States, detailed questions may be included. Conversely, authorities may decide to exclude controversial topics such as citizenship, **ethnicity**, or religion. For instance, the separation of church and state clause in the first amendment to the US Constitution partly explains the lack of questions on religious affiliation in the US census. Similarly, the US census does not require that respondents state their citizenship, because experts worry that those without legal right of residence might be discouraged from completing the form if they feel they could then be targeted. This issue became highly controversial when the Trump administration proposed adding a citizenship question to the US 2020 Census (box 2.3).

Of course, populations change over time and so censuses must evolve with changing socio-demographic conditions. Figure 2.2 illustrates how the questions asked in the US census have evolved. Racial categories provide a particularly powerful example. Questions on **race** have been included in every US census since 1790, but the specific categories used have changed repeatedly. Prior to 1960, census enumerators were required to determine and record the race of every individual. In the earliest census, this involved distinguishing black and white individuals, as well as those who were free or enslaved. By 1890, census enumerators were being asked to make impossible determinations between a multitude of supposed racial groups, as can be seen in instructions given to enumerators at the time:

Write white, black, mulatto, quadroon, octoroon, Chinese, Japanese, or Indian, according to the color or race of the person enumerated. Be particularly careful to distinguish between blacks, mulattoes, quadroons, and octoroons. The word "black" should be used to describe those persons who have three-fourths or more black blood; "mulatto," those persons who have from three-eighths to five-eighths black blood; "quadroon," those persons who have one-fourth black blood; and "octoroon," those persons who have one-eighth or any trace of black blood.

United States Department of the Interior (1890)

BOX 2.3 CITIZENSHIP AND THE US 2020 CENSUS

In March of 2018 the US Justice Department requested that all individuals be asked to provide their citizenship status and whether they obtained citizenship through birth or naturalization on the 2020 Census. The Trump administration argued that citizenship was an obvious piece of information to collect and that many other national censuses have citizenship questions. Supporters also argued that it was important for collecting detailed and accurate data about the US population for purposes such as drawing electoral districts and to avoid counting those who do not actually reside in the United States (Wallace 2019).

The announcement was strongly criticized by many scholars, politicians, immigrant advocacy groups, and others. Opponents believed that the citizenship question could lower the response rate for the entire census if respondents from vulnerable communities (particularly recent immigrants and ethnic minorities) feared that their responses could be used against them. This could produce population undercounts or inaccurate information for already vulnerable groups. The Census Bureau estimated that almost one-third of respondents to a related survey—the American Community Survey—had not answered or had provided false information to a citizenship question in recent years. Considering the political climate of the time and low levels of public trust in the federal government, the no-response rate to the census question could have been even higher.

Normally, the Census Bureau starts preparation for each census years in advance and conducts tests of new or potentially controversial questions. Preparations for the 2020 Census had begun in 2007, with final testing completed in 2018, just before the request for the addition of the citizenship question, making it impossible to assess its potential impact. Critics argued that the Trump administration had ulterior motives for proposing the last-minute addition (Levitt 2019). An undercount of immigrants could be used as evidence of the effectiveness of Trump-era immigration policies. An undercount could also affect political representation and funding to the potential benefit of Republicans if recent immigrants were deterred from submitting their census forms.

In subsequent years, the number of racial categories increased further—the Asian category, for instance, expanded from one in 1870 (Chinese) to seven in 2000. Native Americans were not properly enumerated until the 1890 census, with Native Americans living outside of white communities originally excluded. The biases and racist attitudes of the day are thus clearly recorded in the census.

From 1960 onwards, people were asked to self-report their race. Originally, individuals were required to select just one racial category, forcing people of mixed race to ignore the complexity of their heritage, but from the year 2000 individuals could select more than one racial category and provide more information about their ethnic heritage (figure 2.3). Although it is clearly important for racial categories to change in step with evolving understandings of ethnic and racial identity, changing categories make comparative studies of race and ethnicity over time from census data challenging.

Questions related to household and family characteristics are now also beginning to change in some countries as acceptance of alternative lifestyles increases. Several countries now include questions that acknowledge non-traditional household structures (e.g., Australia has recognized same-sex couples on its census since 1996

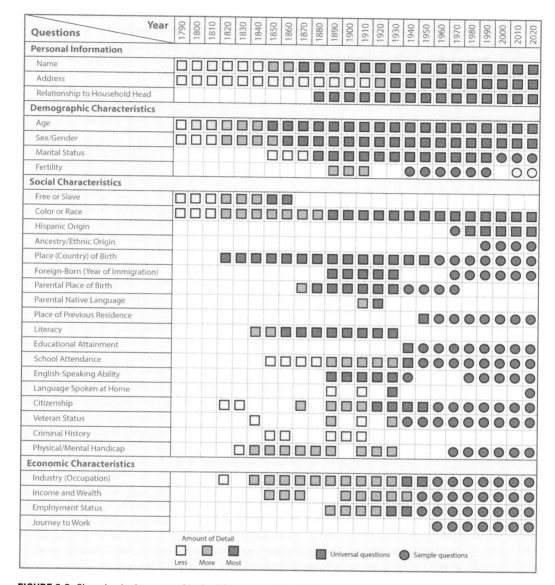

FIGURE 2.2 Changing topics covered in the US census, 1790–2020
"Universal questions" are asked of all those taking the census; "sample questions" are reserved for a sub-set of respondents who receive a more detailed form.
Source: United States Census Bureau (2021).

and Canada since 2001). Today, as more fluid concepts of gender are becoming increasingly accepted, options beyond the traditional male/female choice are also being added (box 2.4).

Census errors

No census has ever been error free. There are two main types of census error: coverage error and content error. **Coverage error** is the omission or double counting

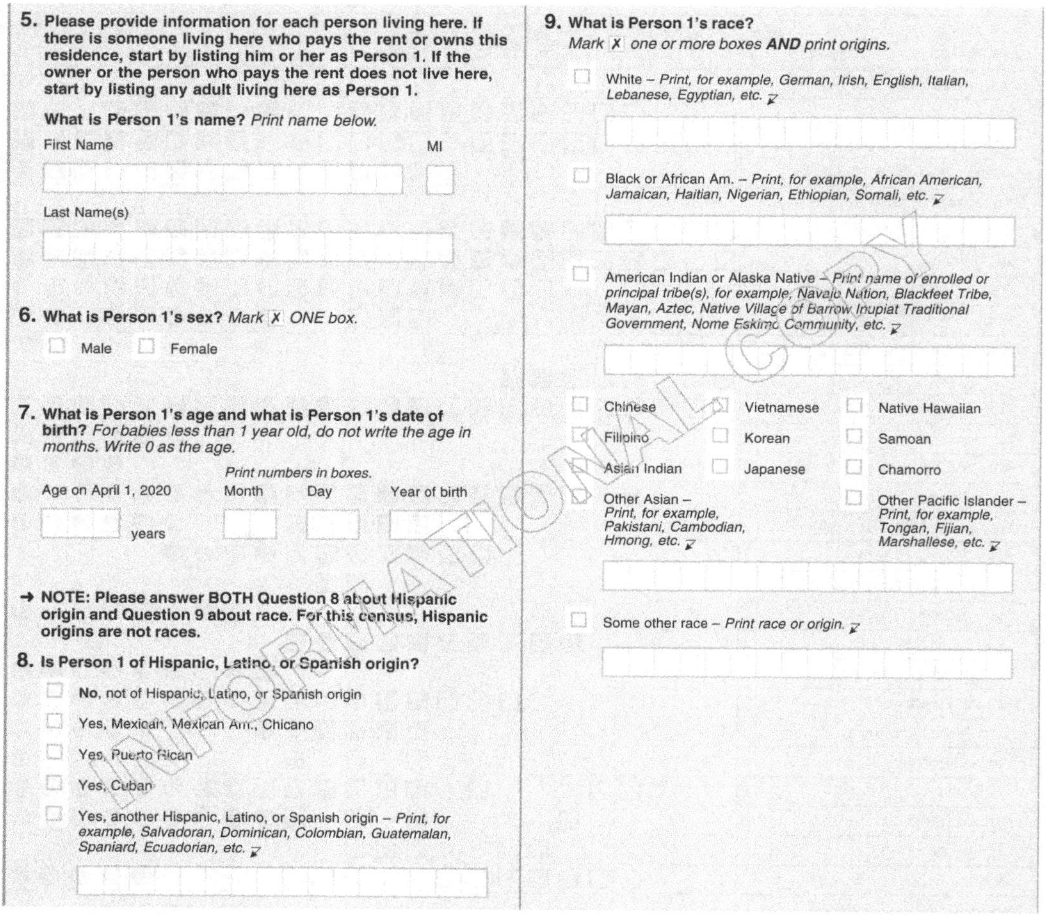

5. Please provide information for each person living here. If there is someone living here who pays the rent or owns this residence, start by listing him or her as Person 1. If the owner or the person who pays the rent does not live here, start by listing any adult living here as Person 1.

What is Person 1's name? *Print name below.*

First Name   MI

Last Name(s)

6. What is Person 1's sex? *Mark* X *ONE box.*

  ☐ Male   ☐ Female

7. What is Person 1's age and what is Person 1's date of birth? *For babies less than 1 year old, do not write the age in months. Write 0 as the age.*

Print numbers in boxes.

Age on April 1, 2020   Month   Day   Year of birth

        years

→ NOTE: Please answer BOTH Question 8 about Hispanic origin and Question 9 about race. For this census, Hispanic origins are not races.

8. Is Person 1 of Hispanic, Latino, or Spanish origin?

  ☐ No, not of Hispanic, Latino, or Spanish origin

  ☐ Yes, Mexican, Mexican Am., Chicano

  ☐ Yes, Puerto Rican

  ☐ Yes, Cuban

  ☐ Yes, another Hispanic, Latino, or Spanish origin – *Print, for example, Salvadoran, Dominican, Colombian, Guatemalan, Spaniard, Ecuadorian, etc.* ⤵

9. What is Person 1's race?

Mark X *one or more boxes AND print origins.*

  ☐ White – *Print, for example, German, Irish, English, Italian, Lebanese, Egyptian, etc.* ⤵

  ☐ Black or African Am. – *Print, for example, African American, Jamaican, Haitian, Nigerian, Ethiopian, Somali, etc.* ⤵

  ☐ American Indian or Alaska Native – *Print name of enrolled or principal tribe(s), for example, Navajo Nation, Blackfeet Tribe, Mayan, Aztec, Native Village of Barrow Inupiat Traditional Government, Nome Eskimo Community, etc.* ⤵

  ☐ Chinese   ☐ Vietnamese   ☐ Native Hawaiian
  ☐ Filipino   ☐ Korean   ☐ Samoan
  ☐ Asian Indian   ☐ Japanese   ☐ Chamorro
  ☐ Other Asian – *Print, for example, Pakistani, Cambodian, Hmong, etc.* ⤵   ☐ Other Pacific Islander – *Print, for example, Tongan, Fijian, Marshallese, etc.* ⤵

  ☐ Some other race – *Print race or origin.* ⤵

FIGURE 2.3 Questions pertaining to race and ethnic origin from the US 2020 Census
The entire census questionnaire is eight pages long and can be downloaded from the US Census Bureau's website.
Source: United States Census Bureau (2021).

of some individuals during the enumeration process. In the United States, under-counting was first acknowledged as a problem in the 1940 census when "more young black males registered for the draft than the census bureau thought were in the country" (Holden 2009, 1008). The Census Bureau has been estimating the undercount for every census since then. Racial and ethnic minorities, some immigrant groups, people experiencing homelessness, and nomadic populations are frequently undercounted in censuses. This may be related to having less access to the form or because marginalized groups are more likely to be skeptical of the benefits of censuses to their communities or even outright distrustful of how the data might be used. On the other hand, white and more affluent households are most likely to be counted twice in the United States, related to factors such as second home ownership. Undercounting children is another common problem. The US 2010 Census failed to count almost 1 million young children, many of them in female-headed households living in poverty, households with limited English-speaking abilities, and immigrant families. It is thought that some households may not understand

BOX 2.4 SEX, GENDER, AND THE CENSUS

Sex is a standard question on any census, with male and female used as default binary categories. This simple classification of individuals into males and females has begun to be challenged, however, by activists who argue that biological sex is actually a spectrum and that individuals may identify with a gender that is not necessarily aligned with their sex assigned at birth. In response to these ideas, some survey forms are incorporating additional gender options. Although we often associate this movement with progressive Western countries, India and Pakistan were among the first countries in the world to provide alternative sex/gender options on their censuses. The 2011 Census of India and 2017 Census of Pakistan had "other" and "transsexual" options for the sex/gender category respectively (UNECE 2019); Bangladesh and Nepal were due to follow suit in their 2021 Censuses (Gurubacharya 2020; The Daily Star 2021). The addition of a third category in these cases is recognition of long-established Hijra communities—whose members identify as a third gender that includes elements of both male and female—in these South Asian countries.

Among Western countries, Canada had two questions on sex/gender in its 2021 Census; one asked for sex at birth ("male/female"), the other for gender ("male/female/please specify in writing"; Statistics Canada 2020). The UK also added questions on gender identity and sexual orientation in its 2021 Census. Individuals in Australia had an option to identify their sex as male, female, or other on the 2016 Census form. However, those checking the "other" option had to complete an additional form to be counted in that category. Many respondents did not fill out this additional form, and so only 1,260 individuals (out of a population of 23 million) were identified as gender diverse. To avoid similar undercounts in the future, the 2021 Census has eliminated the additional form (Truu 2021). New Zealand is planning to add questions on gender and sexual identity in its 2023 Census.

that young children should be counted, and children living in joint custody arrangements or non-traditional households are also at greater risk of being missed (Jacobsen, Mather, and Yorke 2020). In other cases, non-reporting may be deliberate. In China's 2000 Census, for instance, it has been estimated that 37 million children were not reported to avoid penalties for violating China's strict fertility policies (Goodkind 2004).

Content error is the incorrect reporting or non-reporting of responses to census questions. It can occur for a variety of reasons, including poor questionnaire design, errors in asking the question or recording the response by the census enumerator, erroneous information provided by respondents, or fear of government action against the respondent (UNSD 2020). In some countries, there is growing public distrust in the security of personal data or sometimes even the perception that state-sponsored censuses are unnecessary government intrusion into the private lives of citizens (Coleman 2013). Although age has been the most frequently misreported fact (especially in the past), census results on religious affiliation, race, and ethnicity have been widely questioned in some countries as inaccurate or simply false (Lundquist, Anderton, and Yaukey 2015).

A growing number of countries have been switching from traditional censuses to other methods of population enumeration, owing to the rising costs of censuses and the significant time gaps between enumerations (usually 10 years or more) and between data collection and publication (often 1 to 3 years). In an era of continuously updated information, the "snapshot" nature of census data can seem dated. France is now using a **rolling census**, for example. A rolling census enumerates a country's population over a period of time, rather than on a specific day, and is a combination of traditional census and sample survey methodology (Roux 2020). In France, a fifth of small municipalities (under 10,000 population) are enumerated every year using traditional census methods, and a sample survey is administered to 8 percent of households in larger municipalities each year. This method spreads census costs more evenly over time and provides annual population estimates.

Vital statistics

Vital statistics comprise data on births, deaths, marriages, divorces, and other events and related characteristics within a population on a continuous basis. Whereas the census collects information on the entire population at a certain date, vital statistics are collected continuously but only on individuals experiencing particular life events (figure 2.4). Most of the data come from national civil registration systems, which record vital events for the country's entire population as legally mandated. The associated records are important legal documents in most societies. An official birth certificate might be necessary for establishing proof of age or citizenship to be eligible for a driver's license or retirement benefits. An official death certificate is often required for settling an estate or collecting on a life insurance policy.

Registration of vital events has a long history. For example, it was compulsory for ancient Romans to register children within 1 month of their birth. In England,

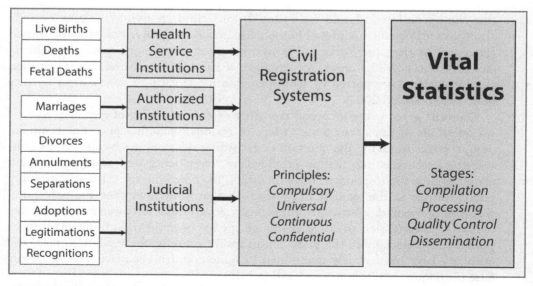

FIGURE 2.4 Basics of a typical national vital statistics system
Source: Adapted from UNSD (2019).

"Bills of Mortality" provided weekly records of burials in London from 1532 until the mid-19th century, and Anglican priests were required to maintain records of weddings and baptisms from 1538. Similarly, the Council of Trent introduced compulsory registration of baptisms, burials, and weddings across Catholic parts of Europe in 1563. Most of these early vital records are of limited value from a demographic point of view because they record religious ceremonies rather than actual events (e.g., baptisms vs births) and were often limited to specific religious communities or parishes. The first examples of modern civil registration of vital events come from America and Europe. The British Colonies of Massachusetts Bay and New Plymouth became the first political entity to have a civil, rather than clerical, registration system when they introduced compulsory registration of births, deaths, and marriages in 1639 (UNDESA 2017). In Europe, the Napoleonic Code of 1804 specified who should report and what should be reported about vital events to the civil authorities. This practice had been adopted across most of Europe by the end of the 19th century.

Vital events were originally recorded primarily for legal purposes but today larger and more complete data sets mean that vital statistics are also useful for demographic analyses. Most countries have formal systems for registering vital events, which can generate very complete data sets in affluent countries. For example, doctors or hospitals may be responsible for ensuring that birth and death certificates are issued, and marriages may have to be documented at a registry office. For poorer countries with disjointed bureaucracies and large rural populations, the data collected may be incomplete, however. A recent report on birth registration in Africa, for instance, found that almost half of all births in the region went unregistered in recent years and about 60 percent of African children under age 5 lack a birth certificate (UNICEF 2022). Those who are not registered are more likely to come from poorer and more remote populations, leading to significant undercounting of certain groups.

At the opposite extreme, some countries have additional registration systems to provide further information beyond simple counts of vital events. **Population registers** are one such system. Population registers record basic information for each individual such as full name, sex/gender, current/previous place of residence, date/place of birth, marital status, nationality, and sometimes even information about employment, educational attainment, income, or health. Residents of countries with population registers have a legal obligation to notify local authorities about any change in their situation (e.g., a change of residence, employment, marriage, or divorce). In Germany, for example, people are expected to register with the Einwohnermeldeamt (registration office) within a week of moving residence, even within the country. Though this system in theory means that the German government knows the number of people in the country at any time and where they live, in practice some people never register or fail to notify authorities when they move. Most European countries have national population registers, and in at least five of them (Austria, Denmark, Finland, Norway, and Sweden), registration systems have replaced census enumeration over the past 2 decades (Valente 2010). The United States notably does not maintain a population register because maintaining records of people's movements is seen as government overreach by many Americans. The United States does register vital events such as births, deaths, and marriages, however.

Sample surveys

A sample survey is a method of collecting data from a sub-set of the total population. Statistical techniques can then be used to consider how results might be generalized to draw conclusions about broader populations. In richer countries, sample surveys are usually funded by government and are designed to provide data to supplement the census and vital statistics to inform public policy, related to topics such as health, employment, disability, migration, and living conditions. Examples include the American Community Survey in the United States, the General Social Survey in Canada, and the Annual Population Survey in the UK.

In poorer countries, with more limited national-scale data, sample surveys are used to gather information on a wide variety of factors, including fertility, health, and economic measures, often funded by non-governmental organizations to assist the local government. One of the best known and most ambitious international sample surveys was the World Fertility Survey, undertaken by the International Statistical Institute in Belgium between 1974 and 1987. It used interviews with 350,000 women in over 60 countries to obtain information on fertility and related topics, including reproductive health, household characteristics, marital status, access to water and sanitation, economic status, religion, and race. The survey results are now available from the Demographic and Health Surveys Program (https://dhsprogram.com), which continues to perform sample surveys in low-income countries to provide data to inform policy and program planning, funded primarily by the US Agency for International Development (USAID).

Challenges of collecting population data

There have always been barriers to enumerating populations, related to practical, political, and ethical issues. Today, data collection is facing mounting challenges, including rising costs, increasing concerns over privacy, reduced respondent cooperation, more complex household structures and racial/gender categorizations, and rising sensitivity to controversial topics. Many countries have responded to these challenges by adapting questionnaire design and data collection methods or introducing new statistical techniques to prevent disclosing personal information (Skinner 2018).

Although we might think of demographic statistics as apolitical, strong ties between census results and political representation and apportionment of resources mean that censuses often become highly politicized. In Nigeria, for example, contested census data were held partly responsible for increasing political and ethnic rivalries between the primarily Muslim north of the country and the predominantly Christian south, culminating in the civil war of 1967–70 (Okolo 1999). More recently, the 2014 Census in Myanmar was contested owing to concerns over questions related to ethnicity (box 2.5).

The Soviet Union provides an example of a situation where the government itself rejected census results. The 1937 census had recorded a population of just 162 million, 8 to 10 million fewer than expected. Elimination of many rich farmers and their families after the Bolshevik Revolution of 1917, as well as forced collectivization and subsequent famine in the early 1930s, had contributed to a slowing of

BOX 2.5 ETHNIC IDENTITY AND THE 2014 MYANMAR CENSUS

Myanmar is one of the most culturally diverse countries in Asia, with at least eight major ethnic groups and numerous sub-groups. The Burmese (Bamar) people comprise the dominant group, accounting for about 68 percent of the population. Other major groups include the Arakanese, Chin, Kachin, Karen, Kayah, Mon, and Shan people. Buddhists comprise about 88 percent of the country's population, but there are also sizeable Christian and Muslim communities. The post-colonial government promoted a policy of ethnic assimilation (one race, one language, one religion) in favor of the Bamar group and the Buddhist religion. The military junta of 1962–2011 continued these policies, often with repressive methods. Consequently, the country's post-colonial history has been characterized by numerous ethnic tensions, frequently resulting in armed conflict between government forces and rebel groups such as the Kachin Independence Army and Shan State Army. The end of military rule in 2011 brought some hope for a better future as the new government began introducing democratic reforms and reestablishing ties with other countries, but that hope was crushed by a coup in February 2021.

The Ministry of Immigration and Population, in cooperation with the United Nations Population Fund, conducted a population census in 2014. Over $75 million (mainly from international grants) was spent on preparing and implementing this large data gathering operation, the first of its kind in 30 years (International Crisis Group 2014). The entire process faced numerous challenges and criticisms from the beginning, including a lack of census expertise, low awareness about the census and its importance among people who had never been enumerated before (about half of the population), difficulties in enumerating people in areas of active conflict, the timing of the census close to the 2015 elections, a lack of consultation with the public during the preparation phase, and the complexity of the census questionnaire.

The major problem, however, was the classification and collection of data on citizenship, ethnicity, and religion. Questions on ethnic identity were particularly controversial (Transnational Institute 2014). The ethnic categorization of Myanmar's population into eight major groups and 135 minor groups in the census was considered outdated because it reflected the colonial system of ethnic classification and did not take into account the lived experiences of many individuals and groups. Critics also claimed that it was designed to diminish the significance of non-Burmese identities and deprive ethnic minorities of adequate political status and representation. Some major ethnic groups were divided into more sub-groups than expected (e.g., the Kachin); some sub-groups would have liked to have the status of major groups (e.g., the Kayan and Palaung); some culturally diverse sub-groups were put in the same ethnic category; whereas other groups—most notably the Rohingya—were denied official recognition at all and were simply recorded in an "other" category. Additionally, only one ethnic category could be reported for each individual on the census form, even though many of Myanmar's inhabitants claim mixed ethnic heritage.

There were particular challenges to census operations in Rakhine State, home to the predominantly Muslim Rohingya and the Buddhist Rakhine people. Centuries-old tense relations between the groups erupted into violent clashes shortly after the end of military rule in 2011. A 1982 nationality law had deprived the Rohingyas of Myanmar citizenship, defining them instead as Bangladeshi Muslims. A proposal to provide self-identification for the Rohingya on the 2014 Census form met with strong resistance from the Rakhine people, and the government made a last-minute decision to remove recognition for the Rohingya after Rakhine protestors threatened to boycott the census. This and other factors aggravated already precarious relations between the groups and led to further conflict, resulting in numerous deaths, the burning of Rohingya villages, and a large-scale Rohingya exodus to Bangladesh.

Most Myanmar census data were released in 2015 and data on religion in 2016. However, by 2021, the sensitive data on ethnicity had not yet been released because, according to government sources, "further negotiations and consultations with ethnic community leaders and representatives, historians, anthropologists and cultural experts were needed to finalize the terminology and classifications of the ethnic groups" (Aung 2018).

population growth—a fact that did not support the Soviet success story the government wanted to portray. The government was also unhappy that less than half of adults identified themselves as atheists in a question on religion. This was much lower than the Soviets had hoped, given considerable efforts to encourage people to be loyal to the state over religious belief systems. Several top census officials were accused of incorrect enumeration procedures and falsifying the results, leading to arrests, imprisonments, and even executions (Thorvaldsen 2019). The 1937 Census results were not published until the breakup of the Soviet Union 5 decades later.

Beyond the challenge of collecting accurate data, there are also issues with respect to how data are used. At one extreme, population data collected by governments have been used to harm or kill people in a list of horrific incidents where governments have harassed, imprisoned, forcefully relocated, or eliminated thousands, sometimes millions, of people because of physical or cultural differences (Seltzer and Anderson 2001). The Nazi Holocaust provides a particularly tragic example: population census and existing or specially designed population register data helped the Nazi government to identify and eliminate members of vulnerable populations, contributing ultimately to the deaths of 6 million Jews and 250,000 Roma people, among others, in concentration camps during World War II. Similarly, the 1994 genocide in Rwanda was facilitated by the population registration system established by the Belgian colonial authorities in the 1930s and used after independence in 1962. This system classified most residents as members of one of two ethnic groups (Hutu or Tutsi), shown on identity cards and in vital statistics records. At the outbreak of hostilities, this information was used to target people based on their ethnicity, resulting in the genocide of some 800,000 people, most of them Tutsi, in just 100 days.

Forced relocations and imprisonment of people have also been triggered by misuse of population data. For example, the US government conducted a series of special censuses to enumerate Native American groups in the 19th century—information that could then be used to forcefully relocate communities (see chapter 11). Data from the 1940 US Census were also used to track and detain Japanese Americans during World War II. The Census Bureau assisted the federal authorities by, among other things, providing census block–level maps showing the number and location of Japanese Americans.

Several safety measures have been recommended to avoid such problems in the future. These include: not collecting particularly sensitive information (e.g., religion, ethnicity), collecting information from a sample group rather than the entire population, introducing some "noise" (random errors) into the data, decentralizing data storage or even storing some data in another country, having strict legal restrictions on the use of data, and putting more emphasis on ethical safeguards and privacy issues. Although some of those recommendations weaken the usefulness of the data for research, this is perhaps a price worth paying given the previous examples.

PART II: VISUALIZATION OF POPULATION DATA

We often say that "a picture is worth a thousand words." This is particularly true in population geography, where graphs and maps can help us identify patterns from the large amount of statistical data in use. Here we describe the use of graphs and maps, their strengths and weaknesses, as well as some common mistakes in making and interpreting them.

Types of population data

Population data can be presented as **raw data**—for instance, a simple count of individuals—or **derived data**, such as rates, ratios, and proportions, where further information is provided to contextualize the data. The number of people in a country or cases of measles are both examples of counts. The advantage of raw data is that it can give an idea of the magnitude of a particular phenomenon—knowing the population of a refugee camp is critical to providing services to that camp, for example. The major problem is that the significance of counts can be hard to estimate unless we know the size of the population they are drawn from—300 cases of measles from the huge population of India is a very different situation from 300 cases occurring in a small town, for instance. Here, the use of a rate or proportion might be preferable.

A **proportion** is the relation of a population sub-group to the entire population of interest, often quoted as a percentage. For instance, we could state that 20 percent of pregnancies are unintended in a particular country. This allows us to understand the frequency of an event in relation to population size. **Rates** show the frequency of a demographic event in a population in relation to the total population subject to that event during a particular time period (usually 1 year). For example, at the end of April 2021, Mongolia's COVID-19 rate of 891 cases per 100,000 was far higher than that of Australia with 116 cases per 100,000 (WHO, n.d.). Even though both countries had approximately the same cumulative number of cases at this point, Australia's *rate* was much lower because these cases were spread across Australia's much larger population.

Ratios are used to indicate the relationship between one sub-group and another in a particular population. Sex ratios (e.g., 105 males: 100 females) and dependency ratios (the dependent population in relation to the economically active population) are frequently used measures of population composition.

Graphical display of demographic data

Population data can be illustrated with a wide variety of different types of graphs. Here we discuss some of the most common ways to display population data, as well as the ways in which graphs can be misinterpreted. Some types of graphs are suitable for showing trends through time; others can show relationships between demographic, economic, and other variables; and still others are useful for showing population composition.

Bar graphs are useful for showing both raw and derived data. Their strength is in allowing a comparison to be made between multiple data points, usually over space, time, or both. Figure 2.5 provides examples of these possibilities using selected population data for Australia. Graph A shows population totals across Australia's eight states/territories, graph B shows population change over time in one state (New South Wales), and graph C illustrates population change over time in each of the eight states/territories.

Stacked bar graphs are similar but display additional information by segmenting each bar. Figure 2.6, for instance, shows the changing composition of Canada's foreign-born population, with each stacked bar divided into segments by immigrants' continent of origin. Here the difference between raw and derived data is significant to how the data are displayed. If we are interested in using counts, in this case the number of immigrants from different regions, the magnitude of the immigrant population is illustrated by the height of each bar (figure 2.6A). If we are more interested in the relative proportions of immigrants coming from different regions, however, we can make the height of each stacked bar the same to emphasize how the *proportion* of immigrants coming from each region has changed over

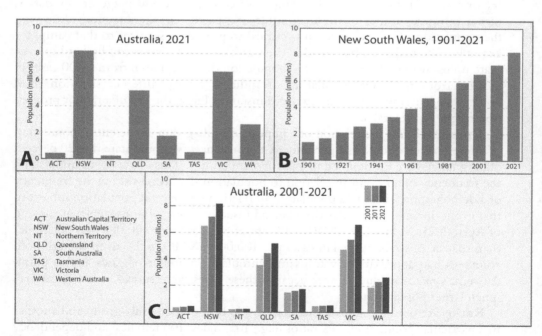

FIGURE 2.5 Bar graphs showing Australia's population across space and time
Data source: Australian Bureau of Statistics (n.d.).

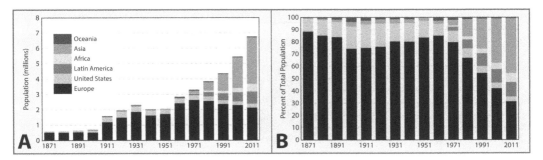

FIGURE 2.6 Stacked bar graphs showing the changing composition of Canada's foreign-born population, as population totals (A) and proportions of the immigrant population (B)
Data source: Statistics Canada (2016).

time (figure 2.6B). The clear message from graph A is that the number of foreign-born people in the Canadian population has increased over this time period. By contrast, graph B better emphasizes changes in the proportion of the immigrant population coming from different regions, particularly the declining importance of European migrants and growing significance of Asian immigrants in recent decades.

Line graphs are well suited for showing changes in demographic variables through time, such as fluctuations in birth and death rates. Because lines occupy less space than bars, they are useful for illustrating long-term trends for one or more variables (figure 2.7A) or trends for several categories of a particular variable (figure 2.7B). When reading a line graph, it is important to pay attention to whether the vertical scale begins at zero, as well as the range of values represented on the graph. Relatively small differences between data points can be greatly exaggerated if a line graph is scaled to show only a small range of values. This can be used to deliberately mislead people by over-emphasizing minor differences such as in crime rates between two groups. Similarly, climate change skeptics sometimes compress the temperature axis of climate graphs to obscure recent increases in average temperatures or extend the range of time shown on the x-axis so that current changes get lost amid longer-term fluctuations.

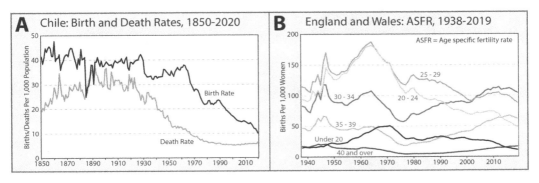

FIGURE 2.7 Line graphs showing birth rate and death rate trends in Chile (A) and age-specific fertility rates for different age groups in England and Wales (B)
Data source: Mitchell and Palgrave Macmillan (2013).

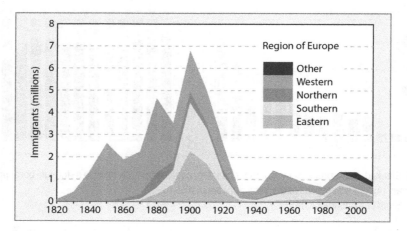

FIGURE 2.8 Area graph showing the changing origin of European immigrants to the United States over time as population totals
Data source: United States Department of Homeland Security (2021).

An **area graph** employs the same principle as a line graph but represents the information as an area rather than just a line. Areas representing different categories are stacked on top of each other. Although similar in principle to stacked bar graphs, area graphs can depict the data for more points in time, showing trends in greater detail. Area graphs are often used to show changes in demographic phenomena over time and can show population counts or proportions (figure 2.8).

Pie charts are used to show the proportion of a whole represented by different sub-categories. Figure 2.9, for instance, illustrates the religious affiliation of people in England and Wales in 2011. At a glance, it is clear that most people identify as Christian. The smaller "zoomed-in" pie chart to the right shows the affiliation of those listed as "other" in the main pie chart. Pie charts could be used to show the ethnic composition of a country, the proportion of a population who voted for various election candidates, or the percentage of women using different kinds of birth

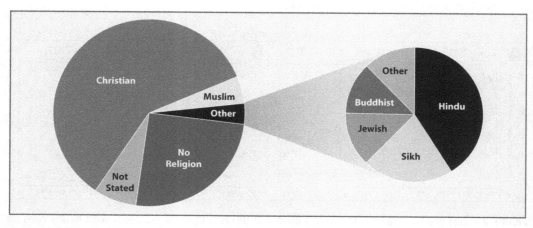

FIGURE 2.9 Pie chart showing religious affiliation of England and Wales, 2011
Data source: United Kingdom Office for National Statistics (2021).

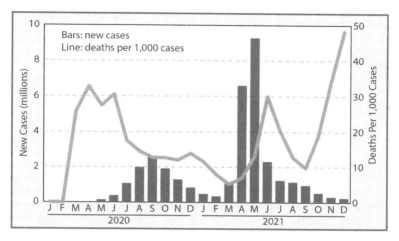

FIGURE 2.10 Bar and line graph showing COVID-19 cases and COVID-related mortality in India, 2020–21
Data source: WHO (n.d.).

control, in each case assuming that the whole pie represents the entire population of interest. Pie charts are not ideal for showing change over time, because this would require a series of charts; here a stacked bar chart might be preferable.

In some cases, two or more graph types can be combined to provide additional information. In figure 2.10, the bars show the number of COVID-19 cases in India over time and the line shows the COVID-related mortality rate. In cases like these it is important to remember that each graph may use a different scale type and range. By convention, one scale is shown on the left-hand axis of the graph and the other on the right. You may be familiar with this principle from climate graphs, where one axis shows the scale for the temperature and the other the scale for precipitation. Combining two graphs can allow us to examine how two phenomena may be related. Indeed, the very construction of such a graph often implies that the two phenomena illustrated are related in some way, and the use of two graph types together is often helpful for generating hypotheses about possible connections. It is very important to bear in mind that just because two events occur together in time and space, this does not mean that one phenomenon has *caused* the other, however.

A more rigorous way to explore relationships between variables is with **scatterplots** (also known as scattergraphs, scattergrams, or scattercharts). In this case, points are plotted on a graph with one variable on the x-axis and another variable on the y-axis, with the deliberate intention of trying to reveal patterns in the data that would indicate that the two data sets are linked in some way. For example, we could use a scatterplot to analyze whether infant mortality rates show a meaningful relationship with affluence at the national scale. Any relationship between the data sets will appear as a spatial pattern on the graph. A **positive correlation** between the data sets will appear as a diagonal line in the data points from bottom left to top right, indicating that as one variable rises, so does the other. A **negative correlation** will show a line from top left to bottom right. The strength of the relationship is indicated by how perfectly the data points form this line. Figure 2.11A shows a weak negative correlation between level of urbanization and population growth rates for countries of the world, indicating that higher rates of natural increase are loosely

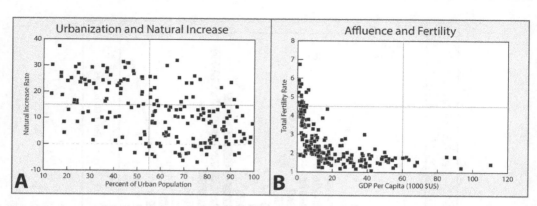

FIGURE 2.11 Scatterplots showing the relationship between two sets of variables for selected countries of the world: a weak relationship between urbanization and natural increase rate (A) and a stronger relationship between affluence and fertility rate (B)
Data source: UNDESA (2019).

correlated with lower levels of urbanization. In this case, we can see that the north-west and southeast quadrants of the graph have more points in them than the south-west and northeast quadrants, indicating that there is a pattern here, albeit a weak one. Figure 2.11B shows a stronger negative correlation, in this case between afflu-ence and fertility rates at the national level. This is illustrated by the tighter clustering of the data points along an imaginary line through the data. In this case, the trend line is curved, providing further information about the relationship between these two data sets—as affluence increases, fertility drops rapidly at lower income levels, but the relationship is much weaker at higher levels of affluence. It is important to note again that we cannot say from this graph whether one phenomenon is causing the other, just that there is some connection between the two data sets.

Bubble graphs are a form of scatterplot that use bubbles instead of points. Here the bubbles provide additional information so that three or four variables can be plotted rather than just two. The two key variables are still plotted on the x and y axes (e.g., national affluence and life expectancy), but a third and fourth variable can be shown by the size and shading/color of the bubbles. This sophisticated form of data representation has been popularized by the Gapminder website (www.gapminder.org), which was designed specifically to make population and health data more accessible to the general public (Gapminder 2021). With the Gapminder soft-ware, the bubbles are used to represent countries and the size of each bubble shows the population size of each country. The color of the bubbles can be changed to divide countries into different world regions, defined by continent or major reli-gion, for instance. The bubble graph in figure 2.12 shows the relationship between life expectancy and national affluence by country. In this case, the size of the bub-bles represents each country's population size, whereas the shading of the bubbles shows infant mortality rates. The clear line formed by the data points indicates a strong positive correlation between life expectancy and national affluence, with infant mortality rate also positively correlated with both variables, as indicated by the fact that the shading also forms a clear pattern—from dark at the bottom left of the graph to light at the top right. Population size is also illustrated on the graph to provide additional information but is not correlated with the other three variables.

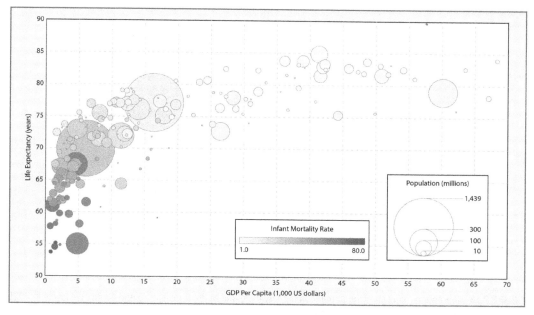

FIGURE 2.12 Bubble graph illustrating the relationship between national affluence, life expectancy, and infant mortality rate
Data source: World Bank (n.d.).

There are many other types of graphs suitable for visualizing demographic data, but space limitations preclude us from discussing them all here. One further specialized graph of note in population studies is the population pyramid, which we discuss in detail in the next chapter.

Cartographic display of demographic data

Although graphs are valuable in population geography, the spatial nature of much of the data being presented lends itself to presentation in maps. Just as with graphs, there are many different types of maps, and we can present only a small selection here. Simple location or **reference maps** are useful to population geographers; for example, to find the location of refugee camps or clinics. Here we focus on **thematic maps**, however, which illustrate a specific data set on a map. Thematic maps use the spatial layout of a map to illustrate spatial patterns in data sets such as infant mortality rates, numbers of migrants, or the spread of a disease over time. Both raw and derived data can be shown on thematic maps.

Choropleth maps are probably the most frequently used maps in population geography. The data are grouped into several data range classes, and a certain shade of gray or color is assigned to each class range. To make the map easy to read, typically no more than about five classes are used, and the shading system should ensure that the lightest shade represents the lowest data category and the darkest the highest. This makes patterns clear to the reader because the places with the highest densities of the phenomenon being shown would be the darkest on the map.

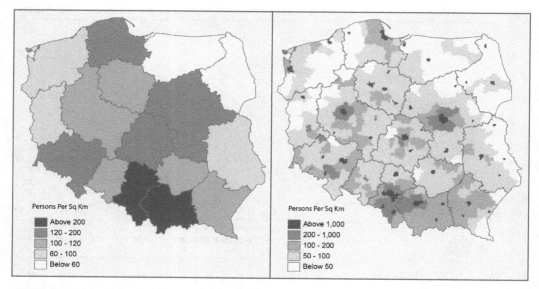

FIGURE 2.13 Choropleth maps showing population density in Poland at the province (A) and county (B) levels in 2020
Data source: Poland Główny Urząd Statystyczny (2021).

Figure 2.13 shows population density in Poland as choropleth maps. Note how the scale at which data are displayed affects the appearance of the maps. Both maps have the same number of classes (although with different value ranges) and the same shading range but use different levels of spatial resolution (provinces vs counties).

It is also important to bear in mind that how data are assigned to classes can lead to very different maps. Typically, data are divided into classes by putting an equal number of cases into each class (quantiles) or by dividing the range of the data into equal parts and making each a class (equal intervals). Unfortunately, this may hide outliers that are highly significant or split cases with very similar characteristics into two different groups. For example, a country where 75.1 percent of women use modern contraception might end up in a different class from a country where 74.9 percent of women do if 75.0 percent becomes the boundary, whereas two countries with rates of 74.9 and 50.5 could end up in the same class. To avoid this problem, mapmakers will sometimes look for "natural breaks" in the data, where the data seem to fall into intuitive categories. A mapmaker can also determine the class intervals that best suit stated research objectives. Figure 2.14 illustrates how different a choropleth map can look using these different techniques for defining classes, despite showing exactly the same data.

A second major type of thematic map is a **dot density map**. In a dot density map, each dot represents a certain number of occurrences of the phenomenon in question such as 10,000 people, 500 students, or five health centers (figure 2.15). The distribution of dots across the map is designed to give the reader an impression of the intensity of the phenomenon across space, providing an intuitive way to illustrate things like population density or the distribution of disease cases. It is important to note that dot density maps can be made in two ways. If the exact location of each case can be identified relatively accurately on the map (often assisted by modern technologies such as global positioning systems, **GPS**), the map

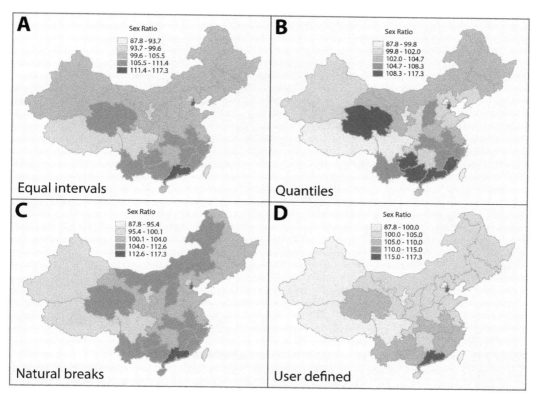

FIGURE 2.14 Choropleth maps showing sex ratios in China by province (2018), illustrating four different systems for creating class boundaries
Data source: China National Bureau of Statistics (n.d.).

will give a very accurate impression of the density of the phenomenon across space. In many instances, this is simply not possible, however, because we often only know the number of cases that occurred in a particular administrative district rather than their exact location. For example, we may know the number of cases of measles in each province of Pakistan during a particular outbreak but have no information about which villages had cases. In such circumstances, dots are placed at random across the administrative districts *at the scale for which data are available*—in this example, across each province. The resulting dot density map is still useful but has limitations—although we know the relative density of cases in one province compared with the next, we cannot use the map to say anything about where cases occurred *within* each province. For this reason, the smallest territorial units for which data are available will usually be used to generate dot density maps, unless this causes privacy concerns.

Proportional symbol maps use symbols (usually circles or squares but sometimes also pictorial images) to show the distribution of a feature. The size of each symbol is proportional to the number of people (or other data) in the respective territorial unit, and the symbol is placed in the geometric center of that unit (figure 2.16). For example, we could use proportional symbols to show the number of refugees accepted by different countries. Proportional symbol maps are intuitive for most readers because a larger symbol simply means "more." The use of pictures as

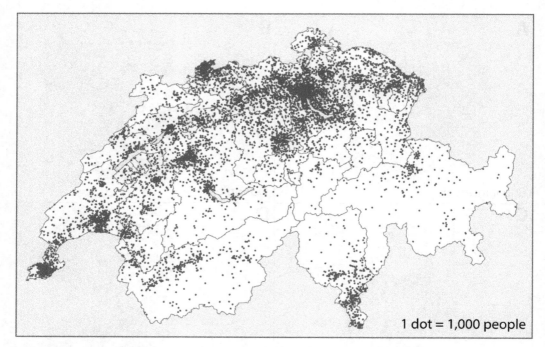

1 dot = 1,000 people

FIGURE 2.15 Dot density map showing population distribution in Switzerland in 2020
Data are at the commune level (one of the finest territorial divisions in Switzerland), with dots distributed randomly across each commune. Commune boundaries are not illustrated on the map because this would impair readability.
Data source: Switzerland Federal Statistical Office (n.d.).

symbols can introduce unnecessary bias into a map, however, if the pictures are chosen inappropriately. For instance, to protest against the supposed health impacts of nuclear power stations, a lobbying group could produce a map that identifies nuclear power stations on a map using a skull and crossbones symbol, prejudicing the audience against them.

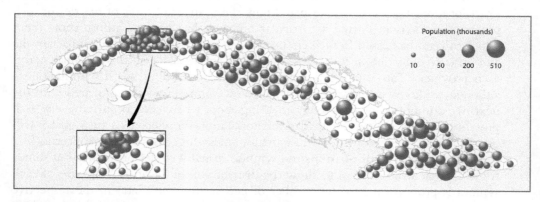

FIGURE 2.16 Proportional symbol map showing population distribution in Cuba by municipality in 2020
Note how proportional symbols can quickly become crowded, obscuring the data. In this case, a "zoomed out" portion of the map has been used to provide further detail for the densely settled part of Cuba around the capital, Havana.
Data source: WorldPop (2022).

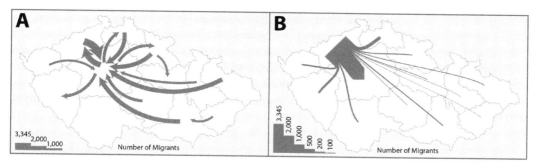

FIGURE 2.17 Flow maps showing the largest inter-provincial migration streams in the Czech Republic (A) and migration from one province to the rest of the country (B) in 2020
Data source: Czech Statistical Office (2022).

Movement can best be shown on **flow maps**, which are frequently used to show migrations as well as other types of movement (e.g., telephone calls, trade; figure 2.17). The direction of movement is shown with arrows connecting origins and destinations (usually the geometric centers of territorial units) and the volume of each flow is illustrated by the thickness of its arrow. The major drawback of flow maps is that multiple flow lines quickly make a map difficult to read, limiting the amount of information that can be displayed in this way.

It is possible to show two or more variables on a map by combining different types of maps (such as choropleth and proportional symbol) and/or assigning colors (or shades of gray) to proportional symbols (figure 2.18). Though such maps show more information about an examined phenomenon and can help the reader to identify associations between two data sets, they can appear cluttered and be harder to read than simpler maps, particularly if symbols obscure the shading underneath.

The idea of overlaying different layers of information is the basis for **GIS** (geographic information systems), which use computer technology to overlay multiple layers of data tied to particular locations. Once the layers have been incorporated

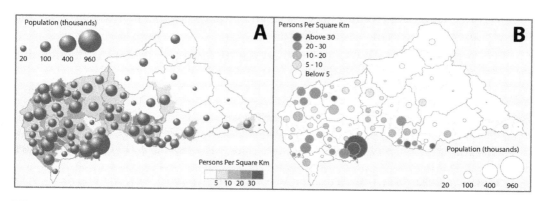

FIGURE 2.18 Population distribution in the Central African Republic in 2015
The two maps show two different ways of showing multiple demographic variables on maps. Map A is a combination of choropleth and proportional symbol techniques. Map B uses only proportional symbols to show both variables. One variable is represented by the size of the symbols and the other by shades of gray assigned to the symbols.
Data source: Humanitarian Data Exchange (n.d.).

into the GIS, the user can then query the database about potential correlations. For instance, in a database that maps population density, the location of health clinics, roads, and bus routes, the user could ask the database to calculate how many people live within a certain distance of each clinic and then go on to calculate approximate journey times for different communities to estimate the physical accessibility of that clinic. GIS is now widely used in population studies to answer spatial questions related to population data.

Conclusion

Population data play a critical role in the organization and administration of modern society. Understanding the strengths and limitations of these data is critical to the study of population geography. Similarly, it is important that we be aware of some of the challenges and limitations of visualizing these data so that we can interpret information correctly. In the chapters that follow, you will see many examples of the different styles of maps and charts that we present here. As you consider them, bear in mind not only what they are depicting but also some of the pros and cons of the different styles of data presentation used.

Discussion questions

1   If you were designing a questionnaire for the next census in your home country and were limited to ten questions, what questions would you include? What questions, if any, do you consider too sensitive to include? Why?

2   Unlike most European countries, the United States does not have a population register. Many Americans see it as a threat to privacy and individual freedoms. Do you agree or disagree with such concerns? Why or why not?

3   Imagine that you are asked to give a presentation on recent demographic trends in your home country. What types of graphs would you use to show changes in population composition (age, race, and ethnicity), fertility and mortality, and immigration at the national level?

4   Which type of map would you choose to show population distribution and change? Justify your choice.

Suggested readings

Coleman, D. 2013. "The Twilight of the Census." *Population and Development Review* 38: 334–51.

Hassan, A. 2022. "The 7 Best Thematic Map Types for Geospatial Data." Built In, January 25, 2022. https://builtin.com/data-science/types-of-thematic-maps

Nagle, N. 2020. "Census 2020 Will Protect Your Privacy More than Ever—But at the Price of Accuracy." The Conversation, April 6, 2020. http://theconversation.com/census-2020-will-protect-your-privacy-more-than-ever-but-at-the-price-of-accuracy-130116

Seltzer, W., and M. Anderson. 2001. "The Dark Side of Numbers: The Role of Population Data Systems in Human Rights Abuses." *Social Research* 68 (2): 481–513.

Glossary

area graph: a graph similar to a line graph where space between lines is filled with colors or shades of gray and stacked if two or more categories are shown

bar graph: a graph that shows categorical data as bars whose height is proportional to the data values

bubble graph: a scatterplot that shows the relationship between two main variables and illustrates additional variables by the color/shading and size of the bubbles

census: a collection of demographic, economic, social, and other information from the entire population in a geographically defined territory at a specific time

choropleth map: a thematic map in which territorial entities are colored/shaded in relation to data values grouped into several class intervals

content error: erroneous information about individuals resulting from accidental or deliberate errors in providing and/or recording the information

coverage error: incorrect counting of a population resulting from missing some people from the census count (undercounting) or counting them more than once (overcounting)

derived data: numerical information obtained from processing raw data (e.g., a birth rate can be calculated from a count of births)

dot density map: a thematic map that shows the distribution of a phenomenon across space, where each dot represents a certain numeric value and reflects the approximate or precise location of the examined phenomenon

ethnicity: belonging to a group of people who share a common origin and culture

flow map: a thematic map that shows the direction of movement of people or commodities by lines of various thickness proportional to the data values

GIS (geographic information system):   a computer-based system for collecting, processing, and displaying various types of spatial data

GPS (global positioning system):   a satellite-based system for determining the precise location of an object on the Earth's surface

line graph: a graph that uses a line(s) to show the data values (vertical axis), often over time (horizontal axis)

natural increase: excess of births over deaths, leading to population increase

negative correlation: relationship between two variables where an increase in the value of one variable is associated with a decrease in the value of the other variable

pie chart: graph showing two or more categories of a phenomenon by pie slices proportional to the relative value of each category

population register: a system of continuous collection of selected demographic information on every individual in a country for legal and statistical purposes

positive correlation: relationship between two variables where an increase in the value of one variable is associated with an increase in the value of the other variable

proportion: a relation of two values where the smaller value is typically quoted as a percentage of the other value

proportional symbol map: a thematic map that uses symbols whose size is proportional to the data values of respective places

race: a group of people considered distinct from other groups, as identified by physical characteristics such as skin color

rate: the frequency of an occurrence of a certain phenomenon in a population (e.g., births per 1,000 population) during a specific time period

ratio: the quantitative relationship between two sub-groups of a population (e.g., 105 males per 100 females)

raw data: data that have not been processed in any way; for example, counts

reference map: map designed to illustrate the location of physical (e.g., rivers) and human (e.g., cities, political boundaries) features

rolling census: continuous and cumulative enumeration of the entire population

sample survey: collection of data from a population sub-group in order to estimate characteristics of the entire population

scatterplot: a graph that displays two numerical data sets with the explicit goal of exploring the relationship between them; each axis of the graph represents a different variable and the position of the points plotted will display a pattern (e.g., a line or group(s) of points) if there is a relationship between the variables

stacked bar graph: a bar graph that shows categorical data as bars whose height is proportional to the data values

thematic map: map that shows the distribution of selected phenomena by symbols of various size, color, or shading related to the data values

vital registration system: a system of recording information about vital events (e.g., births, deaths) in a population on a continuous basis

Works cited

Aung, S. 2018. "Still No Date for Release of Census Findings on Ethnic Populations." *The Irrawaddy*, February 21, 2018. https://www.irrawaddy.com/news/burma/still-no-date-release-census-findings-ethnic-populations.html

Australian Bureau of Statistics. n.d. "Historical Population." Accessed February 24, 2022. https://www.abs.gov.au/statistics/people/population/historical-population/latest-release

Barshad, A. 2019. "In Lebanon, a Census Is Too Dangerous to Implement." *The Nation*, October 17, 2019. https://www.thenation.com/article/archive/lebanon-census/

China National Bureau of Statistics. n.d. *China Statistical Yearbook 2019.* Accessed February 24, 2022. http://www.stats.gov.cn/tjsj/ndsj/2019/indexeh.htm

Coleman, D. 2013. "The Twilight of the Census." *Population and Development Review* 38: 334–51.

Czech Statistical Office. 2022. *Demographic Yearbook of the Czech Republic.* https://www.czso.cz/csu/czso/demographic-yearbook-of-the-czech-republic-2020

The Daily Star. 2021. "Hijras Will Be Included as Separate Gender in National Census 2021: BBS DG." *The Daily Star*, September 17, 2020. https://www.thedailystar.net/country/news/hijras-will-be-included-separate-gender-national-census-2021-bbs-dg-1962849

Gapminder. 2021. "Understand a Changing World: Bubble Chart." https://www.gapminder.org/

Goodkind, D. 2004. "China's Missing Children: The 2000 Census Underreporting Surprise." *Population Studies* 58 (3): 281–95.

Gurubacharya, B. 2020. "Nepal Census Will Add 3rd Gender, Recognizing LGBT Minority." *AP News*, February 6, 2020. https://apnews.com/article/daf0cb11cc4590147c78b8f29e9e70e9

Holden, C. 2009. "America's Uncounted Millions." *Science* 324 (5930): 1008–9. https://doi.org/10.1126/science.324_1008

Humanitarian Data Exchange. n.d. "Central African Republic: Subnational Population Statistics." Accessed February 22, 2022. https://data.humdata.org/dataset/cod-ps-caf

International Crisis Group. 2014. "Counting the Costs: Myanmar's Problematic Census." International Crisis Group, May 15, 2014. https://www.crisisgroup.org/asia/south-east-asia/myanmar/counting-costs-myanmar-s-problematic-census

Jacobsen, L., M. Mather, and L. Yorke. 2020. "Why Are So Many Young Children Undercounted in the U.S. Census?" Population Reference Bureau, March 23, 2020. https://www.prb.org/why-are-so-many-young-children-undercounted-in-the-u-s-census/

Levitt, J. 2019. "Citizenship and the Census." *Columbia Law Review* 119 (5): 1355–98. https://doi.org/10.2307/26650741

Lundquist, J., D. Anderton, and D. Yaukey. 2015. *Demography: The Study of Human Population*, 4th ed. Long Grove, IL: Waveland Press.

Mitchell, B., and Palgrave Macmillan. 2013. *International Historical Statistics: Americas 1750–2010*. London: CLOSER. https://www.closer.ac.uk/data/age-specific-fertility-rates/

Okolo, A. 1999. "The Nigerian Census: Problems and Prospects." *The American Statistician* 53 (4): 321–5. doi:10.1080/00031305.1999.10474483

Poland Główny Urząd Statystyczny. 2021. *Demographic Yearbook of Poland 2021*. https://stat.gov.pl/en/topics/statistical-yearbooks/statistical-yearbooks/demographic-yearbook-of-poland-2021,3,15.html

Roux, V. 2020. "The French Rolling Census: A Census That Allows a Progressive Modernization." *Statistical Journal of the IAOS* 36 (1): 125–34. https://doi.org/10.3233/SJI-190572.

Seltzer, W., and M. Anderson. 2001. "The Dark Side of Numbers: The Role of Population Data Systems in Human Rights Abuses." *Social Research* 68 (2): 481–513.

Skinner, C. 2018. "Issues and Challenges in Census Taking." *Annual Review of Statistics and Its Applications* 5: 49–63. https://doi.org/10.1146/annurev-statistics-041715-033713

Statistics Canada. 2016. "150 Years of Immigration in Canada." https://www150.statcan.gc.ca/n1/pub/11-630-x/11-630-x2016006-eng.htm

———. 2020. "Updated Content for the 2021 Census of Population: Family and Demographic Concepts, and Activities of Daily Living." Statistics Canada, July 17, 2020. https://www12.statcan.gc.ca/census-recensement/2021/ref/98-20-0001/982000012020001-eng.cfm

Switzerland Federal Statistical Office. n.d. "Population." Accessed February 24, 2022. https://www.bfs.admin.ch/bfs/en/home/statistics/population.html

Thorvaldsen, G. 2019. *Censuses and Census Takers: A Global History*. London: Routledge.

Transnational Institute. 2014. "Ethnicity without Meaning, Data without Context. The 2014 Census, Identity and Citizenship in Burma/Myanmar." *Burma Policy Briefing* 13: 1–24.

Truu, M. 2021. "Australian Census to Include a Gender Non-binary Option for the First Time." *SBS News*, January 15, 2021. https://www.sbs.com.au/news/australian-census-to-include-a-gender-non-binary-option-for-the-first-time

UNDESA [United Nations Department of Economic and Social Affairs]. 2017. *Principles and Recommendations for Population and Housing Censuses*, Revision 3. New York: United Nations. https://unstats.un.org/unsd/demographic-social/Standards-and-Methods/files/Principles_and_Recommendations/Population-and-Housing-Censuses/Series_M67rev3-E.pdf

———. 2019. "World Population Prospects 2019." https://population.un.org/wpp/

UNECE [United Nations Economic Commission for Europe]. 2019. "Measurement of Gender Identity." UNECE, June 24, 2019. https://unece.org/statistics/ces/measurement-gender-identity

UNESCO [United Nations Educational, Scientific and Cultural Organization]. 2013. "The 1703 Census of Iceland." UNESCO. https://en.unesco.org/memoryoftheworld/registry/200

UNICEF [United Nations Children's Fund]. 2022. "A Statistical Update on Birth Registration in Africa." UNICEF, Africa-Birth-Registration-Brochure-Oct-2022_Final-LR.pdf

UNSD [United Nations Statistics Division]. 2019. *Handbook on Civil Registration, Vital Statistics and Identity Management Systems: Communication for Development*. UN, September 2019. https://unstats.un.org/unsd/demographic-social/Standards-and-Methods/files/Handbooks/crvs/CRVS-IdM-E.pdf

———. 2020. *Handbook on Population and Housing Census Editing*, Revision 2. New York: United Nations. https://unstats.un.org/unsd/publication/SeriesF/seriesf_82rev2e.pdf

———. 2021. "World Population and Housing Census Programme: Census Dates." https://unstats.un.org/unsd/demographic-social/census/censusdates

United Kingdom Office for National Statistics. 2021. "CT1199_2011 Census." https://www.ons.gov.uk/peoplepopulationandcommunity/culturalidentity/religion/adhocs/13942ct11992011census

United States Census Bureau. 2021. "Decennial Census of Population and Housing by Decades." US Census Bureau, November 23, 2021. https://www.census.gov/programs-surveys/decennial-census/decade.html

United States Department of Homeland Security. 2021. "Yearbook of Immigration Statistics 2020." https://www.dhs.gov/immigration-statistics/yearbook

United States Department of the Interior. 1890. "Eleventh Census of the United States: Instructions to Enumerators." US Census Office, June 1, 1890. https://www.census.gov/history/pdf/1890instructions.pdf

Valente, P. 2010. "Census Taking in Europe: How Are Populations Counted in 2010?" *Population and Societies* 467: 1–4.

Wallace, G. 2019. "Why Does the Trump Administration Want a Citizenship Question on the Census?" CNN Politics, July 6, 2019. https://www.cnn.com/2019/07/06/politics/census-citizenship-question-donald-trump-administration/index.html

WHO [World Health Organization]. n.d. "WHO Coronavirus (Covid-19) Dashboard." Accessed February 24, 2022. https://covid19.who.int/table

World Bank. n.d. "World Bank Open Data." Accessed June 2, 2021. https://data.worldbank.org

WorldPop. 2022. "Mapping Populations." https://www.worldpop.org/methods/populations

3

Population distribution and composition

After reading this chapter, a student should be able to:

1 identify major environmental and social factors influencing population distribution and changes to that distribution;

2 discuss the significance of major aspects of population structure, including race and ethnicity, sex and gender, and age structure;

3 explain why social equity issues frequently arise in connection to population composition.

Population distribution (where people are found across the Earth's surface) and composition (the characteristics of particular populations) are of great importance to addressing geographic questions of *where* and *why*. Issues of population distribution and composition are also of great significance to politicians and administrators entrusted with decision-making tasks, including providing basic services such as healthcare, social support services, and education; determining political representation; and building infrastructure and facilities such as schools, hospitals, and roads. Understanding population distribution and composition can also allow us to identify problems that need to be addressed, such as regional imbalances between population size and resource availability or systematic disadvantages for certain sub-groups of the population.

PART I: POPULATION DISTRIBUTION

Population is unevenly distributed at multiple scales. At the global scale, about half of the population is concentrated on just 3 percent of land (excluding Antarctica), and 90 percent of the population occupies 18 percent of land area. The least densely populated regions, comprising half of global land area, support only 0.5 percent of global population. There is more land north of the Equator and we might therefore expect more people there, but the Northern Hemisphere is more than twice as densely populated as the Southern, with 68 persons per square kilometer (176 per square mile),

DOI: 10.4324/9781003143253-3

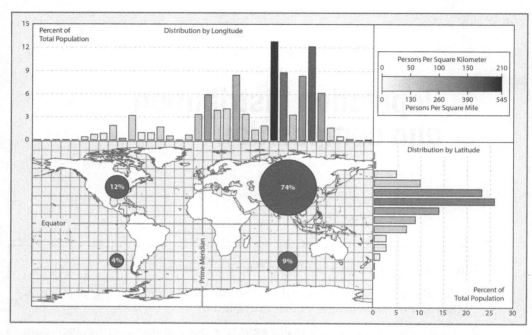

FIGURE 3.1 World population distribution by hemisphere, latitude, and longitude

This map shows the proportion of global population in different quadrants of the Earth. The bar chart on the right of the map shows the distribution of people by latitude, with the majority of people living in the Northern Hemisphere, particularly in subtropical and temperate mid-latitudes. The bar chart along the top shows distribution by longitude, with peaks corresponding to the two demographic giants: India and China.
Data source: WorldPop (2022).

compared with 30 per square kilometer (78 per square mile) in the South. As figure 3.1 shows, fully 85 percent of global population lives north of the Equator. Similar disparities persist between the East and the West, with the huge population centers of Asia apparent in the large proportion of global population focused in the northeastern quadrant of the map. Some of these disparities stem from environmental variations, whereas others can be traced to socio-historical factors such as when particular regions were settled.

Unsurprisingly, people have traditionally settled in regions where climate and land were favorable to agricultural production, leading to particularly high population densities on flat, fertile land and along rivers. Many of the earliest complex civilizations developed in river valleys where the rise of government has sometimes been linked to the need to control water. These so-called hydraulic civilizations include the Ancient Egyptians on the Nile, as well as other civilizations along the Indus, Tigris, Euphrates, and Yangtze rivers. Coastal locations were also popular because they often offered flat land and a more temperate climate, with generally milder temperatures and more rainfall than inland locations. Coastal and river sites also offered good accessibility. Today, about 40 percent of global population still lives within 100 kilometers (62 miles) of the sea and 67 percent in a 500-km (310-mile) coastal belt. Areas below 250 meters (820 feet) above sea level (excluding Antarctica) are home to almost 60 percent of the world's population, even though they comprise just one-third of total land area (figure 3.2). These low-lying regions

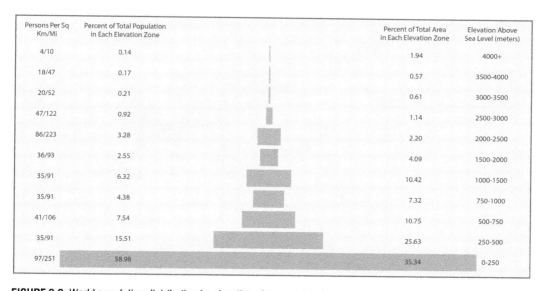

Persons Per Sq Km/Mi	Percent of Total Population in Each Elevation Zone	Percent of Total Area in Each Elevation Zone	Elevation Above Sea Level (meters)
4/10	0.14	1.94	4000+
18/47	0.17	0.57	3500-4000
20/52	0.21	0.61	3000-3500
47/122	0.92	1.14	2500-3000
86/223	3.28	2.20	2000-2500
36/93	2.55	4.09	1500-2000
35/91	6.32	10.42	1000-1500
35/91	4.38	7.32	750-1000
41/106	7.54	10.75	500-750
35/91	15.51	25.63	250-500
97/251	58.98	35.34	0-250

FIGURE 3.2 World population distribution by elevation above sea level
This bar chart shows the proportion of population living at different altitudes. The very wide bar at the bottom represents the large proportion of people living near sea level, with increasingly sparse populations as elevation increases. *Data sources:* OPENDEM (2022) and WorldPop (2022).

often have pleasant climates and high agricultural potential, although in tropical regions areas of higher elevation may be more appealing to human settlement to escape heat and disease in the lowlands. Over half a billion people live above 1,500 meters (4,920 feet) of altitude, over 80 million over 2,500 meters (8,200 feet), and over 14 million over 3,500 meters (11,480 feet), particularly in tropical and subtropical regions such as the Ethiopian highlands and Andean altiplano (Tremblay and Ainslie 2021). Nonetheless, at high altitudes humans (and their food production systems) face multiple environmental stressors, including cold temperatures, low oxygen concentrations, strong winds, and high radiation levels.

The idea of considering the world in terms of its favorability to human settlement has a long tradition. The Ancient Greeks provided us with the terms **ecumene**, or habitable world, and **nonecumene**, the inhospitable part. Although this division has become less clear as technological developments have expanded human ability to live in less hospitable regions, there are still large areas of the world that remain only sparsely populated, particularly cold, dry, and hot–humid regions. Cold climate and permanently frozen ground, in particular, discourage agriculture and settlement—despite their large land areas, Antarctica has no permanent population and Greenland only about 50,000 people. Most people living in very cold regions are found in coastal areas or at mining sites, military bases, or research stations. Arid regions are often sparsely populated owing to a shortage of freshwater. However, dry regions with access to groundwater (e.g., parts of California) or with rivers flowing in from wetter regions (e.g., the Nile Valley) can develop very productive agricultural systems and support high population densities (although not necessarily sustainably if groundwater is used more quickly than it is replenished). In other cases, specific resources, particularly oil, have provided the incentive to settle areas with harsh environments, as illustrated by the Arabian Peninsula. Many hot–humid

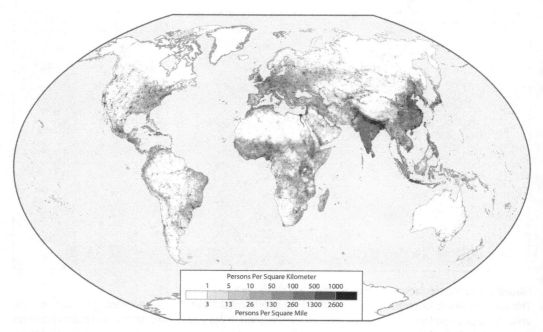

FIGURE 3.3 Distribution of the world's population
Data source: WorldPop (2022).

regions have traditionally been avoided for settlement because of their high risk of disease transmission and uncomfortable climate. Regions covered by tropical rainforest were also challenging for agriculture owing to the labor-intensive task of clearing vegetation, as well as the many pest outbreaks and weed invasions that occur in tropical climates. High temperatures and heavy rain can also leave tropical regions with poor soils once vegetation is cleared, owing to excessive leaching of nutrients. Figure 3.3 illustrates many of these patterns, with low population densities apparent in the cold circumpolar latitudes, the dry conditions of the Sahara Desert and central Australia, and the rainforests of the Amazon Basin.

Figure 3.3 also allows us to identify areas of dense population, including three major population clusters: South Asia, East Asia, and Europe. The South Asian cluster of over 1.7 billion people, focused on India, Pakistan, and Bangladesh, is primarily agricultural. The East Asian cluster, comprising China, Japan, and the Korean Peninsula, has a population of over 1.6 billion and is increasingly industrial and urban. The European cluster is the smallest of the three, with about 600 million people. Other population clusters include West Africa (Nigeria and neighboring areas), Northeast Africa (the Ethiopian Highlands), east-central Africa (the Great Lakes region), the Nile River Valley (Egypt), the eastern United States and Canada, the island of Java (Indonesia), the Philippines, parts of Central America and Mexico, the northern Andes, and coastal Brazil.

A high percentage of global population lives in areas prone to environmental hazards such as floods and hurricanes. In these cases, the threat of a possible future disaster is tempered by the immediate benefits of living in these regions. People have traditionally benefited from nutrient-rich volcanic soils, for instance, explaining high population densities on the volcanic islands of Indonesia. As we have already noted, coastal areas have also been attractive to human settlement, even though they are vulnerable to hurricanes, tsunamis, and sea level rise. Global climate change is posing a particular threat to these regions through more intense coastal storms and rising sea levels. As populations grow, there is also increasing pressure to expand settlement into more hazardous areas, putting larger numbers of people at risk.

Measures of population distribution

From a sustainability perspective, this uneven distribution of population puts greater strain on certain areas of the world than on others. **Population density**— the number of people per unit of land area—is often used to quantify this. **Arithmetic density**, or the number of people per unit of *total* land, is by far the most common population density measure used. It varies widely—from two to three persons per square kilometer (five to eight persons per square mile) in sparsely populated Mongolia and Australia to over 1,200 people per square kilometer (3,100 people per square mile) in Bangladesh and over 2,000 (5,200) in the microstate of Bahrain. A more meaningful measure of population pressure on land is **physiological density**—the number of people per unit of *productive* land (land that could be used for agriculture). Physiological density is always higher than arithmetic density for a particular region, but the two measures diverge most strongly in places with harsh climates and unproductive land (figure 3.4).

Population redistribution

Though the discussion so far has focused on how people have traditionally settled in agriculturally favorable areas, today more than half of the world's population live in cities. Agricultural productivity has increased significantly in the modern era, freeing large numbers of people from having to perform agricultural labor. At the same time, increasingly integrated trade and transportation networks have allowed people to draw resources from greater distances. Industrial lifestyles have thus reduced people's dependence on procuring resources from their immediate environment, allowing population to become more concentrated and enabling the growth of cities.

As recently as 1800, just 5 percent of the global population lived in cities. Since then, **rural-to-urban migration** has moved huge numbers of people from the countryside to urban areas as people sought to escape heavy agricultural labor and rural poverty in hopes of better jobs and higher living standards in cities. By 1950 one-third of global population was urban, by 2020 approximately half, and this figure is projected to rise to two-thirds by 2050. Despite the appeal of cities, urban conditions were initially often appalling, as overcrowding fueled disease and wastes built up. This was evident in the cities of Victorian England in the late 1800s and can be seen in the **megacities** of the Global South today. We expand on the population impacts of urbanization and city living in chapter 12.

In other cases, population movements have been associated with the settling of new land. One way to visualize shifts such as this is by mapping the changing location of a **population centroid**. This is the point from which there is the same

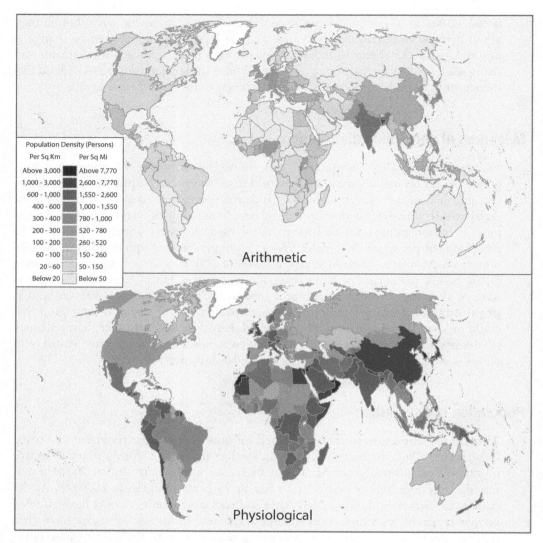

FIGURE 3.4 Arithmetic and physiological densities by country, 2018
Data sources: FAO (2022) and UNDESA (2019).

number of people to the north, south, east, and west within a specified study area. If we consider the movement of the United States' population centroid over the past 200 years, for instance, we can see the influence of settlers' westward expansion from the original 13 colonies in the East (figure 3.5). The centroid also shows a more recent southwards shift, particularly after World War II when increasing availability of air conditioning and industrial developments attracted people to the Sunbelt. More recently, people have moved coastwards due to the availability of jobs and a preference for mild climates and recreational opportunities, whereas the United States' agricultural core has lost population, generating a bicoastal distribution (Rogerson 2021). Only time will tell whether these patterns continue or whether climate change or other factors will change the direction in which the population centroid is moving once again.

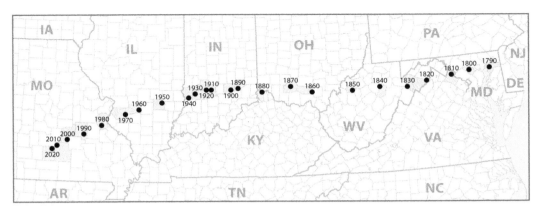

FIGURE 3.5 Population centroids for the United States, 1790–2020
Source: United States Census Bureau (2021).

Many countries are characterized by significant disparities in the spatial distribution of their populations, related to environmental or historical factors. For instance, over 90 percent of Chinese live in the wetter eastern half of China, and over 70 percent of Russians live in the more temperate conditions west of the Ural Mountains. A 2020 United Nations report indicated that some 30 to 40 countries consider uneven population distribution to be a major problem, mostly because of pressure on scarce resources such as water, as well as problems associated with high population densities including pollution, traffic congestion, and squalid living conditions (UNDESA 2020). Several countries have taken active measures toward a more balanced distribution, including Egypt, Indonesia, and Brazil.

Egypt's population of 100 million people is more spatially concentrated than that of any other country in the world, with 55 percent of Egyptians living on just 1 percent of the land and over 95 percent on 5 percent of the land, mainly in the Nile Valley and Delta (figure 3.6). In one of the most ambitious urban development

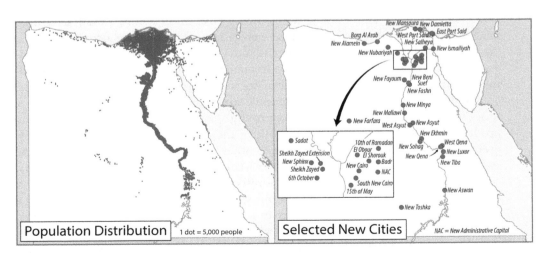

FIGURE 3.6 Population distribution (2020) and the New Cities program in Egypt
Data sources: WorldPop (2022) and Colliers International (2020).

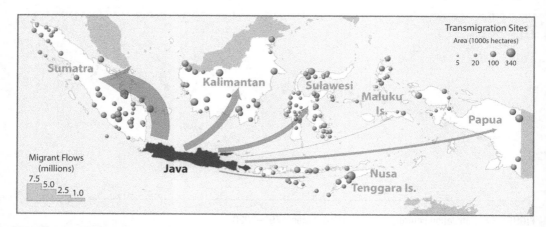

FIGURE 3.7 Transmigration in Indonesia
Data sources: Statistics Indonesia (2021) and SIPUKAT (n.d.).

projects in the world, Egypt developed a program of constructing new cities in the desert in the late 1970s, aiming to relocate some 40 to 50 million people to new towns by the mid-21st century (Ghanem, n.d.). So far, the program has not met expectations because the new cities have not attracted as many people as intended due to low levels of investment in social infrastructure and a lack of diversity in employment opportunities (Tadamun 2015).

Indonesia's transmigration program has also been less successful than hoped (figure 3.7). Indonesia's population of 270 million people makes it the fourth most populous country in the world, but its population is extremely unevenly distributed. The island of Java, including the capital city Jakarta, is particularly densely settled—Java occupies only 7 percent of the country's land area but houses over 55 percent of Indonesia's population. The islands of Bali, Lombok, and Madura are also crowded, with over 1,000 people per square kilometer (2,500 per square mile)—eight times greater than the national average—because of their fertile volcanic soils and favorable climate for growing rice. Even before Indonesia's independence from the Netherlands there were attempts to relocate people from the "inner islands" like Java to the "outer islands" (Hardjono 1977). After independence, a new transmigration policy was implemented to reduce poverty and crowding on the inner islands, exploit the natural resources of the outer islands, and contribute to more balanced demographic and economic development, but this, too, did not achieve its target. Costs were high, and resettled families struggled to produce enough food on the poorer soils of the outer islands. Some people interpreted the relocation of ethnic Javanese to other islands as an attempt to control and assimilate indigenous groups on the outer islands. Furthermore, transmigration has contributed to tropical deforestation as forests were cut to make room for rice farms. Many of these rice farms failed, forcing some transmigrants to switch to illegal logging. More recently, Indonesia has announced that it will be moving its capital city, Jakarta, from densely settled Java, where the capital faces huge challenges associated with sea level rise, to Borneo, prompting concerns that the new city could lead to further deforestation and land degradation in an environmentally vulnerable area (Westfall 2022).

In Brazil, with a population of over 210 million, the concern was not so much uneven population distribution as a desire to "settle" the Amazon to reinforce territorial claims. The Brazilian government recognized the strategic importance of the Amazon region in the late 1930s and encouraged and subsidized migration to the region, although this did not result in large-scale movements. The relocation of the federal government from Rio de Janeiro to the newly built city of Brasília in the interior in 1960, the construction of highways connecting the new capital, and support for mining and agriculture finally helped attract people to the region, although most of Brazil's population remains coastal to this day. By 2020, Amazonia's share of the country's total population had increased to 13.8 percent, but the country's population centroid has shifted westwards only slightly (figure 3.8). Nonetheless, settlement in the Amazon region has resulted in extensive tropical deforestation and disruption of indigenous cultures.

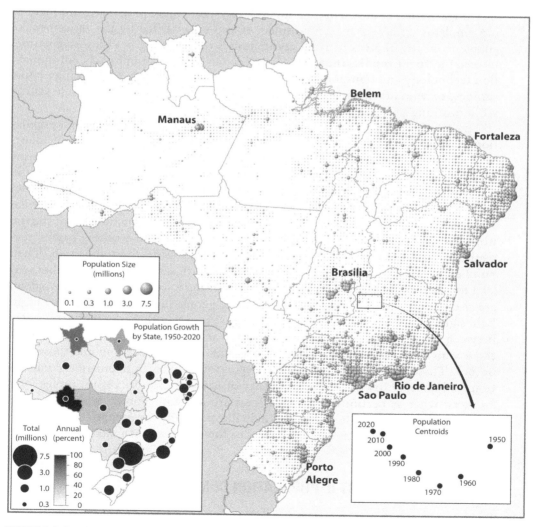

FIGURE 3.8 Population distribution and change in Brazil, 1950–2020
Data sources: IBGE (2022) and WorldPop (2022).

These examples of population redistribution are all the result of deliberate government policies, but more spontaneous population movements are occurring all the time. People may move to regions that offer better jobs or more amenities or be forced to move by political or social upheavals (see chapters 10 and 11). In North America and Western Europe, for instance, we have increasingly seen people moving from declining industrial centers to cities that offer more pleasant living conditions in recent years. Though traditional industries such as car manufacturing were tied to particular locations by the need for raw materials, much of the most dynamic economic development today, such as in computing and communications technologies, is described as **footloose** because it does not need large quantities of raw materials and is therefore not so tied to particular locations. Instead, these high-tech industries are dependent on having a highly skilled labor force. Attracting these skilled workers has become highly competitive, with cities such as Vancouver, San Francisco, and Dublin becoming tech hubs partly owing to the high quality of life they offer in their beautiful natural surroundings. At a more local scale, this increasing emphasis on a pleasant environment is also exemplified by the movement of people from urban areas to rural communities on the edge of cities—sometimes termed the **rural renaissance**. This has been facilitated by improved communication technology and transportation that allow people with remote working opportunities, or who are willing to commute, to live in a pleasant countryside setting while working in well-paid urban jobs.

The COVID-19 pandemic has accentuated many of these tech-related population movements. Though working from home was initially often a difficult adjustment, many people have learned to appreciate its benefits, and some workers are electing not to return to the office. Some attractive small towns with plentiful outdoor recreational opportunities have become known as "Zoom towns" as telecommuters arrive in large numbers (Johanson 2021). Although this particular movement has been triggered by the COVID-19 pandemic, the number of **digital nomads**—people who work remotely from coffee shops, co-working spaces, recreational vehicles, and even abroad—has been increasing for some time. Their numbers thus far are too small to amount to a significant population shift but may lead to meaningful population redistributions in the long term. These movements are also important beyond the small number of people involved because they highlight significant inequalities among population groups, with stark divides emerging between those who are able to work remotely and those who cannot and between those who have the economic means to seek out desirable locations and those who do not. These shifts also contribute to important changes in population composition, as booming towns receive an influx of economically successful professional workers, whereas declining industrial areas are left with poorer, less skilled, and older populations. Aspects such as this help illustrate why the composition of population is such an important issue—it is to this topic that we turn next.

PART II: POPULATION COMPOSITION

Population geographers commonly study age, sex, and ethnic/racial compositions of populations, but many other population characteristics can be considered, including religion, language, and socio-economic status, or even literacy rates and access to clean water. Knowledge about population composition is not just an academic exercise but can also point to real-world problems such as a shortage of schools or inadequate sanitation. In the remainder of this chapter we explore age, sex/gender, and ethnicity/race as primary characteristics of population composition.

Age structure

Age structure provides important information about the characteristics of populations because people of different ages play very different roles in society. Children are dependent on the broader population for support and may spend considerable time in education (box 3.1). Young adults are some of the most dynamic members of society, driving economic migration flows and providing labor and tax revenue to fuel the economy. Many people also become parents in their 20s or 30s, so a large young adult population increases fertility rates. By middle age, women are moving beyond their childbearing years but continue to play a critical role in childrearing. In middle age, many people also have eldercare responsibilities and may find themselves in the **sandwich generation**, concurrently performing childcare and eldercare roles, with clear physical and socio-emotional tolls (Buchanan 2013). Most workers continue in paid employment in their 40s and 50s and have skills to pass on to younger workers; some also take active roles in management or political leadership. By retirement age, people leave the workforce and may become economically dependent on the broader population once again via state pensions and socialized healthcare systems. Older community members may return to playing important roles in childcare too, tending grandchildren while the children's parents are in paid employment. The proportion of people in different age groups in a population therefore has huge implications for fertility, economic growth, migration trends, and health patterns.

One of the key tools used to explore age structure is the **population pyramid**. A population pyramid uses two mirroring bar graphs to show the age and sex structure of a population (figure 3.9). By convention, the male population is shown on the left and females on the right. The vertical axis of the pyramid represents **age cohorts**—groups of people born during the same time period (often 5-year groups)—with the youngest group at the bottom and the oldest at the top. The shape of a population pyramid quickly reveals information about the age structure of the population. Pyramids that are wide at the bottom and narrow at the top (roughly triangular) represent societies characterized by high fertility and rapid population growth but many children dying before reaching adulthood. Population pyramids for more affluent countries are more column-like, with fewer children being born and therefore a narrow base to the pyramid but most people surviving into old age. Long life expectancies mean that even elderly cohorts may still comprise large numbers of people. In the oldest age groups, we expect more females than males, owing to women's slightly longer life expectancies.

At smaller scales, population pyramids can reveal very specific information about a population's age and sex structure. For instance, figure 3.10 shows two US counties—one heavily influenced by college students and the other representing a retirement community. Population pyramids can also reveal unique features about

BOX 3.1 CHILDHOOD

In traditional agricultural societies, farming provided many tasks that could be performed by children from a young age to contribute to the household economy. To this day, children as young as 6 or 7 may be expected to participate in jobs like herding animals or harvesting crops. In villages, direct parental involvement with children may be quite limited, with young children often under the supervision of older siblings and whole communities playing a role in disciplining and caring for children. Shorter life expectancies in traditional communities encourage early marriage and childbearing, often from the mid-teens. In communities with no formal social security system, children may also be responsible for caring for older family members. As a result, childhood in traditional agricultural societies is often short.

The shift to industrial lifestyles brings huge changes. Industrial economies expect people to have multiple years of training before joining the workforce, with students remaining in education for up to 20 years. Due to urbanization, many children are raised in cities. On the positive side, this means access to healthcare and educational opportunities, but it also means polluted air, fewer open spaces, traffic, and often greater perceptions of danger on the part of parents, resulting in less freedom of movement for many urban children. The activities that children participate in also change. Especially in affluent countries, children are likely to spend more of their day indoors than outdoors, to be driven around rather than walking or biking to places independently, to engage in organized after-school activities rather than free play, and to spend significant time in front of screens, with huge implications for physical activity and the development of social skills, creativity, and independence. Increasing aspirations for children associated with competitive education systems and aggressive job markets, particularly in East Asia, have also arguably led to high-stress environments for many children, as well as a growing gap between the experiences of children in rich and poor families.

Household structure has also changed in many industrialized economies, with implications for childrearing. Traditional kin networks were bottom-heavy, with multiple siblings and, as a result, also many cousins, aunts, and uncles—in the recent past, an average individual might have had 5 siblings, 10 uncles/aunts, 30 nieces/nephews, 60 first cousins, and well over 700 second cousins (Jones 2013)! Today, fewer children grow up with large numbers of siblings, focusing more attention on direct lineages, particularly grandparent–parent–child relationships. Though this means a significant focus of resources and attention on children as they grow up, it puts increased pressure on parents, particularly mothers, to perform much of the caring labor for children in the absence of aunts, sisters, and other extended family. Additionally, as these lateral kin networks have shrunk, children may find themselves responsible for caring for several grandparents and parents in adulthood (Jones 2013). Many children also spend much of their family life in the company of adults, particularly the increasing number of only children in affluent societies, with both positive and negative implications for social development. Non-traditional family arrangements are also becoming increasingly common in many countries, including single-headed households, same-sex parents, and blended families, providing a diversity of experiences for children. Though there is nothing inherently problematic with any of these arrangements, they raise different challenges and benefits compared with the extended families that were common until recently.

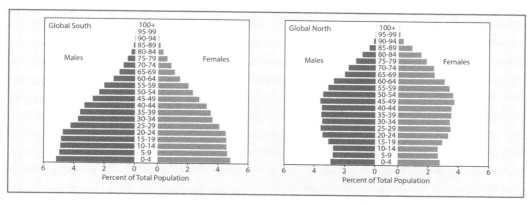

FIGURE 3.9 Population pyramids for economically less and more developed regions, 2015–20
Data source: UNDESA (2019).

the history of a country: a baby boom will appear as a widening of the pyramid, whereas a war may be seen as an indentation in the male side of the pyramid. Over time, bulges and indentations work their way up the pyramid as cohorts age. We may also see an "echo" of a bulge or indentation in the next generation as a particularly large (or small) cohort reaches its childbearing years and goes on to have an unusually large (or small) number of children. For example, in 2021 it was reported that the number of 18-year-olds was rising sharply in the UK as the result of an "echo of an echo" from a baby boom in the mid-1950s (*The Economist* 2021c).

In this way, population pyramids can reflect a country's history. Consider Russia's 2010 population pyramid (figure 3.11). In the oldest age groups, we see diminished population cohorts associated with Stalin's forced collectivization of farming from 1929 until 1933, which took land away from farmers and ultimately resulted in food shortages. We also clearly see the impact of World War II—first the loss of people during the war and then a baby boom after the war, as well as the echoes of these events a generation later. Toward the bottom of the pyramid we see

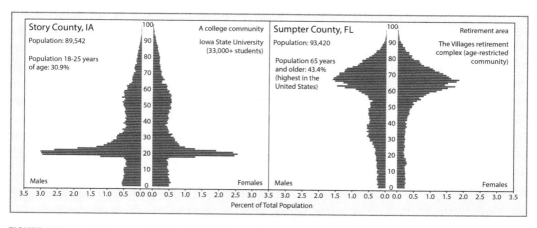

FIGURE 3.10 Population pyramids for selected US counties, 2010
Data source: United States Census Bureau (n.d.).

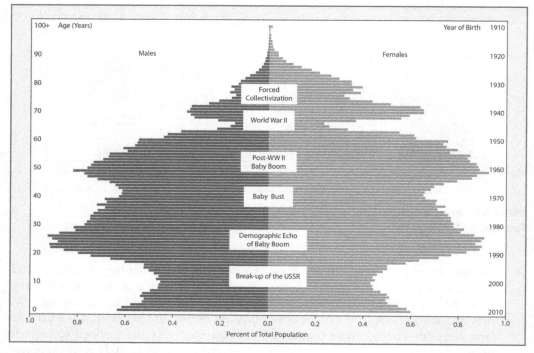

FIGURE 3.11 Population pyramid for Russia, 2010
Data source: Federal State Statistics Service, Russia (n.d.).

the impact of the breakup of the Soviet Union in 1991, which resulted in a rapid decline in the birth rate followed by a gradual recovery.

Aging populations

The major change to age structure over recent years has been steadily aging populations across most of the world owing to rising life expectancies and declining birth rates. This can be seen in a rise in the **median age** (the age at which half of the population is older and half younger) of many populations. In 1970, global median age was just 21.5 years; by 2019 it was over 30 (Ritchie and Roser 2019). In 2018, the global population hit a demographic milestone when the number of people over 64 years surpassed the number of children under 5 for the first time in history. This top-heavy distribution was reached as early as the 1960s and '70s in parts of Europe and North America but is still to occur in much of sub-Saharan Africa—in Nigeria it may not happen until the 2080s (Ritchie and Roser 2019).

As such, today we see different countries at very different stages of their aging trajectory. Populations in East Asia and Europe have aged particularly rapidly, but immigration of young people has reduced the impact on Western European countries somewhat. In North America, immigration has suppressed the aging of the population more significantly. By contrast, Japan, which has largely resisted immigration, has experienced particularly profound aging. In Eastern Europe, outmigration of younger people has intensified demographic aging. In short, immigration tends to soften demographic aging, whereas emigration of workers exacerbates it.

Although aging societies are most often associated with Europe and East Asia, the aging of populations is now becoming significant in the Global South, too. Here, many countries are being confronted with aging populations at lower levels of affluence, lessening the potential for governments to ease the burden of aging via social support measures.

Older population age structures have significant social ramifications. For most of human history, death was a constant possibility, even for young people. Nowadays, the greater likelihood of living to old age has shifted our expectations and changed society. Long life expectancies make it possible to invest a lot of time in the education of young people, which in turn results in more specialized occupations. There is also less rush to start a family, resulting in a longer gap between biological maturity and the formation of a family than at any time in human history. Furthermore, parents can expect to live many years beyond raising their children and achieve other things in life beyond parenthood. In these senses, increasing life expectancies have had far-reaching consequences for how modern societies are organized as well as the aspirations of individuals.

Perhaps most significant, the aging of populations has economic consequences owing to the close ties between age and employment. This is estimated by the **dependency ratio**—the ratio of the dependent population, defined by age (typically those under 15 and over 64 years), compared with the working population (15 to 64 years). Traditionally, high dependency ratios were associated with young populations, but today we are more concerned with the economic implications of large elderly populations. Older populations arguably present an even greater economic challenge than very young ones because the elderly have not only left the workforce but also bring significant costs in terms of healthcare and pensions. Furthermore, in many countries, costs of eldercare are expected to be borne by taxpayers via **welfare state** mechanisms such as socialized healthcare systems, state-run care homes, and government-supported pensions. Most of the costs of raising children, by contrast, are borne by individual families, including huge amounts of unpaid care work on the part of parents. The major public cost of raising children—education—is typically viewed as an investment in the country's economic future and is therefore often willingly borne by a community. As such, the impact of a large elderly population can be particularly profound and often generates disagreements over who should bear these costs. Even in East Asia, which has traditionally benefited from strong family support for the elderly, governments are starting to express alarm over how eldercare should be managed as falling birth rates and people moving for jobs leave many elderly people without support from within the household.

The challenge of caring for elderly populations is therefore on the top of many political agendas. In this respect, we may consider the **potential support ratio**—the ratio of working-age adults (15 to 64 years) to those 65 and over—to estimate the ability of a community to support its elders, whether that burden of care is taken up by professional workers or by family members. The potential support ratio has the same weakness as the dependency ratio in that it assumes that people are dependent just because they have reached a certain age. One suggestion to address this is to replace the fixed age of 65 as a marker of dependency with a dynamic indicator, defined as the point at which remaining life expectancy is less than 10 years. If this approach is used, demographic aging looks less dramatic. For example, in many Western European countries, life expectancy is now over 80, so only people above 70 would be considered dependent through age. This way of defining old age also better reflects healthcare expenditures, which are especially high at the very end of life (Höpflinger 2012).

The negative impacts of demographic aging can also be reduced by policy measures that directly address the new realities of aging populations. For example, by increasing retirement age, some of the economic burden of aging populations could be alleviated. Reforming workplaces in ways that value older workers' experience would also be beneficial. Some commentators note that demographic aging can even have positive effects. For example, though demographic aging in Japan has led to pessimism about the future of Japan's economy, it is also predicted to result in lower housing costs, less crowded living spaces and traffic congestion, reduced ecological impacts, higher wages due to worker shortages, and increased support for family-friendly changes and migration (Kono 2011).

Though many countries struggle with the economic implications of populations of very advanced age, in the early stages of demographic aging communities can actually experience favorable economic conditions. Iran and India can be said to be in this position today. Recent declines in the birth rate mean that their young dependent population has decreased, but short life expectancies until recently mean that their elderly dependent population remains relatively small. During this temporary situation, a country is said to be benefiting from a **demographic dividend** as a population bulge from previously high birth rates moves through its working years, generating a low dependency ratio. It is only a matter of time—usually 20 to 40 years—before these economic benefits are lost, however. The bulge of workers retire and become economically dependent and may live for many years beyond retirement owing to rising life expectancies; meanwhile the birth rate remains low, suppressing the number of workers coming in at the bottom of the pipeline, raising the dependency ratio.

In summary, the aging of populations is a complex issue. Different countries are at different stages of aging, and cultural patterns such as whether there is a strong tradition of eldercare within households can significantly influence the economic and social implications of aging. Furthermore, age is a problematic measure of how economically productive people are, as well as how much support or care they need, as rising living standards have resulted in more years spent in good health. Nonetheless, it is clear that aging societies pose significant challenges and that communities will have to adapt to changing demographic realities. We return to these issues in chapter 6.

Sex and gender

Sex has long been considered another important demographic characteristic, with the sex composition of a population affecting issues ranging from fertility rates to migration patterns to crime rates. Though biological **sex** has traditionally been used in population studies, increasingly scholars are exploring the ways in which our experiences around our sex are socially constructed. Here the term **gender** is often used to reflect how cultural understandings of what it means to be male or female influence our experiences. For instance, how does the culture of the society in which we live influence how men and women are treated differently or the different roles they are expected to play in society? This thinking has been pushed even further by **LGBTQ+** scholars, who have begun to explore the many ways in which sex and gender categories fall short of reflecting the lived experiences of people whose identities do not conform to traditional understandings of male and female or the sex they were assigned at birth.

Sex ratios

Sex ratios have long been used to summarize the relative proportion of men and women in particular societies. In a country with a normal sex ratio, we would expect the number of males and females in each age group to be roughly equal, except for the oldest age groups, which are usually female dominated owing to women's slightly longer life expectancy. Migration is one of the commonest reasons for an unbalanced adult sex ratio. In the United Arab Emirates (UAE), for instance, we see more men in working age groups owing to heavy immigration of South Asian workers for jobs stemming from oil wealth (figure 3.12). The UAE represents an extreme example, with fully 80 percent of the population of the UAE born outside the country.

The dominance of men in military activities can also skew sex ratios. At a local scale, this may mean that towns with military facilities have disproportionately more men of working age (figure 3.13). If countries actually go to war, military losses and refugee patterns can skew the sex ratio of whole countries. Following the Russian invasion of Ukraine in 2022, for instance, men were expected to stay and defend Ukraine while many women and children fled. This changed the demographic composition of both Ukraine and neighboring countries, especially Poland, which took in the bulk of the refugees. It is unclear whether this is a short-term phenomenon until refugees can return to Ukraine or if this will lead to more permanent changes in the population composition of neighboring countries, many of which have aging populations and would benefit demographically from the influx of working-age women and their children. For Ukraine itself, the war is a "demographic disaster" because a quarter of its population has fled the country and life expectancies and birth rates are projected to fall (*The Economist* 2022).

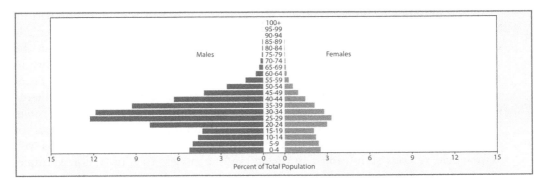

FIGURE 3.12 Population pyramid for the United Arab Emirates, 2015–20
Data source: UNDESA (2019).

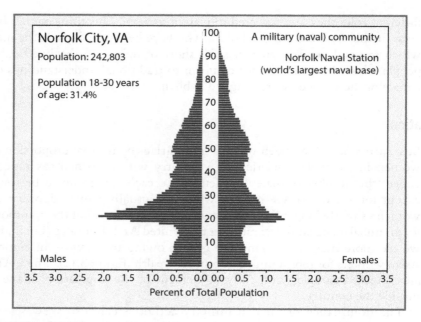

FIGURE 3.13 Population pyramid for a locality dominated by the military
Data source: United States Census Bureau (n.d.).

Sex imbalances can have significant implications for a country's birth rate, social functioning, and even political stability. For instance, large numbers of male migrant workers will likely suppress the birth rate but may generate a thriving sex trade. Large numbers of young men in a population with insufficient employment opportunities are sometimes also associated with political instability. In any situation of unbalanced sex ratios, marriage and birth rates are likely to fall as some proportion of the population fails to find partners. This occurred in Europe after the World Wars when many women were left unable to marry owing to losses during the wars. In the early 1920s, Britain had almost 2 million more women than men at a time when women's role in society was heavily defined by their responsibilities as wives and mothers, leading to public concern over how society would cope with the upheaval.

In the previous examples, skewed sex ratios occurred in the adult population. Where infants and children show a skewed sex ratio, the situation is often considered even more problematic. Although marginally more boys are believed to be conceived under normal circumstances, slightly higher rates of male mortality in infancy usually compensate for this, resulting in approximately even sex ratios through childhood. Where we see very skewed sex ratios in children, we can therefore usually conclude that societal (not biological) influences are responsible. Typically, it is boys who are in the majority, highlighting persistent challenges for females. In **patriarchal** societies, boys are often preferred for philosophical as well as practical reasons stemming from longstanding biases. In China, for example, Confucianist philosophy emphasizes the importance of male heirs, with boys carrying the family name and responsible for taking care of parents in old age. Daughters, by contrast, become part of their husband's family after marriage.

Additionally, in some countries, most notably India, girls have traditionally required a dowry of money and gifts to be given to their family of marriage—in such circumstances, girls may be viewed as an economic burden, with daughters' marriages driving many families into debt. The resulting preference for boys may mean that girls in patriarchal societies are given less care and attention than their brothers, leaving them less well-nourished and less healthy and therefore surviving at lower rates. In some countries, particularly India and China, sex determination of the fetus via ultrasound has also enabled the selective abortion of girls—although the practice is now illegal in both countries, it is believed to persist. There is even evidence that female infanticide (the killing of girls after they are born) has historically occurred in some regions and may persist to this day in extreme circumstances.

Unbalanced sex ratios are most often associated with East and South Asia, but strong traditions of son preference persist or have reemerged also in the Caucasus countries (Armenia, Azerbaijan, and Georgia) and parts of Eastern Europe (Albania and Montenegro; *The Economist* 2013). It has been estimated that 68 million girls and women are "missing" in China, 45 million in India, 4 million in Pakistan, and about 2 million each in Nigeria, Bangladesh, and Indonesia, indicating significant gender inequalities in a wide variety of contexts (Seager 2018). Beyond the obvious challenges this indicates for women, this imbalance also has an impact on men. In China, for instance, the number of prospective grooms substantially exceeds that of prospective brides, with rates of men who cannot find wives expected to peak in the mid-21st century (Guilmoto 2012). This shortage of marriageable women is leading to increased trafficking in women as men desperately try to find partners. To try to address these imbalances, the Chinese government banned sex-selective abortions in 1989 and ultrasounds to determine the sex of the fetus in 2002 (Ebenstein and Sharygin 2009). The abolition of China's controversial "One Child Policy" in 2015 will also likely help to rebalance sex ratios there. Progress has also been made in other countries. In India, recent changes in parental attitudes have been reported, leading to more families wanting a girl in addition to (although notably not instead of) boys (Kaur and Kapoor 2021). The Indian government has even reported that the number of women recently surpassed that of men for the first time, although critics have raised doubts about these numbers (Pandey 2021).

Gender-based violence

Imbalances in the sex ratio are reflective of broader trends in gender-based discrimination and even violence against girls and women (table 3.1). A quarter of all women will endure some sort of violence from their partner in their lifetime. In places like Europe, where such behavior is widely considered unacceptable and laws are in place to punish offenders, about 5 percent of women are abused by their partners, whereas in poorer countries rates are often much higher. In Africa and Asia, it is estimated that 20 percent of women with partners are mistreated annually, and 20,000 African women may be killed by their partners every year (*The Economist* 2021f). In many traditional societies, cultural beliefs that men should dominate women remain relatively widespread, leaving women open to abuse over even minor disagreements as well as circumstances beyond their control. Failure to bear sons has been linked to cases of bride burning in India, and arguments over dowries sometimes escalate to violence, often against young women with little

TABLE 3.1 Selected examples of gender inequality and women's rights violations

	Topic	Facts	Progress
Education and work	Education	In more than a dozen countries in Africa and Asia, girls receive significantly less schooling than boys.	More girls than ever before are in primary education.
	Illiteracy	Two-thirds of the 520 million illiterate people in the world are women. In Mali, 78% of women cannot read; in Niger, that figure is 89%.	Illiteracy rates have been declining for several years, including for women. Female literacy rates are over 90% in Europe, the Americas, and East Asia.
	Wage gap	In most countries women are paid less for their work than men; in poor countries, many more women than men work in the informal sector.	In 2018, Iceland became the first country to make a gender pay gap illegal.
Politics	Voting rights	In most countries, women gained the right to vote much later than men—50 years later in the United States and 123 years later in Switzerland.	In most countries, women now have the right to vote; Saudi Arabia remains an important exception.
	Women in government	The first appointed female head of state did not take office until 1960. Today, just 21% of government ministers are women, and some countries still have no women in government; e.g., Yemen and Papua New Guinea.	In 23 countries (mostly in Europe and southern Africa), women account for over 40% of government; Rwanda was the first country to elect a majority-female government in 2008. In 2021, 26 women served as heads of state.
Women's rights	Child marriage	In Niger, 76% of women are married before they turn 18; 33% of all child brides live in India.	In the 1980s, one in three women married as a child; this has since declined to one in four.
	Rape marriage laws	Laws allowing rapists to marry their victims to avoid punishment exist in several countries, including Syria, Libya, Iraq, Cameroon, and Equatorial Guinea.	Since 2012, eight countries have abolished so-called rape laws that allow rapists to marry their victims to avoid punishment.
	Honor killings	In some Muslim countries, women may be killed (usually by family members) when women are thought to have brought dishonor to their families.	Pakistan introduced stricter laws against honor killings in 2016, but the practice persists.
	Dowry murders	In countries like Bangladesh, India, and Pakistan, women are sometimes killed in dowry disputes.	Dowries have been illegal in India for 60 years, but the practice persists; some communities are banning dowries.
Health	Maternal mortality	Several countries have maternal mortality rates over 700 per 100,000 live births. In countries like Haiti, Nigeria, South Sudan, Bangladesh, and Laos, fewer than half of births are attended by trained personnel.	Maternal mortality is decreasing in most countries in the the world; more and more births are attended by trained personnel.
	Unsafe abortions	In countries where abortion is restricted or forbidden, three-quarters of abortions may be performed in unsafe conditions.	Many countries now provide safe abortions as part of formal reproductive health provision.

Compiled from: Hassan (2021), UNICEF (2020), *The Economist* (2019), Seager (2018), Selby (2016), United Nations Women (n.d.).

power in the household. In some countries such as Iran and Pakistan, men may face only relatively minor punishments if they abuse a wife who steps beyond traditional bounds of expected behavior, and in some very patriarchal communities women may even be murdered in so-called honor killings when they are considered to have brought dishonor to their family (*The Economist* 2021b). Women are often economically dependent on men in poorer countries where there is no welfare system to support them, potentially leaving them trapped in abusive relationships.

It has been argued that gender-based violence hurts not only women but entire societies. For example, there is some evidence that countries with poor women's rights are more likely to be politically unstable (*The Economist* 2021a). Gender inequality is also often cited as a major obstacle to economic development because it hinders the entrepreneurship and economic productivity of half of a country's productive population. The degree to which gender inequalities actively contribute to political instability and economic problems or simply tend to co-occur in societies with broader social problems is hard to tease apart, however.

Beyond outright human rights violations against women, lesser inequalities between men and women persist in even the most affluent and human rights–conscious countries. As a summary measure, the Global Gender Gap Index ranks gender inequality based on a combination of economic, political, education, and health criteria. In 2020, Northern European countries (Iceland, Norway, and Finland) led the world as the most equal countries with respect to gender; inequalities were greatest in the Middle East and parts of Africa (World Economic Forum 2021; figure 3.14).

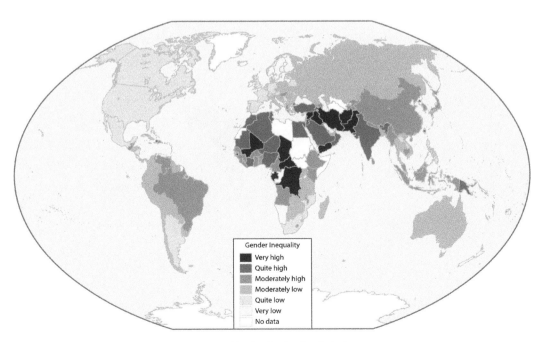

FIGURE 3.14 Global gender inequality (estimated by the Gender Gap Index)
Data source: World Economic Forum (2021).

Sexual orientation and gender identity

From a social justice perspective, gendered issues do not stop at inequalities between men and women. Increasingly, attention is also being directed toward the long-standing marginalization of communities with alternative sexual orientations and/or gender identities. The acronym **LGBTQ+** is often used as shorthand to represent a community of people who identify as lesbian, gay, bisexual, transgender, and/or queer, with the plus sign used to acknowledge that this listing may not capture the full breadth of alternative gender and sexual identities. It is important to note that the sensitivity of questions around gender and sexuality means that different terms may be embraced or rejected by different LGBTQ+ communities around the world.

A variety of social justice issues can be identified with respect to the LGBTQ+ community. First, institutionalized discrimination and victimization persist in many countries for individuals who do not conform to traditional expectations of heterosexual unions. This includes laws such as bans on gay marriage, having an older age of consent for same-sex than heterosexual sexual activities, and homosexual actions being defined as a crime—sometimes even with the possibility of the death penalty (figure 3.15). Today, almost 70 countries criminalize consensual same-sex sexual activity, particularly in Africa and the Middle East. In 11 countries the death penalty remains on the books, with 6 countries still actively implementing it for consensual homosexual activities, including Iran, Saudi Arabia, Somalia, and Yemen (Human Dignity Trust 2022). In other countries, abuses against members of the LGBTQ+ community may simply be overlooked, leaving people at higher risk of mistreatment from law enforcement officers, exposure to so-called

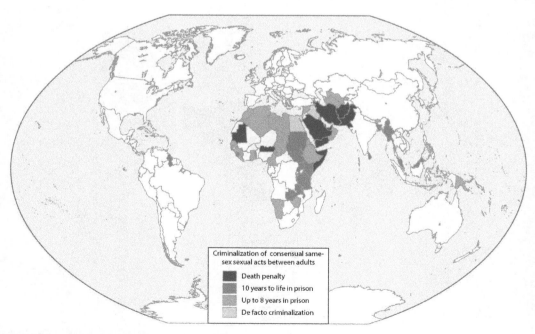

FIGURE 3.15 Criminalization of consensual same-sex sexual activities
Data source: ILGA (2020b).

conversion therapies that attempt to "cure" homosexuality, and even "honor kill-ings" and rapes (Angelo and Bocci 2021). Such abuses are often covered up in the countries where they occur but are increasingly reported as victims of abuse seek asylum in other countries, including a recent surge of applications to the United States from those facing gender-based or sexuality-based persecution in El Savador, Honduras, and Guatemala (Human Rights Watch 2020).

More recently, attention has also been directed toward the discrimination faced by individuals whose gender identity does not match their sex assigned at birth. Fifteen countries criminalize the expression of transgender identities with laws against "cross-dressing," "impersonation," or "disguise" (Human Dignity Trust 2022). Beyond these laws, individuals with alternative gender identities face wide-spread prejudice in many countries and may be forced to either conform to societal expectations of male and female or live in the shadows. This discrimination contin-ues despite mounting evidence that more acceptance of gender diversity is needed. Beyond the growing number of individuals questioning the sex they were assigned at birth is the acknowledgement that a small proportion of babies (possibly up to 1.7 percent, according to United Nations figures; UNFE, n.d.) are born with genitalia or chromosomal characteristics that cannot be definitively classified as male or female at birth—intersex individuals—calling into question the validity of a system that tries to identify all individuals via a binary system of male and female.

Despite evidence such as this, alternative gender identities remain at the margins of public acceptance in many countries and may be subject to sudden swings in policy. In the United States, for instance, after some years of increasing acceptance of alternative gender identities, the Trump administration implemented a ban on transgender soldiers operating in the military and the rolling back of certain healthcare guarantees for transgender patients (Angelo and Bocci 2021). More recently, about 20 US states have enacted or considered legislation that would restrict gender-affirming healthcare to children (Dawson, Kates, and Musumeci 2022). While even doctors remain somewhat divided on the best approaches for working with children experiencing gender identity issues, it is problematic that these discussions have become highly politicized rather than remaining within discussions of medical best practice. Even in countries that have become generally accepting of diverse gender identities, the practical implications of reordering soci-ety around new gender frameworks are complex, and progress has been slow, given how many societal structures currently rely on identifying people as male or female. Debates continue to rage, for instance, around how transgender individu-als should participate in gendered sporting events, how to navigate gendered spaces such as single-sex schools and prisons, and even which bathroom a transgender person should use.

Despite all of these challenges, an overall trend toward greater acceptance of alternative gender and sexual identities has been occurring in many communities for some time, with almost 60 countries now considered to offer "broad protection" against discrimination based on sexual orientation (ILGA 2020a). Although coun-tries in Europe and other parts of the Western world offer some of the strongest protections against sexual discrimination, many Pacific islands, as well as a majority of countries in South America, have also taken steps to protect the rights of indi-viduals identifying as LGBTQ+, and Angola and South Africa stand out in Africa as countries that have broad protection of rights (ILGA 2020a; figure 3.16).

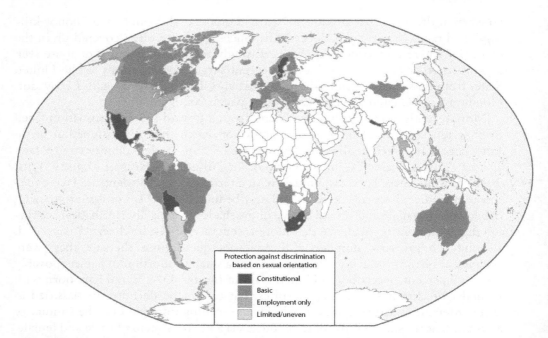

FIGURE 3.16 Legal protections against discrimination based on sexual orientation by country
Data source: ILGA (2020b).

In the most liberal countries, LGBTQ+ communities are now looking beyond acceptance and toward official recognition of genuine *equality* of status. Initially, many countries took the step of recognizing same-sex civil unions, for instance, but refused to acknowlegde them as equal to marriage—some countries are now recognizing gay marriage on an equal footing with traditional heterosexual unions. Another area of lobbying is the ability to obtain official documentation consistent with an indiviual's identity. Some countries now provide alternative gender options on official forms such as the census and passports, including Argentina, Australia, Canada, Denmark, India, Malta, Nepal, the Netherlands, New Zealand, Pakistan, and the United States, which all offered gender-neutral passports by the close of 2021. The situation remains contested, however. For instance, the UK declined to approve gender-neutral passports in December 2021, stating in a high-profile Supreme Court case that gender was important for checking an applicant's identity and that there was potential for confusion across government departments if gender did not match across different documents (Lee 2021). Gender identity remains on the cutting edge of discussions around population-related social norms.

Race and ethnicity

Most people associate the term **race** with physical features such as skin color and hair texture, but race does not clearly reflect genetic differences between different groups of people. Scientists remind us that there is more genetic variation *within* a race than between people of different races. Today race and **ethnicity** (shared ancestry and culture) are usually considered to be largely **socially constructed**, just as gender is increasingly being understood in social terms. This means that definitions and understandings of race and ethnicity are created more by people's understandings of social groups than by biological differences and thus can change over time. For example, today's racial and ethnic categories in the United States differ from those used in other countries and have changed from those used a century ago. Despite the huge diversity of experiences within ethnic groups, ethnic categories still provide us with important information about some of the shared experiences of particular groups. In this respect, we can recognize, for instance, that many **indigenous peoples** are still influenced today by shared histories of land expropriation and cultural suppression, despite huge cultural differences among indigenous groups (box 3.2).

BOX 3.2 INDIGENOUS PEOPLES AROUND THE WORLD

There are about 370 million indigenous people in the world today, accounting for roughly 5 percent of the global population. Known variously as Aboriginal peoples, First Nations, or native peoples, they are the descendants of the earliest known peoples to inhabit a region and continue to maintain at least some aspects of traditional lifestyles. Indigenous groups live from the Arctic to the South Pacific, with the vast majority in Asia (Amnesty International 2021). Though many people are familiar with groups such as the Maya in Central America, Inuit in the Arctic, Saami in Northern Europe, Aborigines in Australia, and Maori in New Zealand, there are over 5,000 indigenous groups globally.

The unique lifeways of indigenous groups set them apart from the dominant society that surrounds them (UNDESA, n.d.), which has historically led to significant marginalization and discrimination that persists to this day for many groups. Indigenous peoples often face similar struggles owing to shared histories of land dispossession and are among the most disadvantaged and vulnerable population groups in the world today. The forcible relocation of Native Americans to reservations in the United States, for instance, not only led to human rights violations but also left many communities disadvantaged because the land they were offered was usually marginal for agriculture and other economic activities. Even today, the land claims of indigenous groups are still being threatened by large infrastructure projects (like the flooding of Kayapo land by the Belo Monte dam in Brazil), natural resource extraction (such as gold in the Amazon or oil in the Arctic), and commercial agriculture. In some places progress has been made by acknowledging and attempting to resolve land claims. Canada, for example, has created a vast territory called Nunavut to give Inuit peoples some political autonomy. However, many groups have not been able to resolve land claims, and providing indigenous people with land does not address all of the social and health issues they face.

Indigenous groups often face problems of poverty, high unemployment rates, and lack of educational opportunities. Many of the social challenges faced by indigenous groups can be traced to past experiences where dominant groups deliberately attempted to destroy traditional cultures in order to speed up assimilation. Indigenous peoples were often restricted from speaking their native language, practicing their culture and religion, and maintaining their political and economic systems. In 19th- and 20th-century Canada, for example, indigenous children were removed from their families and forced into Indian Residential Schools. Similarly, in Australia between 1910 and '70, many children of

Aboriginal heritage were forcibly taken from their birth families and rehoused in institutions or with white parents to try to force their assimilation into white society. Now known as the "Stolen Generations," the extent of the emotional and psychological trauma to these children and their communities is only just being fully recognized (Common Ground, n.d.).

Beyond psychological challenges, many indigenous communities also suffer high rates of chronic diseases such as obesity, heart disease, and diabetes, associated with poverty and the stress of marginalization. Australian Aboriginal populations, for instance, have a life expectancy 8 years shorter than non-indigenous Australians (Packham and Jose 2021). In some indigenous communities, alcoholism and suicide rates are also high, reflecting the huge stresses imposed on these communities. These pre-existing health inequities were accentuated by the COVID-19 pandemic. In Australia, for instance, Aborigines were particularly hard hit owing to a lack of adequate healthcare facilities, high rates of chronic diseases, and living arrangements that made the isolation of infected people challenging (Khalil 2021).

These great injustices are finally being acknowledged, with some national governments issuing apologies for past injustices and granting indigenous people more self-determination. Canada, for example, has created a Truth and Reconciliation Commission report detailing the horrific treatment of Canada's indigenous people. Australia has recently offered financial compensation to the Stolen Generations (Packham and Jose 2021). In recognition of the widespread mistreatment of indigenous people, the United Nations adopted the Declaration on the Rights of Indigenous Peoples in 2007, the most comprehensive document about indigenous rights to date (UNDESA, n.d.). Most important, many indigenous people are now guiding their own communities toward a better future through self-governance, tailored education, and new cultural initiatives.

A subdiscipline of geography, ethnic geography, specifically examines how the distribution of different ethnic (or racial) groups shapes and is influenced by social, economic, and political issues and how ethnic groups interact with one another. In the United States, for instance, the distribution of ethnic groups stems from its history, with populations of African Americans in the South connected to slavery and people of Latino origin in the Southwest reflective of the region's proximity to the current Mexican border and the fact that much of this region was Mexican territory until the mid-19th century. Native Americans are concentrated in the Southwest and states that have Native American reservations, owing to histories of land dispossession. Over time, this distribution has been changing. African Americans migrated north and west, particularly from the 1870s after the abolition of slavery, and the Latino population has both grown and expanded into new regions in the 20th century. The Southwest remains a Latino stronghold, but New York and Miami have also been transformed by Latino migrants. Latinos are now the largest minority in the United States, surpassing African Americans in 2003 (Saenz 2004).

The distribution of ethnic and racial groups has important implications, particularly related to whether ethnic clustering occurs voluntarily or through discrimination. Many immigrants choose to live close to people who are similar to them, at least initially, forming distinctive ethnic neighborhoods. These **ethnic enclaves** can shield people from discrimination and allow communities to preserve their language and culture. Examples include Germantown and Little Italy in New York, formed during the age of massive European migration to the United States, and more recent concentrations such as Little Havana and Little Nicaragua in Miami or Koreatown and Tehrangeles (Little Persia) in Los Angeles. Typically, migrants (or their children) eventually disperse from these ethnic communities. This is often seen as a positive process of gradual assimilation to the new country. In contrast to this largely positive image of ethnic clustering are **ghettos**, where people are forced to live due to discrimination and prejudice. Ghettos are associated with forced **segregation** and often go hand in hand with poverty and other social problems. South Africa's system of Apartheid represents one of the best-known systems of institutionalized segregation. Lasting from 1948 to 1991, the Apartheid system classified all South African residents into four racial categories (Whites, Africans, Coloureds, and Asians) and deprived non-white groups (particularly Africans) of basic political, economic, and social rights. Classifications were made based on appearance, ancestry, socio-economic status, and lifestyle and led to determinations of where you were allowed to live, the jobs you could apply for, and whom you were allowed to socialize with. The system was widely condemned and was countered by anti-Apartheid social movements, both from within South Africa and globally.

Beyond the social implications of segregation, social justice issues are also raised by the physical characteristics of different neighborhoods. For example, polluting industries and other environmental problems are frequently found in close proximity to settlements dominated by ethnic or racial minorities, because they often lack the financial means to move away or the political power to prevent potentially dangerous industries from being set up in their neighborhoods. This disproportionate exposure of minority groups to toxic environments is termed **environmental racism**. US examples include "Cancer Alley" along the Mississippi River, lined with oil refineries and petrochemical plants; the contamination of drinking water in Flint, Michigan; and the proposed storage of nuclear waste at Yucca Mountain, Nevada—all of which place disproportionate health and safety burdens on communities of color. More broadly, health outcomes vary dramatically among ethnic and racial groups, associated with both poverty and the damaging health impacts of discrimination and marginalization.

Despite our increasing understanding of race as a category constructed by people, many people continue to mistakenly believe that there are significant differences in ability or personality that can be defined by race. **Racism** is the practice of discriminating against people because of physical features such as skin color or facial features. The related idea of **colorism** reflects discrimination by skin tone, often from people within the same ethnic group. Although some countries have taken steps to try to outlaw overt racism (e.g., laws requiring that employment decisions may not consider race or laws that define racially motivated attacks as hate crimes), racist attitudes remain deeply embedded in most societies. Additionally, many structures of society were set up in an earlier era when racism was hardly questioned, leading to claims that **structural racism** remains a problem in many countries. Structural racism suggests that racist attitudes are so woven into structures of society that they shape how society functions to this day, leaving minority groups at a systematic disadvantage.

It is important to note that how communities understand race, as well as how racism is perpetuated, varies around the world. In the United States, for example, racial politics draws a strong distinction between those who are considered white and those who are not. This is in contrast to a country such as Brazil, which has multiple terms for a diversity of different biracial and multiracial combinations and where relative skin tone is more important to identity than a simple binary black versus white division. That is not to say that racism is absent in Brazil—it is still a dominant force—just that the way that race is conceived is different.

Today's experiences of race and racism have developed from a long history of racial discrimination in many countries. The idea of **white privilege** has emerged from this problematic history to describe how whiteness has persistently been favored in many societies. To try to redress the balance, some countries such as the United States have experimented with **positive discrimination** (or "affirmative action"), which aims to offer a deliberate advantage to communities of color; for instance, by ensuring that applicants of color are given particularly careful consideration for job openings. For supporters, this represents a reasonable redressing of historic and current disadvantages experienced by minority communities. For opponents, these policies are interpreted as offering certain groups an unfair advantage, generating a backlash against these efforts. Another approach to addressing issues of racism is the idea of being **colorblind**. Here, instead of highlighting inequalities between racial groups, society is charged with not even considering race, in the hope that everyone will then be treated fairly. In France, for example, there is a strong belief that all citizens have equal rights as individuals, rather than because of membership of a particular group. Everyone is viewed simply as French, and the country does not collect any data on ethnicity or race or recognize racial minorities. Unfortunately, inequalities rarely disappear by simply ignoring differences, and President Macron has admitted that racial profiling and white privilege persist in France (*The Economist* 2021e).

Singapore illustrates yet another approach to race: state-enforced **multiracialism**. The country is dominated by three main ethnic groups: about 75 percent of the population are ethnic Chinese, 13 percent Malay, and 9 percent Indian (figure 3.17). In the 1960s, tensions between Chinese and Malays threatened the unity of the newly independent country, and the government stepped in by designating Chinese, Tamil (an Indian language), Malay, and English as official languages and forbidding any form of racism. The government has implemented numerous policies aimed at ensuring racial harmony, most notably racial quotas in public housing, which houses about 80 percent of Singapore's population. Since 1989, this Ethnic Integration Policy (EIP) has required the ethnic makeup of each city block to reflect national ethnic composition. For example, if the Chinese population in a specific block has reached the same percentage as that of residents in the country as a whole, no additional Chinese are allowed to move to that block (Global-is-Asian 2019), resulting in the dispersal of ethnic communities. Today, open discrimination and racism are rare in Singapore, although tensions simmer beneath the surface (*The Economist* 2021d).

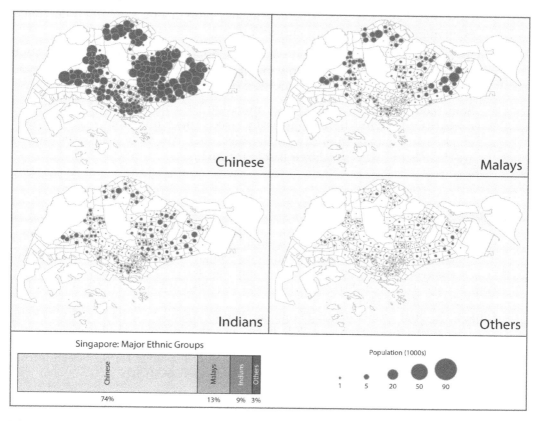

FIGURE 3.17 Population distribution and composition by major ethnic groups in Singapore
Data source: Statistics Singapore (2011).

Conclusion

Knowledge of population distribution and composition is not just an academic exercise but helps us understand a variety of problems and human rights concerns. Although we presented several major demographic issues in separate sections, it is important to note that they often intersect. For example, the age and sex structures of a population are connected to some degree because women tend to outlive men. Furthermore, an elderly woman of color may find herself in a particularly disadvantaged position owing to an intersection of marginalized identities. We have only scratched the surface of topics related to population composition and identity in this chapter, and we will be revisiting many of the issues we raise here in subsequent chapters.

Discussion questions

1   Consider the population distribution of a large country (e.g., Australia, Canada, China). What environmental and human factors can you identify that might explain its uneven population distribution?

2 How useful and accurate is the dependency ratio as a measure of productivity? How might we improve this measure?

3 In recent years, heated discussions have arisen about trans-women's participation in gendered sporting events. How might we resolve some of these controversies, and what other alternatives might there be to improve accessibility for all?

4 Some countries have positive discrimination (affirmative action) policies that favor people who have historically faced discrimination (e.g., through employment quotas, minority scholarships, etc.). Are these policies helpful, or are there better ways to address past injustices?

Suggested readings

Hassan, A. 2021. "'Evil Customs': Why a Kashmiri Village Abandoned Dowries." *The Guardian*, October 12, 2021. https://www.theguardian.com/global-development/2021/oct/12/evil-customs-why-a-kashmiri-village-abandoned-dowries

Kaur, R., and T. Kapoor. 2021. "The Gendered Biopolitics of Sex Selection in India." *Asian Bioethics Review* 13: 111–27.

LaBreck, A. 2021. "Color-blind: Examining France's Approach to Race Policy." *Harvard International Review*, February 1, 2021. https://hir.harvard.edu/color-blind-frances-approach-to-race/

Roan, D., and K. Falkingham. 2022. "Transgender Athletes: What Do the Scientists Say?" BBC Sport, May 11, 2022. https://www.bbc.com/sport/61346517

Glossary

age cohort: a group of people born around the same time period who share certain experiences over their life course

arithmetic density: the number of people per area of land

colorblind approach (to racism): treating everyone equally without consideration of race

colorism: discrimination or prejudice against people with a darker skin tone

demographic dividend: the temporary economic boost an economy receives when a population has a large proportion of workers owing to recent declines in the birth rate

dependency ratio: measure of the number of dependents under age 15 and over 65, compared with the population aged 15 to 64

digital nomad: a person who earns a living working online in a location of their choosing

ecumene: the part of the world inhabited by people or suitable to human settlement

environmental racism: the disproportionate exposure of communities of color to pollution and other environmental hazards

ethnic enclave: an area with a high concentration of people of the same ethnicity

ethnicity: belonging to a group of people who share a common origin and culture

footloose: an industry is footloose when it does not require resources that tie it to a specific location

gender: social groupings related to sex as defined by social and cultural expressions of masculinity and femininity rather than biological aspects

ghetto: a poor urban area typically inhabited by ethnic or racial minorities

indigenous peoples: group of people descended from the first peoples known to have inhabited a country and demonstrating some traditional lifeways

LGBTQ+: acronym representing a community of people who may identify with a diversity of alternative gender identities and sexual orientations

median age: the age at which half of the population is older and half younger

megacity: a very large city, currently usually defined as having at least 10 million inhabitants

multiracialism: a policy that considers racial groups as of equal value and promotes harmony among racial groups

nonecumene: uninhabited or sparsely populated regions of the world

patriarchy: social system dominated by men, marginalizing women

physiological density: the number of people per unit area of arable land

population centroid: the point from which there is the same number of people to the north, south, east, and west within a certain area, usually a country

population density: the concentration of people within a specific area

population pyramid: graphic representation of the age and sex structure of a population

positive discrimination: the practice or policy of favoring people belonging to groups regarded as disadvantaged or discriminated against (e.g., affirmative action in the United States)

potential support ratio: the proportion of people aged 15 to 64 to people aged 65 or older, as a measure of the need for care in a population

race: a group of people considered distinct from other groups, as identified by physical characteristics such as skin color

racism: the practice of discriminating among people of different races, especially in unfair ways

rural renaissance: renewed attraction of rural areas to people, often to avoid problems associated with cities

rural-to-urban migration: migration from the countryside to cities

sandwich generation: people in middle age who have both childcare and eldercare responsibilities simultaneously

segregation: the physical separation of people of different ethnic or racial backgrounds or socio-economic characteristics

sex: biological aspects of being male or female

sex ratio: the number of males in relation to the number of females in a population

socially constructed: a shared idea or belief that has been created by people

structural racism: a system in which policies, practices, and norms perpetuate inequalities among different racial groups

welfare state: a political approach that uses taxes to provide support for individuals via mechanisms such as socialized healthcare and pensions

white privilege: socially constructed advantages white people possess over people of color

Works cited

Amnesty International. 2021. "Indigenous Peoples." https://www.amnesty.org/en/what-we-do/indigenous-peoples/

Angelo, P., and D. Bocci. 2021. "The Changing Landscape of Global LGBTQ+ Rights." Council on Foreign Relations, January 29, 2021. https://www.cfr.org/article/changing-landscape-global-lgbtq-rights?gclid=Cj0KCQjwvZCZBhCiARIsAPXbajutdct-yYigegr JJuT1CoQ5gxjuA2djXQOpttUSXmU1ImsEMKJcttoaAlOtEALw_wcB

Buchanan, A. 2013. "The Impact on Mothers: Managing Competing Needs." In *Fertility Rates and Population Decline*, edited by A. Buchanan and A. Rotkirch, 230–48. London: Palgrave Macmillan.

Colliers International. 2020. "Egypt's Future Cities: New Opportunities." Colliers International, November 4, 2020. https://www.colliers.com/en-eg/research/cairo/egypt-future-cities

Common Ground. n.d. "The Stolen Generations." Accessed February 4, 2022. https://www.commonground.org.au/learn/the-stolen-generations

Dawson, L., J. Kates, and M. Musumeci. 2022. "Youth Access to Gender Affirming Care: The Federal and State Policy Landscape." Kaiser Family Foundation, June 1, 2022. https://www.kff.org/other/issue-brief/youth-access-to-gender-affirming-care-the-federal-and-state-policy-landscape/

Ebenstein, A., and E. Sharygin. 2009. "The Consequences of the 'Missing Girls' of China." *The World Bank Economic Review* 23 (3): 399–425.

The Economist. 2013. "Gendercide in the Caucasus." September 21, 2013. https://www.economist.com/europe/2013/09/21/gendercide-in-the-caucasus

——. 2019. "Love Money: Why Dowries Persist in South Asia." May 18, 2019. https://www.economist.com/asia/2019/05/16/why-dowries-persist-in-south-asia

——. 2021a. "The Cost of Misogyny: Societies That Treat Women Badly Are Poorer and Less Stable." https://www.economist.com/international/2021/09/11/societies-that-treat-women-badly-are-poorer-and-less-stable

——. 2021b. "Honour Killings. Murder, Plain and Simple." February 6, 2021. https://www.economist.com/middle-east-and-africa/2021/02/06/arab-governments-are-doing-too-little-to-end-honour-killings

——. 2021c. "The Number of Young Adults in Britain Is About to Rise Sharply." August 21, 2021. https://www.economist.com/britain/2021/08/19/the-number-of-young-adults-in-britain-is-about-to-rise-sharply

——. 2021d. "Racial Prejudice Rears Its Head in Singapore." July 31, 2021. https://www.economist.com/asia/2021/07/29/racial-prejudice-rears-its-head-in-singapore

——. 2021e. "Racism Tests France's Colour-Blind Model." January 16, 2021. https://www.economist.com/europe/2021/01/14/racism-tests-frances-colour-blind-model

——. 2021f. "A Terrible Toll. Violence against Women Is a Scourge on Poor Countries." March 13, 2021. https://www.economist.com/international/2021/03/11/violence-against-women-is-a-scourge-on-poor-countries

——. 2022. "How the War in Ukraine Is Changing Europe's Demography." April 30, 2022. https://www.economist.com/international/2022/04/30/how-the-war-in-ukraine-is-changing-europes-demography

FAO [Food and Agriculture Organization of the United Nations]. 2022. "FAOSTAT: Land Use." https://www.fao.org/faostat/en/#data/RL

Federal State Statistics Service, Russia. n.d. "Demographic Yearbook of Russia, 2019." Accessed December 14, 2021. https://eng.rosstat.gov.ru/

Ghanem, S. n.d. "New 4th Generation Cities: The New Map of Egypt." Invest.Gate. Accessed June 1, 2021. https://invest-gate.me/features/new-4th-generation-cities-the-new-map-of-egypt/

Global-is-Asian. 2019. "Multiracial Singapore: Ensuring Inclusivity and Integration." https://lkyspp.nus.edu.sg/gia/article/multiracial-singapore-ensuring-inclusivity-and-integration

Guilmoto, C. 2012. "Skewed Sex Ratios at Birth and Future Marriage Squeeze in China and India, 2005–2100." *Demography* 49: 77–100.

Hardjono, J. 1977. *Transmigration in Indonesia.* London: Oxford University Press.

Hassan, A. 2021. "'Evil Customs': Why a Kashmiri Village Abandoned Dowries." *The Guardian*, October 12, 2021. https://www.theguardian.com/global-development/2021/oct/12/evil-customs-why-a-kashmiri-village-abandoned-dowries

Höpflinger, F. 2012. *Bevölkerungssoziologie. Eine Einführung in demographische Prozesse und bevölkerungssoziologische Ansätze.* Weinheim and Basel: Beltz Juventa.

Human Dignity Trust. 2022. "Map of Countries That Criminalize LGBT People." https://www.humandignitytrust.org/lgbt-the-law/map-of-criminalisation/?type_filter_submitted=&type_filter%5B%5D=death_pen_applies

Human Rights Watch. 2020. "Every Day I Live in Fear." Human Rights Watch, October 7, 2020. https://www.hrw.org/report/2020/10/07/every-day-i-live-fear/violence-and-discrimination-against-lgbt-people-el-salvador

IBGE [Instituto Brasileiro de Geografia e Estatística]. 2022. "Censo Demográfico. Séries históricas. População residente, 1872–2010." Accessed May 23, 2022. https://www.ibge.gov.br/en/home-eng.html

ILGA [The International Lesbian, Gay, Bisexual, Trans and Intersex Alliance]. 2020a. "Maps—Sexual Orientation Laws." https://ilga.org/maps-sexual-orientation-laws

———. 2020b. "State-Sponsored Homophobia: Global Legislation Overview Update." https://ilga.org/downloads/ILGA_World_State_Sponsored_Homophobia_report_global_legislation_overview_update_December_2020.pdf

Johanson, M. 2021. "The 'Zoom Towns' Luring Remote Workers to Rural Enclaves." BBC, June 8, 2021. https://www.bbc.com/worklife/article/20210604-the-zoom-towns-luring-remote-workers-to-rural-enclaves

Jones, G. 2013. "The Growth of the One-Child Family and Other Changes in the Low Fertility Countries of Asia." In *Fertility Rates and Population Decline*, edited by A. Buchanan and A. Rotkirch, 44–61. London: Palgrave Macmillan.

Kaur, R., and T. Kapoor. 2021. "The Gendered Biopolitics of Sex Selection in India." *Asian Bioethics Review* 13: 111–27.

Khalil, S. 2021. "Fury as Covid Crisis Hits Australia's Aboriginal Communities." BBC, August 30, 2021. https://www.bbc.com/news/world-australia-58380827

Kono, S. 2011. "Confronting the Demographic Trilemma of Low Fertility, Aging, and Depopulation." In *Imploding Populations in Japan and Germany: A Comparison*, edited by F. Coulmas and R. Lützeler, 35–53. Leiden/Boston: Brill.

Lee, J. 2021. "Gender-Neutral Passports: Campaigner Christie Elan-Cane Loses Supreme Court Case." BBC, December 15, 2021. https://www.bbc.com/news/uk-59667786

OPENDEM. 2022. "Download SRTM Based Contour Lines." Accessed January 12, 2022. https://www.opendem.info/download_contours.html

Packham, C., and R. Jose. 2021. "Australia to Offer Redress Payments to Some of Its 'Stolen Generation.'" Reuters, August 5, 2021. https://www.reuters.com/world/asia-pacific/australia-establish-280-mln-reparations-fund-stolen-generation-2021-08-04/

Pandey, G. 2021. "NFHS: Does India Really Have More Women than Men?" BBC, November 27, 2021. https://www.bbc.com/news/world-asia-india-59428011

Ritchie, H., and M. Roser. 2019. "Age Structure." OurWorldinData. https://ourworldindata.org/age-structure

Rogerson, P. 2021. "Historical Change in the Large-Scale Population Distribution of the United States." *Applied Geography* 136: 102563. https://doi.org/10.1016/j.apgeog.2021.102563

Saenz, R. 2004. "Latinos and the Changing Face of America." Population Reference Bureau, August 20, 2004. https://www.prb.org/resources/latinos-and-the-changing-face-of-america/

Seager, J. 2018. *The Women's Atlas*. Oxford: Myriad Editions.

Selby, D. 2016. "Everything You Should Know about Honor-Based Violence." Global Citizen, July 21, 2016. https://www.globalcitizen.org/en/content/honor-based-violence-killings-women-girls-pakistan/

SIPUKAT [Sistem Informasi Peta Terpadu Kawasan Transmigrasi]. n.d. "Kawasan Transmigrasi." April 10, 2022. https://sipukat.kemendesa.go.id/

Statistics Indonesia. 2021. "Statistical Yearbook of Indonesia 2021." https://www.bps.go.id/

Statistics Singapore. 2011. "Singapore Census of Population 2010." https://www.singstat.gov.sg/publications/cop2010/

Tadamun. 2015. "Egypt's New Cities: Neither Just nor Efficient." Tadamun, December 31, 2015. http://www.tadamun.co/egypts-new-cities-neither-just-efficient/?lang=en#. YLd9IYWSmUk

Tremblay, J., and P. Ainslie. 2021. "Global and Country-Level Estimates of Human Population at High Altitude." *Proceedings of the National Academy of Sciences of the United States of America* 118: 18. https://www.pnas.org/content/118/18/e2102463118

UNDESA [United Nations Department of Economic and Social Affairs]. 2019. "World Population Prospects 2019." https://population.un.org/wpp/

———. 2020. "Policies on Spatial Distribution and Urbanization Have Broad Impacts on Sustainable Development." Population Facts, December 2020. https://www.un.org/development/desa/pd/sites/www.un.org.development.desa.pd/files/undes_pd_2020_popfacts_urbanization_policies.pdf

———. n.d. "Indigenous Peoples." Accessed April 4, 2022. https://www.un.org/development/desa/indigenouspeoples/

UNFE [United Nations, Free and Equal]. n.d. "Factsheet: Intersex." Accessed May 5, 2022. https://www.unfe.org/wp-content/uploads/2017/05/UNFE-Intersex.pdf

UNICEF [United Nations Children's Fund]. 2020. "Gender and Education." https://data.unicef.org/topic/gender/gender-disparities-in-education/

United Nations Women. n.d. "Facts and Figures: Women's Leadership and Political Participation." Accessed December 20, 2021. https://www.unwomen.org/en/what-we-do/leadership-and-political-participation/facts-and-figures

United States Census Bureau. 2021. "Centers of Population." Census Bureau, November 16, 2021. https://www.census.gov/geographies/reference-files/time-series/geo/centers-population.html

———. n.d. "Census 2010 Summary File 1. PCT12 Sex by Age." Accessed April 1, 2022. https://data.census.gov/cedsci/

Westfall, S. 2022. "Indonesia Passes Law to Move Capital from Jakarta to Borneo." *Washington Post*, January 18, 2022. https://www.washingtonpost.com/world/2022/01/18/indonesia-capital-city-jakarta-borneo/

World Economic Forum. 2021. "The Global Gender Gap Index 2020." http://reports.weforum.org/global-gender-gap-report-2020/the-global-gender-gap-index-2020/

WorldPop. 2022. "Mapping Populations." https://www.worldpop.org/methods/populations

4

Population growth and change

After reading this chapter, a student should be able to:

1 describe major changes in population size over time;
2 discuss some of the key factors driving rapid population growth using the example of India;
3 explain major theories of population growth and change associated with Malthus, Marx, and the demographic transition model.

Population growth has been one of the enduring trends of human population dynamics for much of human history. Although populations have fluctuated over short timescales, particularly in early human eras, in more recent centuries the human story has been one of population growth, with global population hitting 8 billion in November 2022. The environmental implications of this are profound, as more and more of the Earth's land surface and energy have been appropriated for human use. In recent decades this trend has begun to slow, however, as many human populations have begun to limit fertility, leading us to a potential turning point—will human population size finally stabilize and even decline in the coming decades?

PART I: POPULATION GROWTH OVER TIME

A short history of population growth

Most of us have seen graphs alerting us to the dramatic growth in human population experienced over the past 200 years. Population growth was indeed dramatic from the late 1700s onwards, with an almost eight-fold increase in global population size between 1800 and 2020 (figure 4.1, inset graph). Prior to the 1700s, the

DOI: 10.4324/9781003143253-4

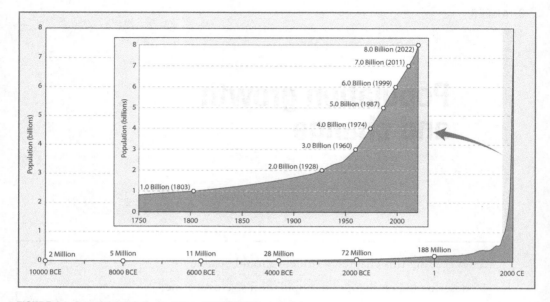

FIGURE 4.1 Growth of global population, 10,000 BCE to present
Note that estimates for prehistoric population sizes vary widely; figures given here provide one estimate but should be considered approximations.
Data source: Manning (n.d.).

story of human population was still of overall growth but in much more gradual and variable terms (figure 4.1, main graph). To understand where we find ourselves today, a history of population growth is instructive.

From an ecological perspective we can see human history in terms of an evolving relationship between people and the natural world, with human societies developing new technologies to capture larger and larger proportions of Earth's energy and resources. For 99 percent of human existence, all human societies lived in the same way—gathering and hunting food from the local environment. The need to gather large enough quantities of starchy roots, fruits, nuts, and other plant foods, along with the desire to supplement this diet with nutrient- and energy-rich foods like meat, fish, and honey, necessarily kept populations small and mobile. Most groups probably consisted of extended families and the total global population may have numbered only 5 million people at the dawn of the **agricultural revolution**, although estimates vary widely (PRB 2020a).

The challenges of this lifestyle cannot be understated. Populations fluctuated significantly in size as climatic variations led to periodic famines, and disease and injury were often life-threatening. Although estimates vary, life expectancy at birth may have been as low as 10 or 12 years during this early period (PRB 2020a). This extremely low life expectancy reflects the many people dying in infancy and childhood, pulling the average down, whereas some of those making it to adulthood may have survived into their 40s, or even longer. Under such circumstances, birth rates might have had to have been as high as 80 live births per 1,000 people just to keep the population stable (for comparison, a birth rate of 30 per 1,000 is considered high today; PRB 2020a).

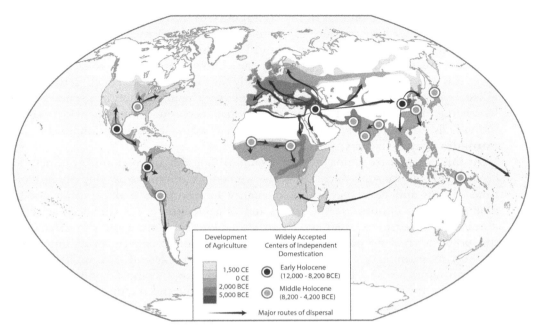

FIGURE 4.2 Agricultural hearths and the spread of early agriculture
Source: Drawn after similar images in Larson et al. (2014) and Bertin, Vidal-Naquet, and HarperCollins (1992).

Circumstances changed dramatically with the development of agriculture around 10,000 to 12,000 years ago. Instead of simply gathering the resources that the local ecosystem had to offer, agriculture allowed people to modify ecosystems in ways that focused a larger proportion of the ecosystem's energy toward producing food. The fact that agriculture was adopted independently in several different regions of the world suggests that it was a relatively intuitive and gradual process rather than a sudden stroke of genius, possibly triggered by regional food shortages or localized climate change (figure 4.2). The move to an agricultural way of life probably involved communities beginning to realize that they could plant or tend favorable species, while concurrently weeding out those of less use. Discarded seeds, as well as human waste containing undigested seeds, may also have led to seeds germinating near the community, leading people to the idea that deliberate cultivation of crops was possible. Communities probably also began to nurture animal herds by killing predators and providing fodder to encourage wild herds to stay close by. Agricultural lifestyles likely first developed in parts of the Middle East (sometimes known as the "Fertile Crescent") and China but subsequently arose independently in a number of different places, including in the Americas, India, and Africa. These so-called **hearths of agriculture** are represented by the circles in figure 4.2. Once agriculture developed in a particular region, it then spread to neighboring areas (represented by the shading on the map), as neighboring communities adopted the new agricultural techniques and as agricultural communities expanded into new territory.

Once communities had turned to agriculture, the amount of energy they could extract from the local ecosystem increased dramatically, allowing more children to survive and populations to increase. The amount of food generated per person also rose, allowing some members of the community to move away from food production toward other activities such as pot making and basket weaving. This produced some of the first consumer goods, as well as vessels to improve food storage and tools to increase productivity. Agriculture also allowed populations to settle because there was no longer a need to roam so widely for food supplies; indeed, settling in a village was critical to being able to tend crops for the many months needed before harvesting them. This settling down also had economic consequences because individuals could now accumulate goods that would have been impossible for **nomadic** peoples to carry.

At this point, there is not only an ecological but also an economic argument to be made. Trade can only begin in earnest when communities develop an agricultural surplus and have the time and energy to devote to making trade goods. Competition to dominate trade routes and valuable territory is enhanced as communities congregate in larger and larger settlements with more assets to defend. At this point, a growing population becomes an economic benefit, providing more workers to manufacture goods and more soldiers to defend community assets. Administrators are also needed to organize increasingly complex systems of food production; for instance, overseeing irrigation systems and the distribution of surpluses. Anthropologists note that these changes probably led to increased social stratification as artisans, soldiers, and government and religious officials began to distinguish themselves from farmers through rituals and symbols of status. This also marked the beginning of *urban* civilization as surplus production allowed people to live at higher population densities and a new class of administrators provided the social organization needed to coordinate the more complex social structures required for large settlements.

The growing economic and political might of these early agricultural communities, in addition to sheer population size, helped them to dominate hunter-gatherer groups, who were increasingly forced off more favorable land (Diamond 1999). Whether individuals actually became happier or healthier as they settled down to agricultural lifestyles has been debated. Diamond (1999) argued that humankind may have paid a heavy price for the conversion to agriculture in terms of an impoverishment of diet (more calories but less diversity of foods and fewer nutrients), increased transmission of infectious disease among higher density populations, longer working hours, and increased social stratification reinforcing inequalities. If we look at agricultural success in terms of population size, however, the impact of embracing agriculture was powerful, triggering a dramatic jump in global population size to perhaps 300 million people by 1 CE (2,000 years ago)—a little less than the current population of the United States—although estimates vary widely (PRB 2020a).

Human population continued to expand rapidly with the abundance of calories provided by this new agricultural way of life, reaching perhaps 500 million people around 1650. Growth may have slowed in this latter period (1 CE to 1650), however, as larger, denser populations became a breeding ground for infectious diseases. This was a particular issue for the interconnected populations of Eurasia and Africa, through which a variety of diseases were known to circulate from historic records—most famously plague, which is believed to have caused millions of deaths in multiple outbreaks during this period (PRB 2020a; box 4.1).

Despite setbacks associated with periodic epidemics and food shortages, global population continued to increase overall, reaching 1 billion by about 1800. It was at

BOX 4.1 PLAGUE

Plague—its very name synonymous with epidemics—produced some of the most infamous disease events in human history. Mortality rates for plague today are 50 to 60 percent if left untreated (WHO, n.d.); death rates may have been even higher in earlier eras. Although some evidence suggests that plague may have infected humans since the Bronze Age (Callaway 2015), the first major recorded outbreaks were the Justinian Plagues of the mid-500s CE, which may have killed 100 million people (WHO, n.d.). More famously, outbreaks in the 14th to 17th centuries, when the disease was known as the "Black Death," killed perhaps 30 to 50 percent of the European population or 75 to 200 million people (UNESCO, n.d.; figure 4.3). This wave of disease is often associated with increased regional integration related to trade routes such as the Silk Road, which likely spread the disease. The disease probably emerged from marmots on the steppes of Central Asia, when fleas from the marmots, carrying the *Yersinia pestis* bacteria that causes plague, may have bitten traders or traveled with them, leading to outbreaks of plague in Europe and China (Suntsov 2015).

The scale of mortality associated with these outbreaks was devastating in terms of loss of human life and disruption to lifestyles. Scholars have also highlighted a host of longer-term impacts. Biologically, the plague acted as a powerful force of natural selection, removing

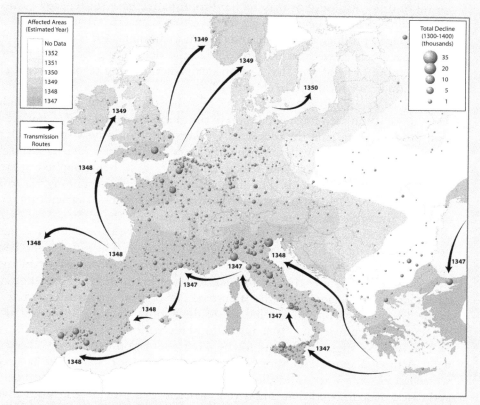

FIGURE 4.3 Spread of the Black Death in Europe
Circle size indicates the decrease in population of cities where population declines were recorded.
Data source: Buringh (2021).

certain individuals and their genes from the population (DeWitte 2014). Socially, the disease reduced the pool of available labor, leading to high demand for workers and increased wages. This may have led to declining economic inequalities as land and assets were redistributed (Alfani 2020). Environmentally, whole swathes of marginal land that had been in agricultural production probably reverted to forest and grassland (van Hoof et al. 2006).

this point that another surge in population was triggered by a new change in humankind's relationship with its environment: the Industrial Revolution. From an ecological perspective, the Industrial Revolution of the late 1700s onwards can be viewed as a further way in which humankind gained control over yet more of the world's energy. The Industrial Revolution was premised on switching from human and animal power to mineral energy in the form of hydrocarbons. The coal, and then oil and gas, that fueled the Industrial Revolution was, in effect, energy stored from previous eras—the compressed tissues of plants and animals that grew from solar energy taken in millions of years ago. Burning these fuels released this stored energy, providing a huge energy boost to human activities over a short period of time. Although it may be hard to visualize how this energy turned into more people, from an ecological perspective that is exactly what happened—more energy helped produce more food and then more humans. Of course, from an ecological perspective, we also now recognize the devastating implications that the rapid burning of fossil fuels has had on our planet, leading to our current climate crisis—a topic we return to in chapter 12.

From an economic perspective, we can also think about how the Industrial Revolution stimulated huge expansion of the global economy, triggering growth in food production to feed the industrial workers who could now spend their wages on food to feed their families. Meanwhile, new technologies such as steam power, mass-produced metal tools, and selective breeding were being applied to farming, increasing agricultural yields. The integration of the global economy via emerging trade routes—an early form of **globalization**—also enhanced agricultural productivity by giving people access to new crops from across the globe and allowing for the redistribution of surpluses (Bähr 2010). It was in this era, for instance, that potatoes were imported from South America to Europe. The potato soon became a staple crop in the cool, wet climate of Northern Europe, allowing for rapid population growth in regions that were marginal for grain crops. These agricultural developments associated with the Industrial Revolution are sometimes referred to as the "second agricultural revolution" to distinguish them from the initial development of agriculture (the "first agricultural revolution").

The result of these combining forces was a population explosion without precedent. Although living conditions remained desperate for many and had arguably worsened in some ways as people moved to overcrowded and disease-ridden towns and cities, overall, conditions improved sufficiently to provide a survival advantage over previous eras and industrial populations grew. It was at this point that Thomas Malthus wrote his famous *An Essay on the Principle of Population* (1798), in which he expressed his concerns that population would outstrip food supply. Given the conditions of destitute poverty that many families lived in at this time, with malnutrition and food shortages common, it is easy to understand how Malthus believed that human populations might be nearing their ecological bounds.

In the more than 200 years since Malthus penned his concerns over food supply, his fears have not been realized, however. Indeed, global population has increased almost eight-fold since Malthus's day and yet today agriculture produces more calories per capita than ever before. The application of scientific methods to farming—the "third agricultural revolution"—is the key to how this was achieved. The discovery that nitrogen from the air could be artificially "fixed" into compounds that could be used by plants was especially significant, paving the way for the development of artificial fertilizers. By the mid-20th century, farming was being revolutionized by artificial fertilizers, pesticides, herbicides, and new high-yielding varieties of crops, dramatically increasing yields. This scientific endeavor began in the laboratories of the Global North, but by the 1960s humanitarian efforts had begun to spread these agricultural technologies to the Global South in a movement termed the **Green Revolution**.

The Green Revolution was an undoubted success in terms of massively increased global crop production, leading to more calories and protein for many, although micro-nutrient deficiencies persisted in many poor communities (Pingali 2012). Despite a doubling of population between 1960 and 2010, the number of people suffering from hunger halved from one in three to one in six over the same period (Wik, Pingali, and Brocai 2008). The Green Revolution's impacts were uneven, however, with some regions benefiting significantly more than others—in particular, Africa and some of the world's poorest communities were largely ignored by early interventions. The Green Revolution also came at considerable ecological cost, related to soil erosion, loss of biodiversity, exhaustion of groundwater supplies, and pollution of waterways, although some landscape conversion was probably avoided owing to the rising yields brought by the new technologies (Pingali 2012). In short, the Green Revolution was a huge success in terms of increasing crop yields but its broader impacts have been more mixed.

Despite the increase in food production, the 1950s and '60s heralded a new era of concern over population growth. By the mid-20th century, the fastest rates of population growth had shifted from Europe to the previously colonized countries of the world and global population size doubled from 3 billion in 1960 to 6 billion by 1999. This trajectory of growth was undeniably unsustainable from an ecological perspective. However, there is considerable disagreement over whether the rhetoric around population growth at this time was fair or helpful, as population growth began to generate explicitly political concerns. Commentators from rich countries expressed fears that rapid population growth in poorer countries could lead to political instability, which would then disrupt the smooth functioning of the global economy. The plethora of new contraceptive technologies being developed at this time (particularly the contraceptive pill) began to be visualized as tools to help "control" population size rather than to increase reproductive freedom for couples.

From a social equity perspective, it is notable that it was when populations of color began to increase rapidly that the Global North began expressing great concern over the need to curtail population growth—with the reproduction of communities in the Global South increasingly being seen as a "problem" that needed "fixing." In 1965, US President Johnson declared that he would "seek new ways to use our knowledge to help deal with the explosion in world population and the growing scarcity of world resources"; in 1969, President Nixon described population growth as "one of the most serious challenges to human destiny" (cited in USAID, n.d.). Population policy at this time was not just race based but also class based, with elite policymakers devising policies aimed at curtailing the fertility of poorer populations. Although considerable funding for these early policies flowed from the Global North to the Global South, many of the policies themselves were created and enacted by local policymakers—a government elite, often Western-educated, that agreed that the speed of population growth in their country was problematic. Many of these early campaigns have since been criticized for being overly focused on demographic targets (i.e., reducing birth rates) at the expense of people's reproductive freedom. For instance, many early programs were heavily focused on promoting sterilization, rather than providing diverse family planning options, which has been interpreted by critics as evidence that the goal was to reduce family size rather than to empower people in their fertility choices.

It was not just political and economic concerns that drove mid- and late-20th-century family planning campaigns, however. Increasingly, attention shifted to the potential ecological implications of growing populations, with the environmental movements of the 1970s onwards popularizing concerns that the human population was approaching global **carrying capacity**. To this day, widespread habitat destruction, water shortages, and pollution are often seen as part and parcel of the burden that growing populations are putting on the Earth. As the environmental movement has become more sophisticated and nuanced in its analyses, however, there has been increased recognition that it is not just population size that is the problem but also consumption patterns. Indeed, the reimagining of the environmental movement toward broader sustainability goals has placed considerable emphasis on ensuring that social justice is not sacrificed for environmental goals. This emphasis on consumption patterns in addition to population size has reintroduced a social equity angle that begins to shift some of the blame for environmental damage back toward affluent populations. This fits into a broader social justice narrative that argues that we need to address a wide variety of social problems—most notably poverty and inequality of access to resources—before we can effectively tackle environmental problems. From this perspective, rising populations are undeniably a significant ecological challenge, but the rich must shoulder their fair share of blame for environmental problems owing to their high-consumption lifestyles. In this view, the poor have the right to raise their own living standards via development projects that try to achieve economic development at minimal ecological cost (see Brundtland 1987).

This social justice narrative has also begun to confront the racist and classist population policies of the past, arguing that family planning campaigns today should only be instituted in ways that enhance reproductive freedom, rather than in ways that place an undue focus on meeting demographic targets. This shift was validated most notably at the 1994 United Nations International Conference on Population and Development in Cairo, which endorsed the idea that population policies needed to move away from "population control" and toward enhancing reproductive freedom (Goldberg 2009). Luckily, this approach of increasing reproductive freedom has also been tremendously successful at concurrently achieving the demographic goal of reducing birth rates, helping to address ecological concerns at the same time.

Across the globe, women have shown huge enthusiasm for reducing their own fertility when given the choice, with the "two-child family" becoming an aspirational goal in many communities. For many parents, small family size is seen as the key to being able to afford to educate children and offer them the opportunity to work their way out of poverty. Additionally, couples all over the world have bought into the idea that small families are critical to the modernization of their countries. Overall, the big demographic story of the early 21st century has been rapidly decreasing global fertility, at rates many experts barely thought possible even 20 years ago, leading some commentators to suggest that, in hindsight, global population growth might well have stabilized even in the absence of many of the coercive population policies that were enacted in the mid- and late 20th century.

Current population growth and population projections

Today, most countries have fertility rates of fewer than three children per woman (PRB 2020b). The one major exception is sub-Saharan Africa, where fertility remains above four in most countries and is over six children per woman in Niger, Mali, and Angola (PRB 2020b). Even here, birth rates are declining, however—just how fast fertility will decline in sub-Saharan Africa is one of the key demographic questions for the next 20 years. Nonetheless, the momentum of the large and very youthful global population at the turn of the 20th century means a still rapidly growing population, with the 4, 5, 6, 7, and 8 billion marks each reached in just over a decade at the peak of population growth in the late 20th and early 21st centuries (table 4.1).

Of course, if family size remains even slightly above the replacement level of about 2 children per woman, populations will continue to grow through **natural increase**, and so global population is predicted to keep growing in the near future despite declining fertility (figure 4.4). The point at which a population just reproduces itself is known as **replacement level fertility** and is a goal for sustainability proponents, who would like to see population size stabilize (and even decline). As you might expect, replacement level fertility is approximately 2 babies per woman (one child to replace each parent) but must always be slightly above 2 to account for

TABLE 4.1 Global population growth

Number of people on Earth	Approximate date	Time taken to add an additional billion
1 Billion	1804	Thousands of years
2 Billion	1927	Approx. 123 years
3 Billion	1960	Approx. 33 years
4 Billion	1974	14 years
5 Billion	1987	13 years
6 Billion	1999	12 years
7 Billion	2011	12 years
8 Billion	2022	11 years

Source: UNDESA (2017).

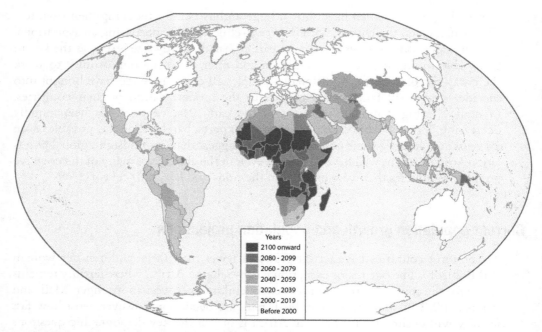

FIGURE 4.4 Timing of projected stabilization of population size by country
Data source: UNDESA (2019).

individuals who die before reaching childbearing age, as well as imbalances in the population's sex ratio—if a population has more men than women, for instance, women will need to have more than 2 children on average to replace the unmarried men who cannot find partners. An estimate of 2.1 babies per woman is commonly assumed to represent the replacement level fertility rate and yet in societies with very low child mortality rates, replacement level fertility might only require a fertility rate of 2.05 births per woman. By contrast, communities with very high infant mortality might require an average of 3 or even more live births per woman to keep the population stable (Gietel-Basten and Scherbov 2020).

By 2020, almost half of the world's population was living in countries with below replacement fertility (figure 4.5). At the other extreme, just nine countries will provide more than half of the population growth projected between 2020 and '50—India, Nigeria, Pakistan, the DR Congo, Ethiopia, Tanzania, Indonesia, Egypt, and the United States (UNDESA 2019). The United States is unusual on this list as the only affluent country projected to see significant growth in the upcoming decades; here growth is premised largely on immigration rather than natural increase, although the United States' relatively young age structure and large number of recent immigrants continue to buoy up its fertility rates compared with Global North averages. By contrast, rates of natural increase are still high in many poor countries, posing significant challenges in terms of meeting development goals such as providing education and sufficient nutrition for all citizens (UNDESA 2019). Africa is projected to be the only world region with strong natural increase for the rest of this century, with its population projected to double by 2050 (UNDESA 2019); Europe and Latin America, by contrast, will experience a decline in population by 2100 (Cillufo and Ruiz 2019).

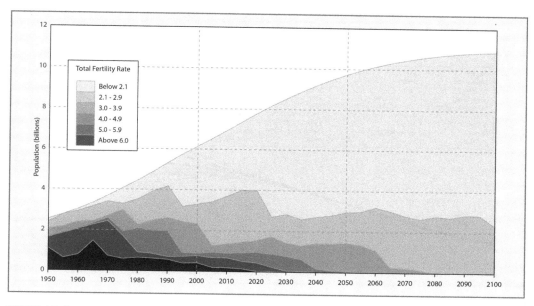

FIGURE 4.5 Proportion of global population residing in countries with different fertility rates
Data source: UNDESA (2019).

Population projections are used to model how population might change in the future. Owing to the large number of variables that influence population, there is always considerable uncertainty inherent in population projections. Will a particular country's fertility trajectory follow that of other countries? Will a political or economic setback lead to people changing their fertility decisions? How rapidly will education improve opportunities for women, with likely declines in fertility? Uncertainties such as these mean that population projections are often given as a range with different probabilities, rather than as a specific number (figure 4.6). In some cases, projections are made for several different scenarios depicting, for instance, how quickly educational attainment might improve (e.g., Lutz et al. 2014). Most recent projections suggest that global population will probably peak at somewhere between 9 and 11 billion before declining, with the demographic future of Africa posing one of the greatest uncertainties. The United Nations, which has a strong track record for accurate projections, estimates with 95 percent probability that population will peak between 9.5 and 13 billion, with a median estimate of just under 11 billion (UNDESA 2019), although this still provides a very wide range of possible outcomes. Most models suggest that population will peak around 2100, although some models suggest that global population may peak as early as the 2060s (e.g., Vollset et al. 2020).

Despite the likelihood of future stabilization of population size, these statistics still pose significant cause for concern for many. Will the Earth be able to feed an additional billion people, let alone 2 or 3 billion more? Will we see more resource conflicts over freshwater and agricultural land? How will we protect the Earth's remaining wildlands as more people need food and space to live? For others, the population crisis has passed, and attention is switching to other looming issues, including regional population decline and population aging. We return to the topic

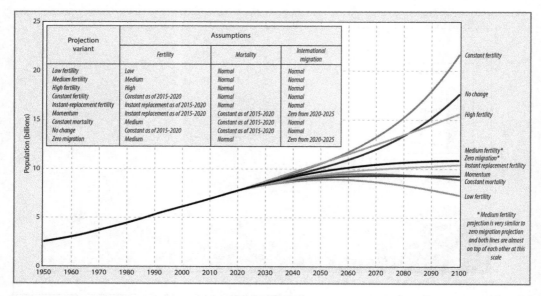

The table within the figure:

Projection variant	Assumptions		
	Fertility	Mortality	International migration
Low fertility	Low	Normal	Normal
Medium fertility	Medium	Normal	Normal
High fertility	High	Normal	Normal
Constant fertility	Constant as of 2015-2020	Normal	Normal
Instant-replacement fertility	Instant replacement as of 2015-2020	Normal	Normal
Momentum	Instant replacement as of 2015-2020	Constant as of 2015-2020	Zero from 2020-2025
Constant mortality	Medium	Constant as of 2015-2020	Normal
No change	Constant as of 2015-2020	Constant as of 2015-2020	Normal
Zero migration	Medium	Normal	Zero from 2020-2025

Constant fertility

No change

High fertility

Medium fertility*
Zero migration*
Instant replacement fertility
Momentum
Constant mortality

Low fertility

* Medium fertility projection is very similar to zero migration projection and both lines are almost on top of each other at this scale

FIGURE 4.6 United Nations' world population projections by 2100
Source: UNDESA (2019).

of population decline in chapter 6. Here, we turn to the case study of India to illustrate many of the patterns of population growth described above.

Population growth in India

India and China are the two global demographic heavyweights, with populations of about 1.4 billion each in 2020—each country comprising approximately 18 percent of the total global population of around 8 billion (PRB 2020b). Although China still has the slightly larger population, India's more rapidly growing population has been catching up for several decades and is projected to surpass China as the world's most populous nation in the mid-2020s (OurWorldinData, n.d.).

Until the late 20th century, India was typical of many pre-industrial countries. A large proportion of the population worked in agriculture, which offered rapid economic benefits from children who could be put to work from a young age, contributing to high birth rates. India's strongly **patriarchal** (male-dominated) society left many women poorly educated and with little authority in the household, with decisions over childbearing often left in the hands of husbands and mothers–in–law. Mothers, who have the strongest incentive to control family size owing to the risk of each pregnancy and the heavy burden associated with childcare, often had little say in limiting their own fertility. In addition, son preference has traditionally been strong across much of India, with cultural expectations dictating that sons carry the family name and support parents in old age. Daughters, by contrast, customarily moved to their husband's new family on marriage and cared for their parents–in–law, representing a significant loss to their family of birth. Daughters were also traditionally expected to bring large dowries of money and goods to their new husband's home, further increasing the economic burden of girls on their birth family. As a result, many women have been pushed by older family members and societal expectations to keep having children in the hopes of having more sons. As infant mortality rates decreased in the mid- to late 20th century, the proportion of children surviving increased rapidly, leading to rapid population expansion.

Finding ways to control this rapid population growth has been on the Indian national agenda since the 1950s, supported enthusiastically in the early years by international organizations such as the US-based Population Council, which promoted family planning in India for reasons ranging from encouraging Indian society to reach its full potential to goals related to geopolitics, conservation, and even **eugenics** (Connelly 2006). Since independence in 1947, at least 36 population control bills have been introduced to the Indian Parliament and in 1951 India became the first country in the Global South to institute a national family planning program (Abbamonte 2019). India's early efforts at population control have been widely criticized as draconian, particularly a period from 1975 to 77 known as "The Emergency" when the Indian government declared a state of emergency and forcibly sterilized at least 8 million people, mostly poor men (Abbamonte 2019; Chandrashekhar 2019). Reports suggested that some government officials were required to meet sterilization quotas, and accounts from poor men in rural villages recounted avoiding government officials for fear they would be coerced into getting sterilized (Harris Green 2018). Some reports suggested that eugenic efforts to sterilize those deemed "subnormal" also took place (Connelly 2006). As one scholar noted, the Indian program "has become emblematic of everything that can go wrong in a program premised on 'population control' rather than on reproductive rights and health" (Connelly 2006, 629).

By the 21st century the situation had changed, with many Indians turning voluntarily toward limiting family size (although reports of human rights abuses related to fertility limitation persist; Chandrashekhar 2019). Increasing access to education has led to a burgeoning middle class in India, with many urban Indians now living at levels of affluence comparable to those in the West and later marriage and small family size becoming the norm for this group. The two-child family is becoming normalized even among the poor, with many poorer Indians seeing fertility limitation as the primary way to provide their children with opportunities that they did not have—education, in particular. The empowerment of women, including through widespread family planning education, has also been pivotal in enabling this transition. The significance of education and empowerment is suggested by regional statistics, with high-fertility states such as Bihar and Uttar Pradesh typically having much poorer socio-economic indicators, particularly for women, than low-fertility states. At the other end of the spectrum, the state of Kerala—whose strongly left-leaning government famously made early strides in providing education and basic healthcare to the general populace—now boasts a literacy rate of 99.3 percent and a fertility rate of 1.8, the same as Sweden's (Chandrashekhar 2019; PRB 2020b). Today, Kerala is beginning to experience problems associated with an aging population.

India's fertility rate has recently been reported to have dropped below replacement level, with the government announcing to much fanfare a total fertility rate of just 2.0 in the period 2019–21 (*The Economist* 2021). This may be significantly below replacement level, given India's unbalanced sex ratio and relatively high infant mortality rate (Chandrashekhar 2019). Indeed, one of the unfortunate side effects of the transition to smaller family size was the unbalancing of the sex ratio, as parents tried to ensure that they still had a son despite having fewer children. This imbalance has been reported since the 1970s, with one recent estimate suggesting that there will be a deficit of a further 6.8 million females between 2017 and 2030, owing largely to sex-selective abortions (Chao et al. 2020). Even though it has been illegal in India since 1994 for doctors to tell parents the sex of their child when they perform an ultrasound, male births still significantly outnumber female births, suggesting that the practice of female feticide is probably still widespread.

The number of children in India's population began to decline almost a decade ago, but it will take several more decades before India reaches its peak population size as a large youth bulge moves through its childbearing years into adulthood, temporarily buoying up the fertility rate—a phenomenon known as **demographic momentum**. The momentum associated with a population the size of India's is indeed huge; India's population is projected to keep growing until 2060, peaking at perhaps 1.7 billion (Chandrashekhar 2019). Experts suggest there is still a huge unmet demand for contraception in India, with an estimated one in seven unplanned pregnancies worldwide occurring in India (UNFPA 2021). It is figures such as these that keep population size on the political agenda, with India's Prime Minister Narendra Modi calling in 2019 for population control policies to be implemented once again and referring to having fewer children as "an act of patriotism" (Abbamonte 2019). Commentators argue that it is more than just population size that is on many politicians' minds, however. Critics have pointed out that there is also a desire to try to control the size of India's Muslim population among India's political elite, who are primarily Hindu. Longstanding tensions between Hindus and Muslims have been heightened by higher birth rates in many Muslim populations, stoking fears that Muslims' strengthening numbers could have political implications in India's democratic political system. Critical scholars also note that the marginalization of Muslim populations in India may at least partially explain their higher birth rates (Shih 2021). As the case of India illustrates, population growth and limitation are often politically charged.

PART II: THEORIES OF POPULATION GROWTH

If we are to fully understand different perspectives on population growth, some theoretical analysis is important. The potential challenges of population growth have been discussed since antiquity—in general, however, the concerns of antiquity were focused around population *decline* rather than expansion (Riddle and Estes 1992). It was only with the Industrial Revolution that rapid population growth and urban crowding really brought the topic of population growth to the fore.

Malthus, Marx, and population

Thomas Malthus, an English cleric, was one of the first scholars credited with calling serious attention to potential problems associated with population growth. Published in 1798, Malthus's *An Essay on the Principle of Population* was written at a time of widespread poverty and growing concern over social conditions. Malthus famously argued that population increases *geometrically* whereas food supply increases only *arithmetically*, with the result that expansion of the food supply can be quickly outpaced by rapid population growth. When this happens, Malthus believed that "positive checks"—war, famine, and disease—would come into play, raising the death rate and decreasing population size. Malthus later acknowledged that "preventive checks," through which people voluntarily constrain fertility, also exist, but he was doubtful that the poor would be able to control themselves adequately for these checks to be sufficient. As a man of the Church, Malthus could only countenance celibacy and sexual restraint as appropriate "preventive checks"; other mechanisms that could reduce the birth rate, including birth control and abortion, were condemned by Malthus on moral and religious grounds (Horner 1997). As such, Malthus was not hopeful that a stable population could be achieved.

Though the basic logic of Malthus's argument—that population can grow faster than food supply—is undeniable, from a social justice perspective Malthus's ideas quickly become problematic. In particular, Malthus's work has been read as an effort to maintain the beneficial position of the ruling classes, which were beginning to see rapid population growth among the poor as a potential revolutionary threat (Horner 1997). Indeed, Malthus argued that Britain's poor laws were actually likely to worsen the problem of poverty. He claimed that giving people handouts of money does nothing to increase agricultural production but instead simply raises the price of food because people have more money to buy it. Even if it does eventually stimulate greater demand for food and increase production, Malthus believed this would simply lead the lower classes to have yet more children, negating any possible benefits (Shermer 2016). This idea was quickly seized upon by politicians to support the idea that the poor were responsible for their own poverty through their lack of moral restraint. In response, efforts were made to restrict subsidies for those in need and make assistance to the poor such as workhouses as punitive as possible.

It is from an ecological perspective that Malthus's views have perhaps left their strongest mark, however. Although Malthus framed his discussion in largely economic terms, he raised legitimate ecological concerns about whether population size will exceed agricultural production, or what we might today refer to as the land's carrying capacity. In the modern era, these ideas have been extended by a **neo-Malthusian** movement that took off in the 1960s. In this era, rapid population growth in the Global South was leading to renewed fears that global food production would not be able to keep pace with population expansion (Ehrlich 1968). The neo-Malthusians also extended Malthus's original ideas to emphasize their environmental message that potentially irreparable damage was being done to Earth's ecosystems by growing populations. Unlike Malthus, the neo-Malthusians embraced artificial methods of birth control and became major proponents of early family planning campaigns, which often proceeded with the explicit goal of slowing population growth. Though it is undeniable that growing populations have diverted more and more land for human use, with detrimental environmental implications, neo-Malthusian ideas have been questioned from a social justice perspective because, at their most extreme, they too have been used to argue that attempts to alleviate famine and hunger simply exacerbate problems of **overpopulation**.

Another significant challenge posed to Malthusian ideas is that the food supply has not yet run out, despite the fact that population has increased dramatically since

Malthus's day. Indeed, today, thanks to the Green Revolution, famine is now rare and is often attributed to political failings rather than ecological limits. For instance, war may reduce certain communities' access to food, or poverty prevents people from buying food, even though it is available in local markets. The work of economist Amartya Sen has been particularly influential in this regard. Sen (1981) argued that hunger is a result of individuals not having *access* to food because of poverty or other barriers, rather than because food is not available owing to ecological limits. The fact that we have enough food to feed everyone at the global scale is often cited as evidence of the fact that it is maldistribution of resources, rather than outright food shortages, that are the problem.

Sen's work is an extension of **Marxist** thinking on population, which strongly emphasizes social inequalities as the root cause of global problems. Working in the mid-1800s, Karl Marx was a philosopher and radical thinker who argued that **capitalism** involves constant class conflict between workers and the ruling classes who control production. With respect to population, he rejected the idea that people were hungry because there were too many mouths to feed and instead promoted the idea that it was poor distribution of resources that was to blame. To this day, many scholars use Marxist approaches to understanding the world—within the realms of population studies, this approach emphasizes unequal patterns of consumption rather than overall population size as key to explaining outcomes such as poverty and malnutrition. Additionally, many Marxist scholars point out how political and economic *structures* of society—global trade networks, colonial relationships, political inequities, etc.—constrain the opportunities of certain groups within society, leaving them more vulnerable to negative outcomes such as hunger and famine.

This is not to suggest that ecological constraints will *never* be reached. It is quite possible that we may be nearing our ultimate potential to feed growing populations. Ecologically, we can also argue that growing human populations have already done significant environmental damage and that the pressures of further human population growth could be devastating. However, these arguments must always be tempered to avoid reaching the unacceptable conclusion that "excess" populations should somehow be removed. A historical overview of thinking on population emphasizes a key division in how we perceive the problem. For Malthus and the neo-Malthusians, the problem is simple: too many people are overstretching Earth's natural limits—an approach that persists to this day in ecological frameworks around carrying capacity. For Marxists and those emphasizing a social justice framework, the problem is instead maldistribution of resources and patterns of consumption. We pick up these threads again in the next chapter where we explore issues of population and resources in greater detail.

Modeling population growth and decline: the demographic transition model

Though philosophical approaches to population growth are instructive, further insights can be gained if we model actual changes in population over time. One of the most influential models in this respect is the **demographic transition model**, which outlines how birth and death rates change over time and the impact of these changes on population size (figure 4.7). Based on the experiences of Western Europe

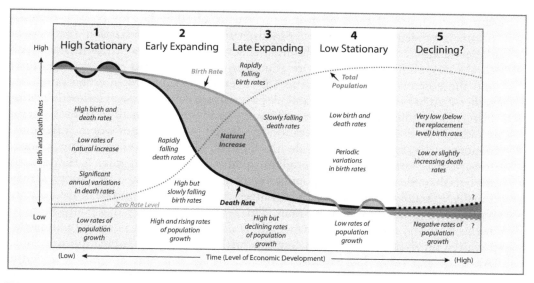

FIGURE 4.7 The demographic transition model
The demographic transition model describes changes in fertility and mortality over time in four stages, moving from a population with high birth and death rates to one with low birth and death rates. Recently, a fifth stage has been suggested for the model to reflect new demographic developments associated with very low fertility and declining population size, as well as the possibility that death rates might rise slightly as populations age.

from the 1700s onwards, the model suggests that populations begin with high birth and death rates and mature demographically by experiencing a decline in their death rate, resulting in a period of dramatic population growth, followed by a decline in the birth rate, before population size stabilizes once again, this time with low birth and death rates. The social implications of the demographic transition are profound— comparable to other major societal developments such as the rise of democratic institutions of government, the Industrial Revolution, and urbanization:

> The transition transforms the demography of societies from many children and few elderly to few children and many elderly; from short life to long; from life-long demands on women in raising young children to the concentration of these demands in a small part of adulthood ...
>
> (Reher 2013, 24)

In the first stage of the model, reflecting a pre-industrial society, both birth rates and death rates are high. Mothers have multiple births and yet population size remains relatively stable owing to high death rates, with many dying young from malnutrition and infectious diseases. Most of us would probably agree that this is not an ideal situation and so the demographic transition is often viewed as reflecting progress from this undesirable baseline toward a preferable demographic situation where people have fewer children but can expect most of those children to live long lives. In stage 2, circumstances are undoubtedly improving as death rates begin to fall, largely in response to improved nutrition and better sanitation. The population grows rapidly as birth rates remain high. In stage 3, death rates continue to decline but now birth rates also decrease as parents begin to choose to have smaller families as more of their offspring survive to adulthood. This process typically occurs in tandem with processes of industrialization and urbanization that increase the economic costs of having children: as societies move away from their agricultural roots, children require more education before they can enter the workforce and the cost of living increases because everyday needs must now be purchased rather than harvested from the land. Aspirations also tend to rise as industrialization brings more consumer goods, encouraging families to stay small to improve living standards. At the same time, education and the empowerment of women offer more opportunities for women outside the household, further encouraging a move toward smaller family size. Although birth rates decline during stage 3, they are still higher than death rates, so population continues to grow, albeit more slowly. Finally, in stage 4, population size stabilizes once again, as birth rates have declined sufficiently to mirror low death rates. This represents a society that many aspire to, focused around the idealized "two-child family"—low death rates, with almost all children surviving to adulthood, and low birth rates keeping population size stable. Although not originally part of the model, a stage 5 is now commonly added in response to recent demographic changes, reflecting a situation where very low birth rates dip below death rates, which rise slightly owing to the rising proportion of elderly people in the population. Now the population growth rate becomes negative and population size decreases unless immigration makes up the shortfall.

Although the model was originally based on the experiences of countries in Western Europe, it has been applied worldwide to explore whether countries of the Global South are experiencing a similar demographic transition. It is problematic to assume that one country will necessarily follow the same demographic trajectory as another, given different social, economic, and political contexts, but the model has nonetheless proved relatively reliable. One notable exception has been the speed of the transition, with the demographic transition occurring more rapidly in the more recent transitions in the Global South, where communities have been able to import technologies that sped up the decline in death rates (e.g., antibiotics, vaccinations, agricultural technologies), as well as effective contraceptives to reliably reduce birth rates. Additionally, fertility decline appears to occur earlier in communities that are geographically and culturally close to communities that have already experienced a transition to smaller family size in a cultural process termed "demographic contagion" (Delventhal, Fernández-Villaverde, and Guner 2021). Another potential critique of the model is whether it over-emphasizes the idea that societies moving from one demographic stage to the next represents "progress"—most people agree that falling death rates are desirable, but critical scholars have pointed out that there is nothing inherently wrong with large families and so we should be more cautious in suggesting that a move to low birth rates is synonymous with progress. Finally, the model implies that countries can only move in one direction demographically, but it is quite possible for countries to "regress" in terms of either birth or death rates. Some countries of sub-Saharan Africa experienced rising death rates during the HIV/AIDS pandemic, for example, and Afghanistan is predicted to see rising birth and death rates associated with the social and political upheavals of the recent Taliban takeover in 2021.

By the year 2020, very few, if any, countries could be said to be in stage one of the demographic transition model. Today, even the poorest countries—including parts of sub-Saharan Africa and war-torn countries like Afghanistan—are really better modeled by stage 2, owing to progress with respect to lowering infant mortality. If we consider Niger, for instance, which currently has the highest birth rate in the world (48 births per 1,000), the infant mortality rate has dropped dramatically over the past 10 years from 108 per 1,000 in 2010 to 69 per 1,000 in 2020 (PRB 2010, 2020b). As infant mortality rates drop, life expectancy increases and populations enter a period of rapid population growth. In the 1940s and '50s, when many countries of the developing world were entering stage 2, some countries achieved life expectancy increases of 10 to 12 years per decade (Reher 2013). In such situations, a community's **doubling time**—the time required for its population to double in size—may be as little as 20 years. Countries also begin to see a youth bulge in stage 2 as large numbers of children are being born but declining mortality means that more and more survive infancy. A youth bulge can be a huge economic stimulus for a country once these children become workers. However, it can also be problematic if a country's economic situation does not provide sufficient jobs for these individuals as young adults, and youth bulges can be associated with political unrest in countries with weak political institutions (Beehner 2007).

We can argue that most countries of the Global South today are experiencing declining fertility *and* declining mortality, putting them in stage 3 of the model. As early as the period 1955 to '90, an estimated 150 countries had already begun to see declines in fertility (Reher 2013)—more recently, even very poor countries such as Niger have begun to see declining birth rates (from 52 per 1,000 in 2010 to 48 per 1,000 in 2020; PRB 2010, 2020b). Indeed, many middle-income countries (including most notably India) now have birth rates around replacement level, suggesting that they are transitioning to stage 4.

The phenomenon of demographic momentum complicates matters at this point, however. **Demographic momentum** refers to the fact that countries whose birth rates have only recently declined may continue to see significant population growth for some years to come as a youth bulge from previously high birth rates makes its way through its childbearing years. This helps to explain why India's population may not peak until the 2060s, despite already having replacement level fertility in 2020 (Chandrashekhar 2019). This same time period may also see a temporary economic boom associated with a **demographic dividend**—as a youth bulge enters its working years, it provides abundant workers to stimulate the economy, and a new pattern of lower birth rates concurrently suppresses the size of the dependent population. The ideas of youth bulges and demographic dividends can perhaps best be seen by looking at how population pyramids change as countries move through the demographic transition (figure 4.8).

Some countries of the Global South are now already in stage 4, joining much of the affluent world. Indeed, several countries that most people still think of as part of the "developing world" have had fertility rates of no more than replacement level for over a decade now, including Thailand, Vietnam, Brazil, Cuba, Costa Rica, and Tunisia (PRB 2010). Furthermore, in recent years, a new and unexpected demographic trend has developed—birth rates dropping *below* death rates, including in some countries of the Global South (table 4.2). By 2010, almost half of the global population was living in countries with below replacement level fertility, and projections suggest that this proportion will increase (UNDESA 2010). In such cases, in the absence of immigration, population size will decline, although this may not occur immediately if life expectancies are increasing. Many scholars describe this as stage 5 of the demographic transition model—low death rates but even lower birth

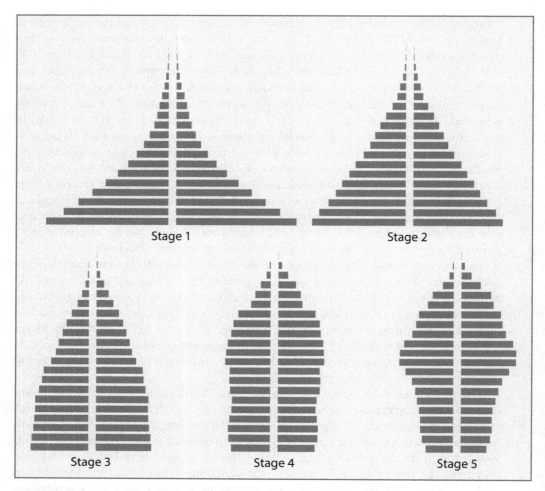

FIGURE 4.8 Population pyramids typical of each stage of the demographic transition model
Note the traditional pyramid shape of stage 1, where high birth rates create a wide base but then individuals are lost to disease and other causes in every age cohort, giving the stepped appearance. In stage 2, the pyramid begins to move toward a more column-like shape as births remain high but fewer people die before old age. By stage 3, the base of the pyramid narrows even further, reflecting fewer births, but a bulge of workers from previously high birth rates is now entering adulthood. In stage 4, the pyramid takes on a narrow profile with few births but few deaths in most cohorts until old age. By this point, a sex imbalance is likely to be apparent in older age cohorts, associated with women's longer life expectancy. By stage 5 the base of the pyramid is very narrow, reflecting a very small number of births and that most of the dependent population is now in elderly age cohorts.
Data source: PopulationPyramid.net (n.d.).

rates. Very low birth rates are the primary driver of such circumstances, but death rates also rise slightly as an increasing proportion of the population is in elderly age groups—life expectancy remains high in such cases, but the large proportion of the population over 65 years means that the *rate* of people dying increases.

The major demographic impact of low fertility rates has been a dramatic change in the age structure of populations (figure 4.9), with increasing proportions of the population in elderly cohorts. The resulting **demographic deficit** is projected to have significant economic impacts associated with the declining proportion of

TABLE 4.2 Examples of countries with below replacement level fertility in 2020

Country	Income level (2020)	Total fertility rate (2010)	Total fertility rate (2020)
Japan	High	1.4	1.3
Spain	High	1.4	1.3
Thailand	Upper-middle	1.8	1.5
Brazil	Upper-middle	2.0	1.7
Ukraine	Lower-middle	1.5	1.3
Bhutan	Lower-middle	3.1	1.7
Global average		*2.5*	*2.3*

Data sources: World Bank (2021) and PRB (2020b).

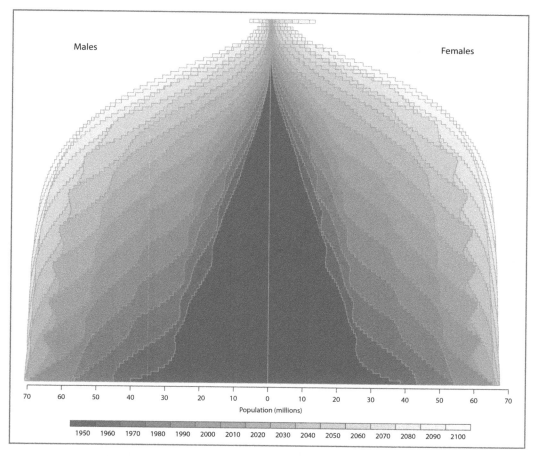

FIGURE 4.9 Changing global population structure by age and sex, 1950–2100
Note the more column-like structure of the pyramid over time, reflecting more children reaching adulthood, as well as the widening of the elderly cohorts and increasing predominance of women in these older age groups.
Data source: UNDESA (2019).

workers in the population. By 2050, we will reach the historically unprecedented situation where more of the global population is over 60 than under 15, with an estimated 22 percent of the global population aged 60 and above by 2050 (in 1950, this figure was just 8 percent)—this situation has already been reached in the Global North (Harper 2013).

One of the great uncertainties of our global demographic future revolves around whether fertility rates will rebound to replacement level or whether sub-replacement birth rates will become the norm. Some evidence suggests that the traditional relationship of increasing affluence being associated with declining fertility reverses at very high levels of development, offering the suggestion that we may eventually see a small rebound in fertility toward replacement level. In one study, researchers used the United Nations' Human Development Index (a measure that summarizes measures of health, education, and economy to estimate overall levels of "human development") and found a modest fertility increase across countries at the highest levels of human development, although some countries did not respond in this way—most notably, Japan, Canada, and South Korea, which saw persistently low fertility despite very high levels of human development (Myrskylä, Kohler, and Billari 2009; figure 4.10). Even among the mostly European countries that did see

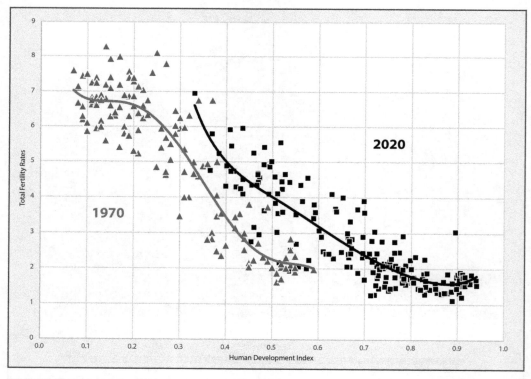

FIGURE 4.10 Relationship between human development index and fertility, 1970 (136 countries) and 2020 (189 countries)

Data from 1970 illustrate the traditional pattern of increasing human development associated with declining fertility. This pattern is also evident in the 2020 data, except at the very highest levels of human development, where we see an indication of a modest uptick in fertility toward replacement level fertility.

Data source: UNDP (2020) and UNDESA (2019).

modest increases in their fertility, this increase was not generally enough to push fertility above replacement level, although the researchers suggest that it would have the potential to offset some of the costs associated with population aging. At this point, we are on the very edge of what we know about population growth and change, and only time will tell how the next chapter of global demographic patterns will develop.

Conclusion

The human population has experienced dramatic growth over the past 200 years, but signs suggest that we may finally be nearing the end of this increase. The ecological costs of this growth have been significant in terms of landscape conversion, consumption of resources, and pollution. Although the rate of population growth is now slowing, growth is projected to continue, and legitimate questions remain over whether additional population can be supported given finite resources and already high levels of contaminants in the environment. Hope for the future rests on a combination of stabilizing population growth as well as developing new technologies and policies that promote lower-consumption lifestyles. We pick up these threads in the next chapter.

As population growth slows and, indeed, becomes population decline in some countries, concerns are turning to focus also on the potential impacts of aging populations and how we can encourage families to once again put faith in the future and create another generation. Here social justice concerns become significant as evidence suggests that those living in more just and empowering societies may be more likely to maintain fertility rates at just about replacement level. Whether this will finally lead to global populations stabilizing and a final conclusion to the demographic transition or just usher in further demographic changes remains to be seen.

Discussion questions

1   Why do you think we are still uncertain of population projections and at what point global population might stabilize? What factors can you list that represent key uncertainties related to when population will stabilize and the population size at that point?

2   Where would you place India in the demographic transition model? How might we consider India's efforts to control its population size in a sustainability framework? What critiques might we put forward from a social justice perspective?

3   Some critical scholars argue that it is problematic to suggest that the demographic transition model indicates "progress." To what extent do you agree? In what ways is it unhelpful to see the demographic transition as progress? What aspects of the transition might we argue are genuinely desirable?

4   Given the benefit of hindsight, what do you believe Malthus and Marx got right about population change? What did they get wrong?

Suggested readings

Cillufo, A., and N. Ruiz. 2019. "World's Population Is Projected to Nearly Stop Growing by the End of the Century." Pew Research Center, June 17, 2019. https://www.pewresearch.org/fact-tank/2019/06/17/worlds-population-is-projected-to-nearly-stop-growing-by-the-end-of-the-century/

Pappas, S. 2011. "7 Population Milestones for 7 Billion People." LiveScience, October 11, 2011. https://www.livescience.com/16489-7-population-milestones-7-billion-people.html

Pingali, P. 2012. "Green Revolution: Impacts, Limits and the Path Ahead." *PNAS* 109 (31): 12302–8. https://doi.org/10.1073/pnas.0912953109

PRB [Population Reference Bureau]. 2020. "How Many People Have Ever Lived on Earth?" PRB, January 23, 2020. https://www.prb.org/howmanypeoplehaveeverlivedonearth/

Glossary

agricultural revolution: the transition from hunting and gathering to sedentary agricultural activities

capitalism: economic system premised on trade and industry, designed toward profit and controlled by private ownership of means of production

carrying capacity: the maximum number of individuals that can be supported by an ecosystem without causing degradation

demographic deficit: the decreased economic productivity associated with having an aging and shrinking population; the opposite of a demographic dividend

demographic dividend: the temporary economic boost an economy receives when a population has a large proportion of workers owing to recent declines in the birth rate

demographic momentum: the potential for a population to continue to grow into the future, despite a recent decline in fertility, because of the population's young age structure

demographic transition model: model that describes changes in birth and death rates as countries develop economically

doubling time: the number of years needed for a population to double in size

eugenics: the premise of trying to improve a human population through selective breeding; the idea is now widely discredited owing to its discriminatory implications

globalization: the growing interconnectedness of the global economy and associated cultural and economic shifts

Green Revolution: efforts to introduce scientific farming techniques to the Global South, especially high-yielding seed varieties, fertilizers, pesticides, and improved irrigation

hearths of agriculture: locations where agriculture arose independently

Marxism: philosophical belief system, following Karl Marx, that focuses on inequality as the root cause of many social and economic problems

natural increase: excess of births over deaths, leading to population increase

neo-Malthusianism: extensions of Malthus's ideas; argues that population growth is leading to environmental collapse, as well as potential resource shortages

nomads: people who follow a lifestyle with no fixed abode, settling in places for only short periods of time

overpopulation: too many individuals to be supported sustainably in a particular region

patriarchy: social system dominated by men, marginalizing women

population projections: estimates of the way in which population will grow (or shrink) in the future

replacement level fertility: the number of babies per woman that keeps the population stable; replacement level fertility is typically around 2.1, but an unbalanced sex ratio or high child mortality will raise this figure

Works cited

Abbamonte, J. 2019. "India's Prime Minister Calls for Population Control, Says Small Families Are an Act of 'Patriotism.'" Population Research Institute, August 26, 2019. https://www.pop.org/india-prime-minister-calls-for-population-control/

Alfani, G. 2020. "The Economic Consequences of Plague: Lessons for the Age of Covid-19." History and Policy, June 29, 2020. http://www.historyandpolicy.org/policy-papers/papers/the-economic-consequences-of-plague-lessons-for-the-age-of-covid-19

Bähr, J. 2010. Bevölkerungsgeographie. Stuttgart: Ulmer UTB.

Beehner, L. 2007. "The Effects of 'Youth Bulge' on Civil Conflicts." Council on Foreign Relations, April 13, 2007. https://www.cfr.org/backgrounder/effects-youth-bulge-civil-conflicts

Bertin, J., P. Vidal-Naquet, and HarperCollins. 1992. The Harper Atlas of World History. New York: HarperCollins.

Brundtland, G. 1987. "Our Common Future: Report of the World Commission on Environment and Development." UN-Document A/42/427. Geneva. https://sustainabledevelopment.un.org/content/documents/5987our-common-future.pdf

Buringh, E. 2021. "The Population of European Cities from 700 to 2000." Research Data Journal for the Humanities and Social Sciences 6 (1): 1–18. https://brill.com/view/journals/rdj/6/1/article-p1_3.xml

Callaway, E. 2015. "Bronze Age Skeletons Were Earliest Plague Victims." Nature News, October 22, 2015. https://www.nature.com/news/bronze-age-skeletons-were-earliest-plague-victims-1.18633

Chandrashekhar, V. 2019. "Why India Is Making Progress in Slowing Its Population Growth." Yale Environment 360, December 12, 2019. https://e360.yale.edu/features/why-india-is-making-progress-in-slowing-its-population-growth

Chao, F., C. Guilmoto, K. Samir, and H. Ombao. 2020. "Probabilistic Projection of the Sex Ratio at Birth and Missing Female Births by State and Union Territory in India." PLOS One 15 (8): e0236673. https://doi.org/10.1371/journal.pone.0236673

Cillufo, A., and N. Ruiz. 2019. "World's Population Is Projected to Nearly Stop Growing by the End of the Century." Pew Research Center, June 17, 2019. https://www.pewresearch.org/fact-tank/2019/06/17/worlds-population-is-projected-to-nearly-stop-growing-by-the-end-of-the-century/

Connelly, M. 2006. "Population Control in India: Prologue to the Emergency Period." Population and Development Review 32 (4): 629–67.

Delventhal, M., J. Fernández-Villaverde, and N. Guner. 2021. "Demographic Transitions across Time and Space." https://www.sas.upenn.edu/~jesusfv/Demographic_Transitions.pdf

DeWitte, S. 2014. "Mortality, Risk and Survival in the Aftermath of the Black Death." PLOS One 9 (5): e96513. doi:10.1371/journal.pone.0096513

Diamond, J. 1999. "The Worst Mistake of the Human Race." Discover Magazine, April 30, 1999. https://www.discovermagazine.com/planet-earth/the-worst-mistake-in-the-history-of-the-human-race

The Economist. 2021. "Why the Demographic Transition Is Speeding Up." December 11, 2021. https://www.economist.com/finance-and-economics/2021/12/11/why-the-demographic-transition-is-speeding-up

Ehrlich, P. 1968. The Population Bomb. New York: Ballantine Books.

Gietel-Basten, S., and S. Scherbov. 2020. "Exploring the 'True Value' of Replacement Rate Fertility." Population Research and Policy Review 39: 763–72.

Goldberg, M. 2009. "Cairo and Beijing." In The Means of Reproduction, 103–20. New York: Penguin.

Harper, S. 2013. "Falling Fertility, Ageing and Europe's Demographic Deficit." In Fertility Rates and Population Decline, edited by A. Buchanan and A. Rotkirch, 221–9. London: Palgrave Macmillan.

Harris Green, H. 2018. "The Legacy of India's Quest to Sterilise Millions of Men." *Quartz India*, October 5, 2018. https://qz.com/india/1414774/the-legacy-of-indias-quest-to-sterilise-millions-of-men/

Horner, J. 1997. "Henry George on Thomas Robert Malthus: Abundance vs Scarcity." *The American Journal of Economics and Sociology* 56 (4): 595–607.

Larson, G., D. Piperno, R. Allaby, M. Purugganan, et al. 2014. "Current Perspectives and the Future of Domestication Studies." *Proceedings of the National Academy of Sciences* 111 (17): 6139–46. doi:10.1073/pnas.1323964111

Lutz, W., B. Butz, K. Samir, W. Sanderson, and S. Scherbov. 2014. "9 Billion or 11 Billion? The Research behind New Population Projections." IIASA World Population Program, September 23, 2014. https://blog.iiasa.ac.at/2014/09/23/9-billion-or-11-billion-the-research-behind-new-population-projections/

Malthus, T. 1798. *An Essay on the Principle of Population.* London: J. Johnson.

Manning, S. n.d. "World Population Estimates Interpolated and Averaged." Accessed December 12, 2021. https://www.scottmanning.com/archives/World%20Population%20Estimates%20Interpolated%20and%20Averaged.pdf

Myrskylä, M., H.-P. Kohler, and F. Billari. 2009. "Advances in Development Reverse Fertility Declines." *Nature* 460: 741–3. doi:10.1038/nature08230

OurWorldinData. n.d. "Historic and Projected Population." Accessed March 3, 2022. https://ourworldindata.org/grapher/historic-and-projected-population?country=IND~CHN

Pingali, P. 2012. "Green Revolution: Impacts, Limits and the Path Ahead." *PNAS* 109 (31): 12302–8. https://doi.org/10.1073/pnas.0912953109

PopulationPyramid.net. n.d. "Population Pyramids of the World from 1950 to 2100." Accessed March 12, 2022. https://www.populationpyramid.net/

PRB [Population Reference Bureau]. 2010. *World Population Datasheet 2010.* Washington, DC: PRB. https://www.prb.org/wp-content/uploads/2010/11/10wpds_eng.pdf

———. 2020a. "How Many People Have Ever Lived on Earth?" PRB, January 23, 2020. https://www.prb.org/howmanypeoplehaveeverlivedonearth/

———. 2020b. *World Population Datasheet 2020.* Washington, DC: PRB. https://www.prb.org/wp-content/uploads/2020/07/letter-booklet-2020-world-population.pdf

Reher, D. 2013. "Demographic Transitions and Familial Change: Comparative International Perspectives." In *Fertility Rates and Population Decline*, edited by A. Buchanan and A. Rotkirch, 22–43. London: Palgrave Macmillan.

Riddle, J., and J. Estes. 1992. "Contraceptives in Ancient and Medieval Times." *American Scientist*, May–June 1992. https://www.jstor.org/stable/29774642?seq=2#metadata_info_tab_contents

Sen, A. 1981. *Poverty and Famine: An Essay on Entitlement and Deprivation.* Oxford: Oxford University Press.

Shermer, M. 2016. "Why Malthus Is Still Wrong." *Scientific American*, May 1, 2016. https://www.scientificamerican.com/article/why-malthus-is-still-wrong/

Shih, G. 2021. "In India, a Debate over Population Control Turns Explosive." *The Washington Post*, August 29, 2021. https://www.washingtonpost.com/world/2021/08/29/india-population-hindus-muslims/

Suntsov, V. 2015. "On the Origin of *Yersinia pestis*, a Causative Agent of the Plague: A Concept of Population-Genetic Macroevolution in Transitive Environment." *Zhumal Obshchei Biologii* 76 (4): 310–8. https://pubmed.ncbi.nlm.nih.gov/26353398/

UNDESA [United Nations Department of Economic and Social Affairs]. 2010. "World Population Prospects: The 2010 Revision." http://esa.un.org/unpd/wpp/Analytical-Figures/htm/fig_8.htm

———. 2017. "World Population Prospects, the 2017 Revision." https://www.un.org/development/desa/publications/world-population-prospects-the-2017-revision.html

———. 2019. "World Population Prospects 2019." https://population.un.org/wpp/

UNDP [United Nations Development Programme]. 2020. "Human Development Report 2020." https://hdr.undp.org/en/2020-report

UNESCO [United Nations Educational, Scientific and Cultural Organization]. n.d. "The Spread of Disease along the Silk Road." Accessed November 3, 2022. https://en.unesco.org/silkroad/content/spread-disease-along-silk-roads

UNFPA [United Nations Population Fund]. 2021. "My Body Is My Own." https://www.unfpa.org/sites/default/files/pub-pdf/SoWP2021_Report_-_EN_web.3.21_0.pdf

USAID [United States Agency for International Development]. n.d. "Family Planning Timeline." Accessed November 10, 2022. https://www.usaid.gov/sites/default/files/documents/1864/timeline_b.pdf

van Hoof, T., F. Bunnik, J. Waucomont, W. Kürschner, and H. Visscher. 2006. "Forest Re-growth on Medieval Farmland after the Black Death Pandemic—Implications for Atmospheric CO_2 Levels." *Palaeogeography Palaeoclimatology Palaeoecology* 237: 396–411. doi:10.1016/j.palaeo.2005.12.013

Vollset, S., E. Goren, C.-W. Yuan, J. Cao, et al. 2020. "Fertility, Mortality, Migration, and Population Scenarios for 195 Countries and Territories from 2017 to 2100: A Forecasting Analysis for the Global Burden of Disease Study." *The Lancet* 396: 1285–306. https://www.thelancet.com/article/S0140-6736(20)30677-2/fulltext

WHO [World Health Organization]. n.d. "WHO Report on Global Surveillance of Epidemic-Prone Infectious Diseases." WHO/CDS/CSR/ISR/2000.1. Accessed November 20, 2021. https://www.who.int/csr/resources/publications/surveillance/plague.pdf

Wik, M., P. Pingali, and S. Brocai. 2008. *Global Agricultural Performance: Past Trends and Future Prospects.* Washington, DC: World Bank. https://openknowledge.worldbank.org/handle/10986/9122

World Bank. 2021. "World Bank Country and Lending Groups." https://datahelpdesk.worldbank.org/knowledgebase/articles/906519-world-bank-country-and-lending-groups

5

Population and resources

After reading this chapter, a student should be able to:

1 consider how a growing population can be a sustainability challenge;

2 use a social justice lens to consider consumption patterns as a further threat to sustainability;

3 apply these theoretical ideas to the issues of finite resources and food/agriculture.

A central theme in population studies is the relationship between people and resources. As biological beings, we cannot escape the fact that we consume resources and emit wastes, with impacts on natural systems. Every additional individual increases this impact, leading to concerns that growing populations cause environmental damage and that we may eventually reach ecological limits associated with finite resources. On the other hand, the *way* in which we interact with the environment has huge implications for how great a burden a community places on ecological systems, with high-consumption lifestyles arguably posing greater burdens on the environment than high populations, raising critical social justice questions. In this chapter, we think explicitly about how population size and patterns of consumption intersect with environmental issues.

PART I: THEORIES OF POPULATION AND ENVIRONMENT

Large population as a problem

As early as the late 1700s, commentators were expressing concerns over growing populations and finite resources, most notably Thomas Malthus, who argued that population size would eventually outstrip food supply (Malthus 1798). As noted in

DOI: 10.4324/9781003143253-5

chapter 4, we have not yet reached this situation, thanks largely to the development of new agricultural technologies and the globalization of food production networks, although concerns are resurfacing over whether we can keep expanding global food supplies given a potential future population of at least 9 billion people (Godfray et al. 2010). Additionally, emphasis is increasingly placed on the broader impacts of large human populations on the natural environment.

For many, the 1960s represent the beginning of the era of modern environmentalism. Rapid population growth at the time led to renewed fears that Earth's ecological systems might collapse, expanding Malthus's original ideas on food supply to also emphasize the environmental impact of growing populations. Proponents of this movement have been labeled **neo–Malthusians** ("new followers of Malthus"). Slowing population growth was the primary focus of the neo-Malthusian movement, and new methods of family planning were embraced, even though Malthus himself considered artificial birth control immoral. Among the most famous neo-Malthusians are Paul and Anne Ehrlich, whose book *The Population Bomb* documented fears that rapid population growth would lead to famine as well as social and ecological collapse (Ehrlich 1968). Although the Ehrlichs were not enthusiastic about the title "The Population Bomb" that the publisher requested, the idea of a **population explosion** took hold in the popular imagination and has been used to justify global efforts to control population growth. Indeed, perhaps the key long-term impact of the Ehrlichs' book was to make the idea of population control acceptable, even necessary (Mann 2018).

Critics of the Ehrlichs' book considered the tone of the book to be overly doom-laden and noted that it focused heavily on population size rather than patterns of consumption, although the Ehrlichs addressed consumption in subsequent publications (e.g., Ehrlich and Ehrlich 2009). Critics also noted that much of the fearful language in the book focused on images of growing populations in the Global South, especially India, fueling the belief that poor populations of color were responsible for **"overpopulation"** (Mann 2018). Although the scenarios outlined in the book have not played out exactly as described, the Ehrlichs have stood by their foundational points that: "it can be a very bad thing to have more than a certain number of people alive at the same time, that Earth has a finite carrying capacity, and that the future of civilization was in grave doubt" (Ehrlich and Ehrlich 2009, 63).

In the same era, the notion of **Spaceship Earth** became a popular metaphor for the limits of our planet. The globalization of communication networks and improved transportation allowed people to visualize the Earth as one self-contained unit, focusing attention on the issue of **finite resources**. Additionally, images of Earth taken from space have been credited with emphasizing Earth as a bounded system and for increasing the scale at which environmental issues were envisaged. New scientific methods began to be used to model how growing populations might interact with ecological systems given natural limits. Notably, in a 1972 report titled *The Limits to Growth*, a group of researchers associated with the Club of Rome (a group of policymakers and thinkers who came together in 1968 to seek holistic solutions to humankind's interconnected environmental and social challenges) used emerging computer technologies to simulate the interaction of exponential human and economic growth given finite resources (Meadows et al. 1972). The report's conclusions were startling, suggesting that human populations would indeed meet limits to growth in terms of food, population, and industrial capacity by the end of the 21st century given the trajectories of the time, with the authors proposing an urgent need to move toward more sustainable modes of production.

Meanwhile, an ecologist, Garrett Hardin, was publishing his ideas about overpopulation. Hardin used the example of a population of herders to explain how self-interest encourages herders to put more and more animals on common grazing land because each herder benefits from the extra animals they can graze. The unfortunate outcome is that the common resource of the pasture becomes degraded, penalizing the whole community. He termed this the **Tragedy of the Commons** (Hardin 1968). The Tragedy of the Commons is still referenced today in discussions over shared resources such as the atmosphere, oceans, and freshwater. Climate change provides a particularly glaring example—as individuals we use global energy reserves to make our own lives more comfortable but at the cost of changes to the atmosphere that negatively affect everyone. Other scholars have questioned Hardin's work, pointing out circumstances where communities have developed sustainable common resource management systems.

Hardin's work soon also began to draw criticism from a social justice perspective. Hardin expanded on the idea of finite resources by arguing that a closed system inevitably leads to a competitive struggle for resources. He suggested that **"lifeboat ethics"** provide a good analogy for how people in rich countries are currently afloat with comfortable lifestyles but could be overwhelmed if those from poorer countries try to "climb aboard" by demanding a greater share of global resources. For Hardin, the use of food aid and the expansion of the welfare state were problematic if they supported the growth of struggling populations, further increasing the number of poverty-stricken individuals and threatening to destabilize richer societies (Hardin 1974). Hardin's ideas have since been denounced by social justice scholars for implying that we should deny resources to poor communities. Critical scholars also argue that Hardin ignored the economic reality of unfair trade systems that do not allow poorer countries and individuals to compete for resources on a fair playing field (Potter 2011). More broadly, critical scholars reacted to extreme neo-Malthusian ideas by noting that they put overwhelming emphasis on the ecological framing of the situation at the expense of recognizing the humanity of populations. In this respect, social justice scholars have decried the very use of the term "overpopulation," arguing that the prefix "over-" implies that some portion of the global population is excess or undesirable.

Although these social justice issues represent important critiques, neo-Malthusian thinkers have nonetheless played an important role in raising awareness of Earth's ecological bounds and the potential problems of rapid population growth. Many commentators agree that we have to acknowledge that food supplies will become overstretched and the Earth's ecological systems damaged if population growth continues (Union of Concerned Scientists 2022). The size of the Earth's sustainable population remains a matter of intense debate, with some estimates falling as low as 2 to 3 billion or even less than 1 billion people, if we are to allow space for natural ecosystems to thrive as well as humans (e.g., Tucker 2019). The logic of identifying an optimal population size becomes unacceptable for many, however, when it is interpreted in ways that lead to coercive population control campaigns or even policies that deny assistance to the poor on the grounds that this might lead to further population growth. A social justice critique also forces us to engage with the reality that consumption patterns are just as critical as population size when it comes to environmental outcomes.

Consumption and the role of technology

Many scholars have pointed out that environmental impact is a result of not just population size but also consumption patterns. This is often linked to **Marxist** ideas that emphasize uneven allocation of resources as a root cause of global problems, redirecting the blame for environmental problems from poor families with many children to the rich who consume many resources. Here we can argue that an affluent Canadian couple poses a greater strain on ecological resources than a large Cambodian peasant family, if the Canadian couple has a lifestyle that might include two cars, frequent air travel, a centrally heated house, abundant consumer goods, and high meat consumption—all of which impose significant environmental costs. On the other hand, some scholars have argued that remaining childless is one of the best things that can be done for the planet, because this means fewer future children and grandchildren, effectively projecting the ecological benefits through future generations (Hamity et al. 2019). The relative contribution of population size versus consumption patterns remains highly contested!

One method for estimating the significance of consumption is the notion of an **ecological footprint**. In its original manifestation, the ecological footprint estimated the land area needed to support an individual, community, or product, in terms of providing all required resources as well as disposing of wastes (Rees and Wackernagel 1994). If we envisage a city, we can think about all of the food and resources pouring into that city, as well as the large area needed to dispose of the pollutants it produces—taken together, these requirements mean that the city has an environmental impact much larger than the city limits: its ecological footprint. Ecological footprints highlight the challenges posed by affluent lifestyles, with countries such as the United States, Canada, and Australia having much higher per capita footprints than Angola, Pakistan, or Madagascar (figure 5.1). If you would like to see what factors influence your personal ecological footprint, several online sources provide ecological footprint calculators that consider factors such as the size of your house, whether you regularly drive or fly, whether you eat meat and dairy, and even whether you have a pet cat or dog!

Ecological footprint estimates suggest that, globally, we are using far more resources than the Earth can regenerate. In 2020 people were using the equivalent **biocapacity** of perhaps 1.5 Earths, or 50 percent more resources than can be regenerated sustainably (WWF 2020). The unsustainable proportion of this resource use is referred to as **ecological overshoot**. In the short term, overshoot leads to diminishing resources (e.g., declining fish stocks) and pollutants left circulating in the environment; sustained biocapacity deficits can eventually lead to the collapse of ecological systems (Global Footprint Network 2021b). Figure 5.2 shows which countries are currently operating an ecological deficit and which have biocapacity sufficient to support their current population and consumption patterns. The United States, with its very high per capita demands and large population, is in significant deficit, as is China with its large population and

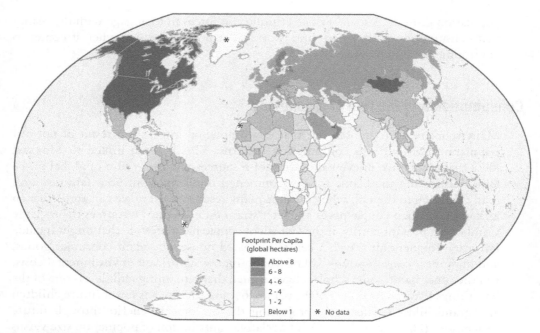

FIGURE 5.1 Ecological footprint by country, 2017
The shading on this map clearly shows those countries with high per capita consumption, with affluent countries typi-cally having higher footprints than poorer countries. Among affluent countries, the United States, Canada, and Australia have especially high footprints, associated with less efficient transit and larger houses than the higher den-sity countries of Europe and East Asia. Outside the Global North, Mongolia and Oman stand out as having especially high ecological footprints—in these cases hydrocarbon reserves (oil in Oman and coal in Mongolia) are likely partly responsible.
Data source: Global Footprint Network (2022).

aggressive industrialization. By contrast, sparsely populated Australia and Canada are revealed to have biocapacity reserves, despite high-consumption lifestyles, owing to their huge territories. Scale is highly significant in this respect: on one hand, we could argue that Australia and Canada can sustainably continue using resources as they currently do; on the other hand, we could argue that Australians and Canadians, with their high per capita resource use, should share their resources with people from other countries that are in deficit to achieve better equity at the global scale. A national-scale analysis also encourages us to explore the ways in which pollutants and natural resources are moved around the globe via international trade networks, further complicating the picture. At the global scale, the picture is clear, however—the Earth is estimated to have been in biocap-acity deficit since the 1970s, and this deficit has been growing ever since (WWF 2020). "Earth Overshoot Day" is designed to publicize the issue by identifying the point in the year when Earth's population has already consumed the resources that can be sustainably regenerated in a year. This date has been moving steadily forward—in 2020, it was calculated to be August 22; by 2022, it had moved to July 28. If we consider a scenario in which the whole global population lived at the standards of the affluent Swiss, the date would be May 11 (Global Footprint Network 2021a).

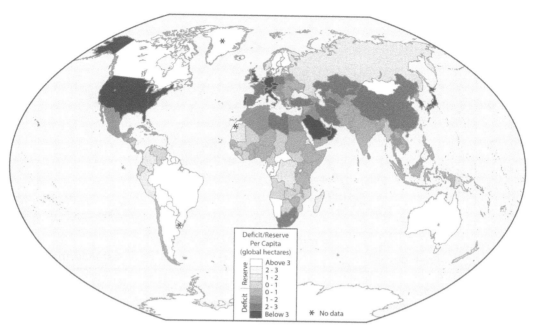

FIGURE 5.2 Ecological reserve and deficit by country, 2017
Data source: Global Footprint Network (2022).

The significance of population size must therefore be balanced by consideration of consumption patterns, leading to a profound philosophical divide between those who see too many people as unsustainable and those who view affluent lifestyles as more troublesome. In reality, of course, we must recognize that both population growth *and* high-consumption lifestyles contribute to environmental degradation. This combination of impacts is recognized in the formula I=PAT, which outlines how impact on the environment (I) is a combination of population size (P), consumption patterns or affluence (A), and technology (T). Devised in the 1970s, this equation is generally credited to have arisen from conversations between Paul Ehrlich, John Holdren, and Barry Commoner—three environmental thinkers of the day—although the emphasis they placed on different aspects of the equation differed significantly (Chertow 2001). Ehrlich was most concerned with the impact of growing populations (consistent with his neo-Malthusian views), whereas Commoner argued that it was consumption patterns and technologies associated with capitalist modes of production that were the greater problem. Commoner believed that low-income countries were disadvantaged by historical inequalities that placed poor countries in marginalized positions in the global economy, perpetuating their poverty and contributing to rapid population growth. He argued that alleviating poverty through technological and social developments could slow population growth in the long run (Commoner 1971, 1976).

Beyond social justice concerns, the introduction of technology to the debate is also significant because it suggests that *how* we navigate the relationship between people and the environment is important. Although the impact of technology on the environment has historically been perceived as largely negative (chimneys bellowing out toxic smoke or water fouled by industrial runoff), increasingly we are

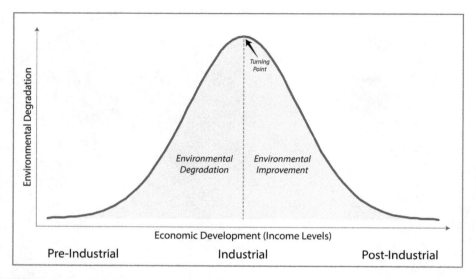

FIGURE 5.3 Environmental Kuznets curve
Source: Drawn after a similar illustration in Sarkodie and Strezov (2018).

focusing on technology to provide solutions to environmental problems. Many clean air acts, for instance, rely on technological fixes to remedy pollution problems caused by earlier technological developments (Chertow 2001). Today, many environmentalists hope that a shift to **green technologies** such as renewable energy generation and efficient recycling will alleviate environmental problems. In this respect, it is often hoped that a community's influence on the environment can be reduced significantly, without any decrease in living standards, given the application of the right technologies.

This idea is consistent with a theory from economics that suggests that the environmental impact of a particular community changes over time as that community develops economically, as represented by the **environmental Kuznets curve** (figure 5.3). As a country industrializes, rising pollution and loss of natural resources initially have detrimental environmental impacts (the rising limb of the curve). At some point, however, rising affluence leads people to begin to tackle the deteriorating environmental situation, and the community begins to invest in pollution-remedying technology and other environmental measures. At this point, environmental conditions begin to improve (the falling limb). This curve has been shown to fit quite well with respect to the emission of some pollutants such as sulfur dioxide and lead into the atmosphere. The curve is also consistent with the fact that newly industrializing middle-income countries such as India and China have some of the highest levels of airborne pollutants today, with very poor as well as affluent countries generally having better air quality—poor countries because they have not yet industrialized and rich countries because they employ pollution mitigation strategies. The model has been criticized for being overly simplistic, however, because pollution responds to policy measures as well as changes in income and the interplay of the two is complex. Furthermore, rich countries may export some of their environmental damages—for instance, domestic deforestation can decline in affluent countries if they turn to importing timber. Empirically, too, not all pollutants seem to follow the curve—solid waste and carbon dioxide emissions, for example, have so far seemed to just keep rising as societies become more affluent, although perhaps they too will finally peak and then decline.

Policy is also brought to the fore when we consider the relative importance of reducing consumption versus technological solutions to environmental degradation. The idea that new technologies can solve environmental problems *without* the need for high levels of environmental regulation or reductions in consumption is popular with many politicians, who want to be seen to be tackling environmental problems but without having to ask their voters to accept lower living standards. In many ways, green technologies provide the ultimate win–win situation—they address environmental concerns without requiring that consumers reduce consumption and are often touted for creating jobs along the way! Critics counter that there is no escaping that we will eventually have to curb our consumption in an era of growing populations and finite resources, even if technology can help alleviate some environmental problems in the short term.

This division between those who see technology as providing a solution to environmental problems and those who believe that consumption must be reduced as well reflects further philosophical divides. For a very optimistic group—sometimes referred to as **cornucopians**—the Earth provides huge abundance, and technological progress and human ingenuity will always keep humankind ahead of the curve—as one resource runs out, there will be another to turn to; as populations grow, farmers will devise new approaches to raise yields. In such a worldview, reducing consumption is unnecessary. Although largely a historical perspective with its roots in religious philosophy, traces of cornucopianism can be found in some recent commentators' writings, particularly those using economic frameworks to explore issues of population and resources. For one such group, sometimes called the **optimistic economists**, larger populations are interpreted to mean more workers, consumers, and innovators, and therefore economic growth, even in situations of finite resources. An example of this approach was provided by Ester Boserup, a Danish economist, who argued in the 1960s that population growth stimulates innovations in agriculture, leading to higher yields, keeping food supply one step ahead of population growth (Boserup 1965). Similarly, Julian Simon, an American economist working in the latter half of the 20th century, became well known for his view that population growth stimulates the economy. Although he rejected cornucopian views, he believed that human ingenuity would provide substitutes to resources that were running out and that there was little cause for concern over finite resources.

A further concern related to population and limited resources considers the source of their scarcity. Here we can distinguish between "demand-induced scarcity," whereby population growth and increased consumption lead to fewer resources per capita, and "supply-induced scarcity," whereby environmental degradation reduces the resources available. A further concept is "structural scarcity," representing circumstances where uneven access to resources leaves certain populations under-served, raising important social justice issues. In all three cases, resource scarcity can increase tensions and even generate conflict (Kennedy 2001).

Although these philosophical standoffs may seem dated today, their key aspects continue to be debated. Is population size or consumption the major problem? Will technology allow us to decrease our impact on the environment while maintaining current living standards, or are lifestyle changes essential? Are we nearing the Earth's limits, or will innovative humans continue to find solutions and substitutions for finite resources? Does economic development automatically lead to improving environmental conditions, or is policy required to guide this transition? How can we address social justice concerns within this framework? This last question leads us to one final theoretical thread before we turn to consider several case studies in light of these philosophical foundations.

Population, resources, and social justice

As an environmental movement developed in the West from the 1960s onwards, a challenge quickly arose related to how to balance global environmental needs with the need for economic development for communities living in poverty. Poorer countries were urged to avoid the environmental mistakes that the rich countries had made through activities such as deforestation and pollution, and yet many countries of the Global South argued that they needed the income from resource extraction and other environmentally damaging activities to raise their populations out of poverty.

The 1972 "Conference on the Human Environment" was the United Nations' first attempt to try to iron out these differences. As the Global North pressed for greater efforts to address an ever-increasing list of environmental issues (deforestation, acid rain, desertification, biodiversity loss, etc.), many in the Global South argued for the need to address poverty before environmental goals could be met. To try to achieve consensus between these diverging viewpoints, a United Nations independent commission, the Brundtland Commission, was convened in 1983 to develop proposals for reconciling economic and environmental goals. The Commission's final report, *Our Common Future* (often referred to as the **Brundtland Report**), argued that environmental goals could not be met unless the global community addressed poverty at the same time (World Commission on Environment and Development 1987), heralding in a new era in which social justice was seen as pivotal to environmental progress.

The Brundtland Report introduced the idea of **sustainable development** or "development that meets the needs of the present without compromising the ability of future generations to meet their own needs," as a way to reconcile environmental and economic goals (World Commission on Environment and Development 1987). Though this may seem like a reasonable compromise, critics argued that the report was too idealistic. In particular, "development" implies consumption of resources, which is counter to efforts to protect environmental assets and ecosystems, and so exactly how these twin goals would be achieved was left unclear. Nonetheless, the Brundtland Report was an important philosophical turning point in recognizing the needs of poor populations alongside environmental concerns.

Over the coming years, the concept of **sustainability** emerged from these earlier discussions. Today, efforts toward sustainability are generally accepted as having to consider three main aspects: environment, economy, and social equity, representing a genuine embrace of the need to balance multiple goals. Originally envisaged as the straightforward intersection of goals in these three realms (figure 5.4), more recently some scholars have argued that the natural environment should be

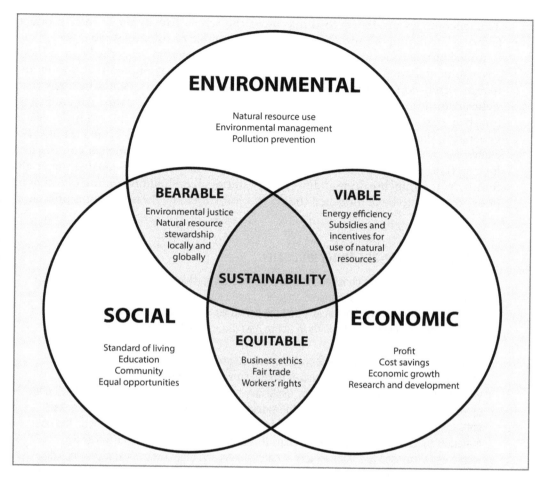

FIGURE 5.4 The three pillars of sustainability
Source: Drawn after a similar image in Hafizyar and Dheyaaldin (2019).

seen as encapsulating both society and the economy to reemphasize the environmental concerns that are supposed to be central to sustainability debates. For some, sustainability simply requires that we inject ethical goals (both environmental and social justice) into traditional economic and political systems; others endorse far more radical approaches, suggesting the need to overturn current economic systems to rebuild them from the ground up.

One such effort to rethink traditional practices has been the development of **environmental economics**, which argues that current economic systems cause environmental problems through not acknowledging environmental costs in economic decisions. For example, we know that factories produce pollutants and yet the cost of disposing of pollutants is rarely included in the cost of products. Instead, the economy assumes that the natural environment will provide raw materials and environmental services for free. Environmental economists attempt to quantify the value of these so–called **environmental externalities**, requiring that the producer or consumer cover the cost of cleaning up pollution, sustainably sourcing materials, and safely disposing of the product at the end of its useful life, among other things.

This encourages producers to think about the whole **life cycle** of their product, favoring the design of products that are more durable or fully recyclable.

Social justice issues are also raised when we consider who bears the brunt of poor environmental conditions. Many social justice scholars have noted that poorer and marginalized communities often experience greater environmental burdens than richer communities, particularly in terms of exposure to pollutants. An additional injustice exists in the fact that the poorer community may have played little role in generating the pollutants affecting them. For instance, there has been a tradition of electronic waste (old computers, cellphones, etc.) being sent from rich countries to low-income countries where they are broken down to retrieve valuable materials. Similarly, the ship-breaking industry has traditionally relied on sending old ships to poor countries to be demolished (box 5.1). Unfortunately, these processes often lead

BOX 5.1 THE SHIP-BREAKING INDUSTRY

The global shipping fleet consists of around 55,000 cargo ships. Their average life span is 20 to 25 years, with several hundred ships scrapped each year. Most of this dirty and dangerous work is carried out in low-income countries to take advantage of cheap labor and lax environmental laws, particularly in South Asia (Bangladesh, India, and Pakistan), China, and Turkey (ILO 2021).

Ships consist of many different components, so taking them apart is difficult and time-consuming. On the positive side, ship-breaking provides employment for thousands of workers and recycles many materials, especially huge amounts of steel. In Bangladesh, one of the world leaders in ship-breaking, the industry supplies a high percentage of the steel consumed in the country and has been a catalyst for other industries dependent on the raw materials that can be recovered from ships (Hossain et al. 2016). On the negative side, workers are exposed to a wide variety of carcinogens and other toxic substances (including PCBs, asbestos, cadmium, mercury, lead, and sulfuric acid), endangering worker health as well as the environment.

The International Labour Organization (ILO) has identified ship-breaking as one of the most dangerous types of work due to high numbers of fatalities, injuries, and diseases contracted at work. The most common causes of injuries and fatalities are falls, being hit by ship parts, and explosions; most long-term health issues occur through toxic exposures. Data on accidents are hard to find, however, because most workers are employed informally and authorities may not keep records. Workers typically have very little education or training, leaving them vulnerable to exploitation. Worksites often lack regulations or safety procedures and workers may be offered little protective equipment. Living conditions are often also poor, with a lack of adequate housing, food, sanitation, and health services (Greenpeace-FIDH 2005).

To be scrapped, ships are brought close to the coast at high water and beached. Because they are not taken apart in shipyards, there are often no facilities to properly treat or dispose of hazardous waste (ILO 2021). Toxic substances may be released directly into coastal waters or contaminate soils, with detrimental impacts on marine ecosystems and biodiversity. Beyond human rights concerns, ship-breaking is a prime example of environmental injustice because it transfers hazardous waste to poorer communities, while the money saved through improper disposal of wastes benefits large corporations.

to the receiving communities being exposed to toxic chemicals, as well as high levels of pollutants left in the local environment, with attendant health threats. In another powerful example, many commentators have pointed out that the environmental costs of climate change are likely to be borne disproportionately by low-income communities, even though the bulk of the responsibility for climate change lies with emissions made by the affluent. Circumstances such as these have been highlighted as instances of **environmental injustice**.

The environmental justice movement can be traced to the early 1980s when African American communities in the US South protested against dumping of hazardous waste in their neighborhoods. Although similar grassroots protests had occurred since the 1960s associated with the US civil rights movement, it was only when academics and policymakers began collecting data on the issue that the scale of the problem was realized (e.g., Bullard 1990). Since then, numerous articles and books have been published that demonstrate the disproportionate burden of environmental pollution borne by poor and minority populations. Many early reports quantified how hazardous facilities were more likely to be sited in marginalized neighborhoods (Reichel 2018). The notion of environmental justice (as well as the related idea of **environmental racism**) has now been extended to explore a wide variety of issues, including higher rates of air pollution in minority neighborhoods (Grineski, Bolin, and Boone 2007), high levels of heat stress in communities of color (Hsu et al. 2021), and the disproportionate impact of mining on indigenous health (Lewis, Hoover, and MacKenzie 2017).

One final social equity issue linking population and environment revolves around indigenous land ethics. Today, indigenous communities are often celebrated by environmentalists for their close connections to the land and roles as **environmental stewards**. Many indigenous communities retain a strong **land ethic** consistent with low-consumption lifestyles that prioritize protection of natural resources (Berkes 1999). This has been demonstrated in a variety of ways, from efforts to protect sacred forests to protests against environmentally damaging practices. In parts of the Brazilian Amazon, for instance, many indigenous reserves are proving to be strongholds against deforestation, with several recent studies showing them to be as effective as national parks at preventing deforestation (Walker et al. 2020). In another example, some Andean communities in Peru have protested mineral extraction on traditional lands, arguing that it is leading to pollution of water and soil and destroying sacred landscapes (Hufstader 2009). It is important not to over-generalize indigenous beliefs, however. There is a huge diversity of belief systems, both among and within indigenous groups, including in terms of interactions with the environment. Whereas some indigenous groups have embraced the label of environmental steward, others have argued that expectations that indigenous groups must protect resources or be participating in "traditional" lifestyles as proof of their indigeneity provide just one more way in which dominant groups attempt to restrict their activities.

PART II: APPLYING THEORIES OF POPULATION AND ENVIRONMENT

In the remainder of this chapter, we use these theories as a starting point for analyzing two pressing environmental issues: (1) finite resources and unsustainable resource extraction and (2) food and agriculture, as representative of some of the key themes in the population and environment literature. We address the global climate crisis in chapter 12, providing a further example of the ways in which environmental issues can be viewed from a population perspective.

Finite resources and unsustainable resource extraction

One clear intersection between population and the environment concerns **finite resources**. Finite or non-renewable resources are those that are limited by the total amount currently available on Earth, because the regeneration of these resources takes too long for them to be considered renewable. Given that some of the minerals that we use today are found in only minute quantities, the idea that we are likely to run out of them causes considerable concern. Though an optimistic economist might argue that as one mineral runs out human ingenuity will find a replacement, many scientists are skeptical of how easily such transitions might be made and note that we are already at a point where supplies of some commonly used minerals such as copper are approaching a shortage (Ali et al. 2017). Of particular concern are some of the rare minerals that have become integral parts of our digital society. Indium, for instance, was critical to the development of touch screen technology, but scientists worry that supplies are insufficient to meet demand (Wong 2015). Scientists also note that our rush to decarbonize society (reduce CO_2 emissions) via new climate-friendly technologies is likely to pose further demands on rare minerals that the world's current trade structures are ill-equipped to address (Wong 2015). The mining of these minerals also takes a hefty toll on the environment, leading to complex environmental trade-offs—the mining of lithium and cobalt, used in electric car batteries, has been linked to a variety of toxic accidents as well as environmental degradation, for instance (Katwala 2018).

The environmental costs of mining have always been high, related to landscape clearance, toxic by-products contaminating land and water, as well as the impact of opening up remote areas to further development. The increasing value of some minerals as they become scarcer makes it economically viable to mine increasingly low-grade ores (rocks with very little of the target mineral in them), as well as in more remote locations, each with significant environmental implications. Low-grade ores require the extraction of more ore for the same amount of mineral, resulting in more landscape destruction and greater energy costs. Large-scale operators have the technology to literally move mountains in large open-pit operations in the search for so-called invisible gold, for instance. In addition, processing low-grade ores often requires the use of toxic chemicals to extract the traces of valuable mineral. In the Yanacocha gold mine in the Peruvian Andes, for instance, this has led to pollution of the local environment with chemicals such as cyanide and mercury (Hufstader 2009). In other cases, mining is pushing farther into remote areas. A gold rush has been going on in parts of West Africa, for instance, destroying some of the region's remaining forest and polluting water courses. This effort is driven by small-scale miners working for unscrupulous middlemen, who pay little heed to worker health or environmental protection. Small-scale mining operations such as these are also known to involve child labor, with the International Labor Organization estimating that as many as 1 million children may work in small-scale mines and quarries (Neff 2013; Bales 2016).

Soil provides another example of a resource that is not easily renewable on human timescales. Unsustainable farming techniques, such as removing all crop cover from a field and leaving the soil open to the elements, allow topsoil to be washed and blown away—half of global topsoil may have been lost over the past 150 years (WWF 2021). Some longstanding areas of human cultivation show extremely impoverished soils after thousands of years of unsustainable agricultural use—the so-called Fertile Crescent that was an early hearth of agricultural activity in the Middle East is now far from fertile, for instance (Diamond 2003). This reflects a huge loss of agricultural potential, particularly because many soils suffer from additional problems such as soil compaction, nutrient loss, and increasing salinity. Furthermore, the lost soil can cause the sedimentation of nearby water courses and leave the thin soils that remain less able to hold onto moisture, raising the risk of both drought and flooding (WWF 2021).

Freshwater provides yet another intriguing case of renewability and scarcity. Of all of the water on Earth, over 97 percent is saline and about 60 to 70 percent of freshwater is locked up as ice, making freshwater a very limited resource (figure 5.5). It has become common to note that 20th-century wars were often driven by oil, whereas future conflicts will likely focus around water. Water is critical to life and is also non-substitutable for most purposes: we may choose to eat apples if bananas become scarce, but there is nothing to substitute for water. One of our few options is desalination of saltwater into fresh, but this comes at high energy cost. We may think first of drinking and washing as key water uses, but agriculture accounts for 70 percent of global water consumption, and industry also uses more water than domestic consumption (Khokhar 2017). Estimates of water scarcity vary widely, but commentators usually agree that the problem is growing. For example, Mekonnen and Hoekstra (2016) estimated that half a billion people around the world face water scarcity year-round, and two-thirds of the global population experience severe water scarcity for at least 1 month per year, especially in China and India. As figure 5.6 illustrates, the drylands of the Middle East and North Africa, as well as the dense populations of Asia, put particularly high demands on water supplies.

Rising populations clearly create growing demands for water—an example of demand-induced scarcity. Even if we consider only increasing agricultural requirements, we may see a 15 percent increase in water withdrawals by 2050 (Khokhar

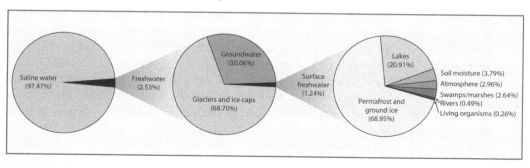

FIGURE 5.5 The Earth's freshwater resources
Data source: USGS (2018).

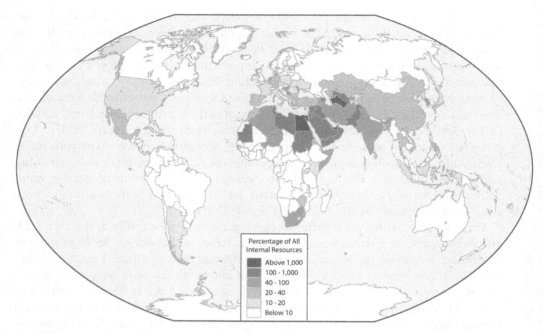

FIGURE 5.6 Freshwater withdrawals as a percentage of national water resources, 2017
Countries in North Africa, the Middle East, and West Asia experience some of the greatest pressure on water resources, given generally dry climates as well as dense populations in places. Where countries are reported to withdraw more than 100 percent of national water resources, they are either extracting water unsustainably from large aquifers or producing a large amount of water from desalination.
Data source: Ritchie and Roser (2017).

2017). Increasing affluence is projected to put further demands on already over-stretched supplies. Rising meat consumption will likely be a key factor because of the water needed to grow grain for feed (Myers and Kent 2003). Supply-induced scarcity is also likely to increase as climate change alters precipitation patterns in ways that make drought more common. Many climate change models predict that a larger proportion of rain will fall in strong storms, potentially leading to short-term flooding as large quantities of water run quickly off the land, but then drought between these strong storms. Additionally, sea level rise has begun to contaminate coastal groundwater supplies.

Water use has a significant social justice component, too, illustrating "structural scarcity," with affluent communities using significantly more water than poorer ones (figure 5.7). In 2013–17, per capita freshwater withdrawals totaled 1,367m³ per year in the United States, 673m³ in Australia, 408m³ in France, and just 51m³ in Mozambique (FAO 2021a). Global trade practices add yet more complications. Globalized trade networks mean that many low-income countries are heavily dependent on agricultural produce for foreign exchange, and yet the export of these products puts heavy demands on water supplies. Some particularly thirsty cash crops such as cotton are now being grown extensively in drylands using irrigation, competing with water use for domestic food production. Many of these crops are exported, effectively representing the export of water from a dry country. One study estimated that approximately 11 percent of non-renewable groundwater used

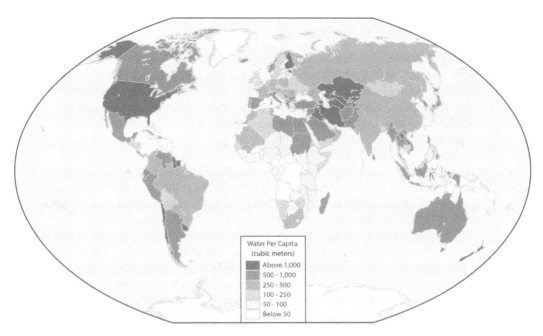

FIGURE 5.7 Freshwater withdrawal per capita by country, 2017
Data source: Ritchie and Roser (2017).

for irrigation is embedded in this "virtual water trade," with the United States, Mexico, and Pakistan generating two-thirds of water exports (Dalin et al. 2017).

Aquifers (underground rock formations that hold water) store perhaps 30 percent of all of the freshwater on Earth and so groundwater represents a further important source of freshwater, particularly in arid regions with erratic or sparse rainfall. Although aquifers can be refilled when rainfall percolates into the ground, many sources of groundwater are, in all practical terms, finite owing to the extremely long timescales needed to recharge them. Many aquifers were filled in previous geological eras, sometimes over thousands of years or when climatic conditions in the region were much wetter. As such, a large proportion of aquifers are being drained more rapidly than they can be recharged. This **groundwater depletion** not only leads to dwindling supplies but can also lead to land subsidence and incursion of saline water into coastal aquifers. Groundwater depletion is a particular problem in places where drylands have been developed for agriculture, with wheat and rice production being the primary drivers of this water use (Dalin et al. 2017).

Even completely renewable resources like timber and fish may be used at unsustainable rates (box 5.2). The process of deforestation reflects trees being removed more quickly than they can grow back, for example. This represents not only a loss of resources but also the degradation of huge areas of habitat, with associated biodiversity loss. Currently, tropical deforestation is one of the key drivers of global forest loss, with 420 million hectares (1,040 million acres) of forest lost to other land uses since 1990. Although the rate of forest loss has slowed in recent years, deforestation continued at about 10 million hectares (25 million acres) per year between 2015 and 2020, with agricultural expansion persisting as the main cause (FAO 2020c). Efforts to restore forests are beginning to bear some fruit, with several recent

BOX 5.2 FISHERIES AND SUSTAINABILITY

Sustainable Development Goal 14 is to "conserve and sustainably use the oceans, seas and marine resources for sustainable development." Unfortunately, fisheries have often been used in unsustainable ways, with research pointing to a potential fisheries crisis (Worm et al. 2006). Although the extent of the crisis is debated, the proportion of fish harvested from stocks considered biologically sustainable has declined from 90 percent in 1974 to 66 percent in 2017 (FAO 2020a).

Traditionally, fishing has been largely restricted to coastal waters, where it could be monitored by governments and local communities. New fishing technology has allowed people to go farther offshore, allowing fishing fleets access to the "high seas"—areas not subject to national jurisdictions. The global fish catch has increased rapidly from 15 million tonnes in 1938 to 86 million tonnes in 1989 and nearly 180 million tonnes in 2018 (FAO 2020a). Today, almost two-thirds of global commercial fish stocks are estimated to be fully exploited or overfished (Arthur et al. 2019). Whether fish catches will continue to increase in the future remains open to question, with the UN's Food and Agriculture Organization predicting continued growth in the fisheries sector, whereas other estimates suggest that wild catches may already have peaked and that some fisheries are subject to imminent collapse (Pauly and Zeller 2016). Another key issue will be the role of **aquaculture**, which could pick up shortfalls in wild catches (figure 5.8). Although aquaculture could relieve pressure on wild stocks, it brings other environmental challenges, including pollution from pesticides, antibiotics, and excess feed; the conversion of coastal environments into fish farms; and the potential introduction of non-native fish species.

Significant social justice issues also surround fisheries. Approximately 3 billion people receive at least 20 percent of their protein from aquatic sources, rising to well over half in some poor countries of West Africa and Asia (HLPE 2014). Given the importance of aquatic sources of protein and micro-nutrients, declining catches have been identified as threatening

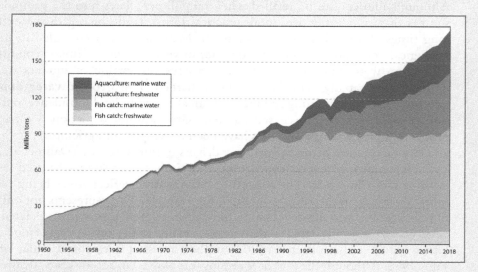

FIGURE 5.8 Global fish catch and aquaculture, 1950–2018
Data source: FAO (2019).

human health (Golden et al. 2016). Today, fish is a globalized commodity, however, with small-scale local fishing fleets often outcompeted by better-funded, more efficient commercial fleets (Pauly and Zeller 2016). Recently, the top seven fishing nations (China, Indonesia, Peru, India, Russia, the United States, and Vietnam) caught almost half of all fish taken from the wild (FAO 2020a). Though the sale of fishing rights provides quick cash for a government, it means that profits from the fishing industry leave the country, as do the fish themselves. This may limit access to a protein-rich food source for local people, as well as prevent the development of supplemental jobs in fish processing.

analyses emphasizing the economic benefits that can accompany forest restoration—one estimate suggested that for every dollar spent on forest regeneration, $9 of economic benefits can be realized (Taylor 2020).

Food and agriculture

As the fisheries example in the previous section illustrates, resource issues are not just an environmental challenge but also have significant ramifications for human communities. Running out of food has long been a source of concern with respect to growing populations. Although huge increases in production achieved by **agricultural extensification** (converting new land to crops) and **agricultural intensification** (increasing yields through more intensive farming methods) have kept food production ahead of population growth so far, there is no guarantee that this will continue. Global population is projected to increase by at least 1 billion people by the mid-21st century, calling into question whether food production will continue to keep up.

It is not just population size that is at issue. Consumption patterns are also critical. Despite sufficient food at the global scale, almost 9 percent of the global population is currently undernourished (FAO 2021c; figure 5.9). Indeed, maldistribution of food currently leaves some populations undernourished, even as others consume more than is necessary. Although these social equity implications are critical, in this chapter we focus on the environmental impacts of growing populations and changing consumption patterns with respect to food.

Today, agricultural and pasture lands consume about half of Earth's ice-free land surface (FAO 2013), with **landscape conversion** for farming continuing to destroy wildlands in many parts of the world. Although some clearance occurs as peasant farmers develop farms for their own family's subsistence, large areas of land are also being cleared for commercial agriculture. Illegal clearing for commercial agriculture was probably responsible for almost half of recent deforestation worldwide, with the tropical forests of Brazil and Indonesia responsible for large proportions of this total (Forest Trends 2014). In recent years, the Brazilian government has also actively encouraged the conversion of one of the last remaining wild grasslands, the *cerrado*, to large-scale export-oriented soybean farms.

Agricultural intensification brings with it a suite of other environmentally damaging impacts. As humans have intensified agricultural systems, pushing for ever higher yields in more "efficient" systems, demands on natural ecosystems have

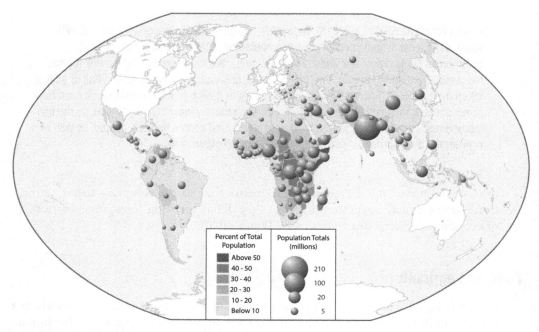

FIGURE 5.9 World's undernourished population, 2020
Undernourishment remains a significant problem in sub-Saharan Africa, where around half the population is under-nourished in many countries. In most of the remainder of the low-income world, rates of undernutrition are lower, but the large populations of South and East Asia mean that India in particular has many of the world's undernourished people.
Data source: FAO (2021d).

increased, leading in many cases to land degradation. Plants have three basic needs: water, nutrients, and sunlight. Intensive agriculture typically focuses on providing irrigation water and nutrients from fertilizers to make up for shortfalls from the natural environment. In addition, heavy pesticide use is common to protect the rows of identical crops that are vulnerable to pest and disease outbreaks. Such intensive agriculture requires heavy application of a diversity of agricultural inputs but primarily fertilizer, pesticides, and irrigation water. This highly unbalanced system comes with many detrimental side effects related to over-use of natural resources and emission of pollutants.

With respect to water, for example, irrigation has led to the use of water from rivers and aquifers at unsustainable rates. The US Southwest illustrates this well. This dryland area has relied on rivers such as the Colorado combined with groundwater to provide the water needed to boost its agricultural production. The Colorado River is now well known for failing to reach the sea in many years as so much water is extracted for agriculture. This generates significant social justice concerns as the United States diverts water out of the river before it ever reaches Mexico, in what has been referred to as a "striking example [of] environmental injustice" (Lakhani 2019). Further problems are associated with the over-pumping of aquifers—some of Southern California's aquifers have been so over-exploited that significant land subsidence has been recorded, for instance (USGS, n.d.).

Another major problem with intensive agriculture is pollution from excessive fertilizer and pesticide use. Pesticides frequently kill non-target species or **bio-accumulate** to toxic levels in predatory species that eat other animals contaminated with pesticide. The impact of excess fertilizer is more complex. Although, ideally, each farmer only applies the amount of fertilizer needed by the crop, in practice this is challenging, leading to the runoff of excess fertilizer into local water bodies. This can lead to the overgrowth of algae and simple plants in rivers and lakes. In addition to blocking out sunlight, the decomposition of these plants depletes oxygen in the water, choking out more complex plant and animal life in a process known as **eutrophication**.

Intensive agriculture has also been implicated in contributing to climate change. Globally, agriculture and food production release around one-quarter of all greenhouse gases (Tilman and Clark 2014). Although many people assume that it is emissions associated with the transportation of food that are the problem, transportation is usually only a relatively small contributor to food's overall carbon footprint. A small number of fragile foods are transported by air and do indeed contribute a high carbon footprint from airfreighting (e.g., berries and asparagus), but for most foods it is other processes—including landscape conversion, agricultural practices, food processing, retail, and packaging—that contribute significant quantities of greenhouse gases (Ritchie 2020). In this respect, we need to think about a much wider **life cycle analysis** of the food we eat to understand its full climate impacts (figure 5.10). A large proportion of this impact comes from agricultural processes, including enteric fermentation (digestion in ruminant animals producing methane), emissions from manure and artificial fertilizers spread on fields, decomposition of organic matter, burning of crop residues, and the running of agricultural machinery (Ritchie 2020). When we do these sorts of life cycle analyses, we find that lamb and beef have especially high greenhouse gas emissions—about 250 times the emissions per gram of protein compared with legumes (beans and peas; Tilman and Clark 2014). Estimates such as these have contributed to a growing literature arguing that careful consideration of the diets we eat is critical to improving sustainability (box 5.3).

Given increasing populations and rising affluence, estimates suggest that we may need to increase food production by 50 to 100 percent by 2050 (Royal Society 2009). Many of the products that consumers turn to once they have more choice over their diet, especially meat, are precisely those that put the greatest demands on the environment. As noted in box 5.3, a shift from plant-based diets to ones rich in meat and dairy is particularly detrimental, but increasing demand for luxury products such as chocolate, coffee, and out-of-season produce also brings high environmental costs associated with the complex production and transportation requirements of these finicky and perishable products.

Given that important efforts are underway to protect the few remaining wildlands on Earth, future increases in food production will require intensification of production on existing farmland rather than converting new land to agriculture ("Feeding 9 Billion," n.d.). Many commentators note that this must be done in a sustainable manner, with the term **sustainable intensification** recently coined to reflect this aspiration (Royal Society 2009). Although sustainable intensification is, in many ways, just as much of a contradiction in terms as sustainable development, some existing approaches are promising. Sustainable intensification relies on more

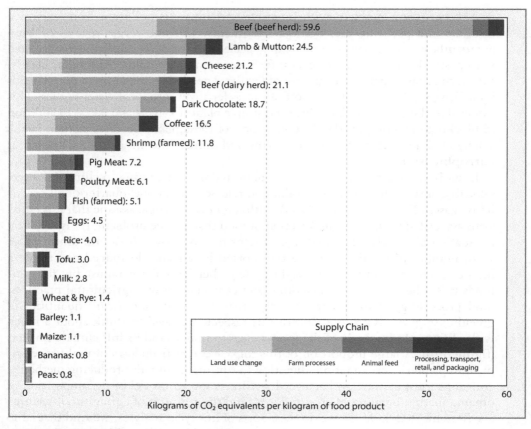

FIGURE 5.10 Greenhouse gas emissions in the food chain
Note the huge significance of animal-based products to greenhouse gas emissions from food compared with many plant-based alternatives, as well as the wide variations between different types of meat. Also, consider how different methods of production can influence emissions of the same food: beef from dairy herds, for instance, has a lower footprint than beef from herds used exclusively for meat, owing to the dairy products that are made from the same production system.
Data source: Ritchie (2020).

efficiently using the agricultural resources we have to obtain higher yields from the same land area but with greater care and attention paid to the pollutants and other side effects that this sort of intensive production typically has. Many see new technological advances such as monitored drip irrigation and carefully measured application of fertilizers as potential approaches. These high-tech systems are often expensive to install, however, and are most viable for large farms that can realize economies of scale. For poorer farmers, considerable hope is being pinned on the potential for genetically modified (GM) crops. We have already dramatically increased food yields with high-yielding crop varieties developed by traditional plant breeding techniques, and many commentators see GM crops as the obvious next step. Genetic modification could both raise yields (with Africa seen as the next frontier for introducing high-yielding crop varieties) and reduce pollution—crops are now being engineered to be more resistant to pests so that farmers can apply

BOX 5.3 DIET AND SUSTAINABILITY

Today, many people are turning to plant-based diets for environmental reasons. The most basic issue revolves around the fact that energy is lost as we move up the food chain and so eating plants is a much more efficient use of energy than meat and dairy (figure 5.11). By contrast, diets high in animal protein need larger land areas for agriculture, as well as greater quantities of the inputs needed to grow crops, which contribute to the higher greenhouse gas emissions associated with animal-based foods. Additionally, fermentation in the guts of some herbivores, particularly cattle and sheep, is associated with the production of methane—a very powerful greenhouse gas—making beef and lamb among the highest agricultural contributors to greenhouse gas emissions.

As such, a general rule of thumb suggests that decreasing our consumption of animal protein and increasing plant-based foods could be an important step toward sustainability. Mediterranean, pescatarian, and vegetarian diets could all reduce land use requirements and greenhouse gas emissions compared with a traditional Western diet (Tilman and Clark 2014). Even small changes could be significant—simply switching from beef to poultry would have significant impacts in terms of greenhouse gas emissions. Many of these changes are consistent with improved health, too. In the Western world, where diets are typically rich in animal protein, people who consume a higher proportion of fruits and

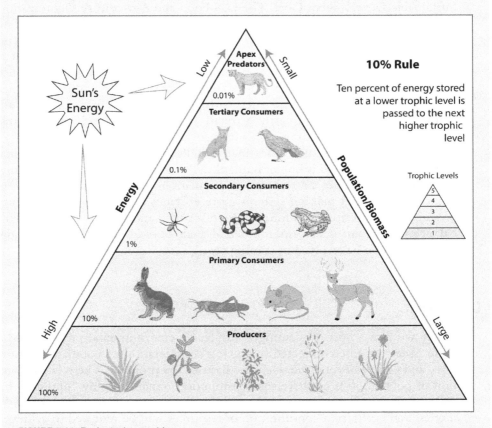

FIGURE 5.11 Ecological pyramid

vegetables in their diets are at lower risk of a wide variety of negative health outcomes, including obesity, certain cancers, and heart disease. In such cases, reducing the consumption of animal products is consistent with both health and environmental benefits (Tilman and Clark 2014).

Here we face a social justice dilemma, however. Although rich consumers can easily decrease their consumption of animal products and experience health benefits, for poor populations with diets deficient in protein and micro-nutrients, increasing consumption of meat and dairy is consistent with improving nutrition. As a result, the UN's Food and Agriculture Organization has noted that some poor communities may need to increase their consumption of animal products, and by extension also their greenhouse gas emissions, simply to meet the basic dietary needs of their populations (FAO 2020b).

pesticides more sparingly, for instance. Efforts are also being made to increase the nutritional quality of crops by genetic engineering to help address food insecurity; "golden rice" has been developed to increase the quantity of vitamin A in rice, for example.

An even clearer win–win situation is to reduce food waste, which is also a Sustainable Development Goal. The UN Food and Agriculture Organization estimates that approximately one-third of food produced globally is lost or wasted (FAO 2021b), with wastage of fruits and vegetables perhaps as high as 50 percent (UNEP, n.d.). To put these losses in perspective, it has been estimated that people in rich countries waste about 222 million tonnes of food every year—almost the entire net food production of sub-Saharan Africa (UNEP, n.d.). In low-income countries, losses are mostly associated with processing and storing crops; in affluent countries, losses occur primarily at retail and consumer levels, with changes in consumer behavior providing a key target for intervention.

In summary, agriculture poses many challenges with respect to sustainability and social equity. Food is one of humankind's most basic needs, so increasing food production is going to be essential as populations grow. How far we can get by reducing waste and making systems more efficient remains to be seen, but these provide win–win situations for both food security and the environment. How we balance food security and environmental goals beyond this is still widely debated but will have to be resolved in coming years.

Conclusion

Today we face a host of seemingly intractable environmental problems. Many of these issues are interconnected, requiring consideration of population, consumption, and equity beyond the scientific basis of the problem. The relative contribution of population growth versus consumption remains highly contested, but the scale of the problem suggests that stabilization of population growth *and* more modest and egalitarian consumption of resources will be required to achieve some measure of environmental sustainability.

Discussion questions

1   Consider the challenges posed by increasing demands for rare metals in the high-tech industry. How might a neo-Malthusian explain the problem? How about an optimistic economist? How do you think that the global community should tackle issues related to finite resources? Is this a scientific problem or a policy concern?

2   How would you propose increasing the amount of food produced to meet growing needs while minimizing environmental impacts? In your discussion, you might think about the pros and cons of organic agriculture, genetically modified crops, and industrial agriculture—for both the environment and human health—as well as social justice goals.

3   What are some of the pros and cons of wild-caught versus farmed fish from environmental sustainability and social justice perspectives?

Suggested readings

Ahmed, S. F. 2016. "The Global Cost of Electronic Waste." *The Atlantic*, September 29, 2016. https://www.theatlantic.com/technology/archive/2016/09/the-global-cost-of-electronic-waste/502019/

Bland, A. 2014. "The Environmental Disaster That Is the Gold Industry." *Smithsonian Magazine*, February 14, 2014. https://www.smithsonianmag.com/science-nature/environmental-disaster-gold-industry-180949762/

"Can the 'Blue Revolution' Solve the World's Food Puzzle?" n.d. *National Geographic*. https://www.nationalgeographic.com/foodfeatures/aquaculture/

"Feeding 9 Billion." n.d. *National Geographic*. https://www.nationalgeographic.com/foodfeatures/feeding-9-billion/

Godfray, C., J. Beddington, I. Crute, L. Haddad, et al. 2010. "Food Security: The Challenge of Feeding 9 Billion People." *Science* 327: 812–18. doi:10.1126/science.1185383

Glossary

agricultural extensification: increasing agricultural production by converting wild lands into farmland

agricultural intensification: increasing agricultural production by raising yields on land that is already being farmed via agricultural inputs such as fertilizers and pesticides

aquaculture: farming fish for food

bioaccumulation: the process of chemicals building up in the tissues of an organism; these chemicals can also accumulate up the food chain as animals feed on other animals that have absorbed them

biocapacity: an estimate of the biological productivity of an ecosystem in terms of its production of natural resources and sometimes also its potential to absorb wastes

Brundtland Report: report that introduced the idea of "sustainable development" in 1987; its official title was *Our Common Future*

cornucopianism: philosophy suggesting that the Earth will always produce a bounty of natural resources for human use

ecological footprint: the impact of one's activities on the environment; in its original manifestation the ecological footprint estimated the land area needed to provide all required resources, as well as dispose of wastes, associated with an individual, community, or product

ecological overshoot: ecological demands exceeding the regenerative capacity of an ecosystem; ecosystems can function for a period of time in this non-sustainable manner but will degrade and could eventually collapse

environmental economics: a recent economic approach that emphasizes quantifying the valuable role of ecological systems in economic processes

environmental externalities: environmental impacts of producing goods that are not incorporated in traditional economic mechanisms; for example, pollution

environmental injustice: the disproportionate exposure of marginalized communities to pollution and other environmental hazards

environmental Kuznets curve: economic model that suggests that a community's impact on the environment is initially low but will increase with economic development up to a point where the community begins to invest in green technologies that can reduce the environmental impacts of their activities

environmental racism: the disproportionate exposure of communities of color to pollution and other environmental hazards

environmental stewardship: a community living in an environmentally sustainable manner is sometimes described as displaying environmental stewardship

eutrophication: degradation of a freshwater ecosystem owing to oxygen depletion associated with overgrowth of algae and simple plants owing to runoff of agricultural fertilizers

finite resources: natural resources that cannot be regenerated over a human timescale; for example, minerals

green technologies: technologies specifically designed to reduce the environmental impact of a particular activity

groundwater depletion: unsustainable use of water from aquifers

land ethic: philosophical ideals of how a community should live sustainably on the land

landscape conversion: changing land use from one activity to another; the term typically refers to the loss of wildlands to human activities, particularly agriculture

lifeboat ethics: theory by ecologist Garrett Hardin that suggests that global resource distribution can be understood using the metaphor of lifeboats, where the rich nations are safely aboard but at constant risk of being overwhelmed by poor individuals trying to claim a share of global resources

life cycle analysis: an effort to assess the impact of the whole life cycle of a product, from where the raw materials come from to how the product will be disposed of at the end of its useful life

Marxism: philosophical belief system, following Karl Marx, that focuses on inequality as the root cause of many social and economic problems

neo-Malthusianism: extensions of Malthus's ideas; argues that population growth is leading to environmental collapse as well as potential resource shortages

optimistic economists: in the framework of population growth, optimistic economists believe that more people lead to more workers and consumers and therefore economic growth

overpopulation: too many individuals in a population to be supported sustainably

population explosion: concept of population growing rapidly and unsustainably, popularized in the 1960s and '70s

Spaceship Earth: concept popularized in the 1960s and '70s that refers to the idea of Earth as a bounded system with finite resources

sustainability: a social movement that strives for ecological balance, alongside economic and social justice goals

sustainable development: economic development designed to bring people out of poverty but at lowest ecological cost; often criticized for being overly idealistic and unattainable

sustainable intensification: raising agricultural productivity through increasing yields but in ways that limit the negative environmental consequences that usually accompany agricultural intensification

Tragedy of the Commons: theory by ecologist Garrett Hardin that states that shared resources are likely to be degraded over time as individuals have an incentive to over-use them for their own benefit

Works cited

Ali, S., D. Giurco, N. Arndt, E. Nickless, et al. 2017. "Mineral Supply for Sustainable Development Requires Resource Governance." *Nature* 543: 367–72. https://www.nature.com/articles/nature21359

Arthur, R., S. Heyworth, J. Pearce, and W. Sharkey. 2019. "The Cost of Harmful Fishing Subsidies." IIED Working Paper. http://pubs.iied.org/16654IIED

Bales, K. 2016. *Blood and Earth: Modern Slavery, Ecocide and the Secret to Saving the World*. New York: Random House.

Berkes, F. 1999. *Sacred Ecology: Traditional Ecological Knowledge and Resource Management*. Philadelphia: Taylor & Francis.

Boserup, E. 1965. *The Conditions of Agricultural Growth: The Economics of Agrarian Change under Population Pressure*. London: Allen and Unwin.

Bullard, R. 1990. *Dumping in Dixie: Race, Class and Environmental Quality*. New York: Routledge.

Chertow, M. 2001. "The IPAT Equation and Its Variants." *Journal of Industrial Ecology* 4 (4): 13–29.

Commoner, B. 1971. *The Closing Circle: Nature, Man and Technology*. New York: Random House.

———. 1976. *The Poverty of Power*. New York: Bantam Books.

Dalin, C., Y. Wada, T. Wastner, and M. Puma. 2017. "Groundwater Depletion Embedded in International Food Trade." *Nature* 543: 700–4. https://doi.org/10.1038/nature21403

Diamond, J. 2003. "The Erosion of Civilization." *Los Angeles Times*, June 15, 2003. https://www.latimes.com/archives/la-xpm-2003-jun-15-op-diamond15-story.html

Ehrlich, P. 1968. *The Population Bomb*. New York: Ballantine Books.

Ehrlich, P., and A. Ehrlich. 2009. "The Population Bomb Revisited." *The Electronic Journal of Sustainable Development* 1 (3): 63–71.

FAO [Food and Agriculture Organization]. 2013. "Food and Agriculture Data." http://faostat.fao.org

———. 2019. *FAO Yearbook of Fishery and Aquaculture Statistics*. https://www.fao.org/fishery/en/topic/166334/en

———. 2020a. "The State of Food Security and Nutrition in the World." http://www.fao.org/3/ca9692en/online/ca9692en.html#chapter-executive_summary

———. 2020b. *The State of the World's Fisheries*. Rome: FAO and UNEP. https://doi.org/10.4060/ca9231en

———. 2020c. *The State of the World's Forests*. Rome: FAO and UNEP. https://www.fao.org/documents/card/en/c/ca8642en

———. 2021a. "AQUASTAT Database." http://www.fao.org/aquastat/statistics/query/index.html

———. 2021b. "Food Loss and Waste Database." http://www.fao.org/food-loss-and-food-waste/flw-data

———. 2021c. "Hunger and Food Insecurity." http://www.fao.org/hunger/en/

————. 2021d. *The State of Food Security and Nutrition in the World 2012.* https://www.fao.org/documents/card/en/c/cb4474en/

"Feeding 9 Billion." n.d. *National Geographic.* Accessed September 21, 2021. https://www.nationalgeographic.com/foodfeatures/feeding-9-billion/

Forest Trends. 2014. "Forest Trends Report Series: Forest Trade and Finance." September 2014. https://www.forest-trends.org/wp-content/uploads/imported/for168-consumer-goods-and-deforestation-letter-14-0916-hr-no-crops_web-pdf.pdf

Global Footprint Network. 2021a. "About Earth Overshoot Day." https://www.overshootday.org/about-earth-overshoot-day/

————. 2021b. "Ecological Footprint." https://www.footprintnetwork.org/our-work/ecological-footprint/

————. 2022. "Ecological Deficit/Reserve." https://www.footprintnetwork.org

Godfray, C., J. Beddington, I. Crute, L. Haddad, et al. 2010. "Food Security: The Challenge of Feeding 9 Billion People." *Science* 327: 812–18. doi:10.1126/science.1185383

Golden, C., E. Allison, W. Cheung, M. Dey, et al. 2016. "Nutrition: Fall in Fish Catch Threatens Human Health." *Nature* 534: 317–20.

Greenpeace-FIDH. 2005. "End of Life Ships: The Human Cost of Breaking Ships." https://www.fidh.org/IMG/pdf/shipbreaking2005a.pdf

Grineski, S., B. Bolin, and C. Boone. 2007. "Criteria Air Pollution and Marginalized Populations: Environmental Inequity in Metropolitan Phoenix, Arizona." *Social Science Quarterly* 88 (2): 535–54.

Hafizyar, R., and M. Dheyaaldin. 2019. "Concrete Technology and Sustainability Developed from Past to Future." *Sustainable Structure and Materials* 2 (1): 1–13.

Hamity, M., C. Dillard, S. Bexell, and C. Graff-Hughey. 2019. "A Human Rights Approach to Planning Families." *Social Change* 49 (3): 469–92. doi:10.1177/0049085719863894

Hardin, G. 1968. "The Tragedy of the Commons." *Science* 162: 1243–8.

————. 1974. "Lifeboat Ethics: The Case against Helping the Poor." *Psychology Today* 8: 38–43.

HLPE [High Level Panel of Experts on Food Security and Nutrition]. 2014. "Sustainable Fisheries and Aquaculture for Food Security and Nutrition." www.fao.org/3/a-i3844e.pdf

Hossain, S., A. Fakhruddin, M. Chowdhury, and S. Gan. 2016. "Impact of Ship-breaking Activities on the Coastal Environment of Bangladesh and a Management System for its Sustainability." *Environmental Science and Policy* 60: 84–94.

Hsu, A., G. Sheriff, T. Chakraborty, and D. Manya. 2021. "Disproportionate Exposure to Urban Heat Island Intensity across Major US Cities." *Nature Communications* 12: 2721. https://doi.org/10.1038/s41467-021-22799-5

Hufstader, C. 2009. "Conflict Surrounds Expansion of Peru Gold Mine." Oxfam. https://www.oxfamamerica.org/explore/stories/conflict-surrounds-expansion-of-peru-gold-mine/

ILO [International Labour Organization]. 2021. "Ship-breaking: A Hazardous Work." https://www.ilo.org/safework/areasofwork/hazardous-work/WCMS_110335t/lang–en/index.htm

Katwala, A. 2018. "The Spiralling Environmental Cost of Our Lithium Battery Addiction." *Wired*, May 8, 2018. https://www.wired.co.uk/article/lithium-batteries-environment-impact

Kennedy, B. 2001. "Environmental Scarcity and the Outbreak of Conflict." PRB, January 1, 2001. https://www.prb.org/resources/environmental-scarcity-and-the-outbreak-of-conflict/

Khokhar, T. 2017. "Chart: Globally, 70% of Freshwater Is Used for Agriculture." World Bank Blogs, March 22, 2017. https://blogs.worldbank.org/opendata/chart-globally-70-freshwater-used-agriculture

Lakhani, N. 2019. "The Lost River: Mexicans Fight for Mighty Waterway Taken by the US." *The Guardian*, October 21, 2019. https://www.theguardian.com/environment/2019/oct/21/the-lost-river-mexicans-fight-for-mighty-waterway-taken-by-the-us

Lewis, J., J. Hoover, and D. MacKenzie. 2017. "Mining and Environmental Health Disparities in Native American Communities." *Current Environmental Health Reports* 4: 130–41. https://link.springer.com/article/10.1007/s40572-017-0140-5

Malthus, T. 1798. *An Essay on the Principle of Population.* London: J. Johnson.

Mann, C. 2018. "The Book That Incited a Worldwide Fear of Overpopulation." *Smithsonian Magazine*, Jan/Feb. https://www.smithsonianmag.com/innovation/book-incited-worldwide-fear-overpopulation-180967499/

Meadows, D., L. Dennis, J. Randers, and W. Behrens. 1972. *The Limits to Growth: A Report for the Club of Rome's Project on the Predicament of Mankind.* New York: Universe Books.

Mekonnen, M., and A. Hoekstra. 2016. "Four Billion People Facing Severe Water Scarcity." *Science Advances* 2 (2): e150323. doi:10.1126/sciadv.1500323

Myers, N., and J. Kent. 2003. "New Consumers: The Influence of Affluence on the Environment." *PNAS* 100 (8): 4963–8. doi:10.1073/pnas.0438061100

Neff, Z. 2013. "Africa's Child Mining Shame." Human Rights Watch, September 11, 2013. https://www.hrw.org/news/2013/09/11/africas-child-mining-shame#

Pauly, D., and D. Zeller. 2016. "Catch Reconstructions Reveal that Global Marine Fisheries Catches Are Higher than Reported and Declining." *Nature Communications* 7: 10244. https://www.nature.com/articles/ncomms10244?dom=pscau&src=syn

Potter, M. 2011. "Lifeboat Ethics." In *Encyclopedia of Global Justice*, edited by D. Chatterjee, 660–2. Heidelberg: Springer.

Rees, W., and M. Wackernagel. 1994. "Ecological Footprints and Appropriated Carrying Capacity: Measuring the Natural Capital Requirements of the Human Economy." In *Investing in Natural Capital*, edited by A. Jansson, C. Folke, M. Hammer, and R. Costanza, 362–91. Washington, DC: Island Press.

Reichel, C. 2018. "Toxic Waste Sites and Environmental Justice: Research Roundup." Harvard Kennedy School, The Journalist's Resource. https://journalistsresource.org/environment/superfund-toxic-waste-race-research/

Ritchie, H. 2020. "You Want to Reduce the Carbon Footprint of Your Food? Focus on What You Eat Not Whether Your Food Is Local." *OurWorldInData*, January 24, 2020. https://ourworldindata.org/food-choice-vs-eating-local

Ritchie, H., and M. Roser. 2017. "Water Use and Stress." OurWorldInData. https://ourworldindata.org/grapher/annual-freshwater-withdrawals

Royal Society. 2009. "Reaping the Benefits: Science and the Sustainable Intensification of Agriculture." https://royalsociety.org/~/media/royal_society_content/policy/publications/2009/4294967719.pdf

Sarkodie, S., and V. Strezov. 2018. "Empirical Study of the Environmental Kuznets Curve and Environmental Sustainability Curve Hypothesis for Australia, China, Ghana and USA." *Journal of Cleaner Production* 201: 98–110. https://doi.org/10.1016/j.jclepro.2018.08.039

Taylor, M. 2020. "Global Reforestation Drive Grows Fast as Governments Grasp Benefits." Reuters, September 2, 2020. https://www.reuters.com/article/global-climatechange-forests/global-reforestation-drive-grows-fast-as-governments-grasp-benefits-idUKL4N2FY2O0

Tilman, D., and M. Clark. 2014. "Global Diets Link Environmental Sustainability and Human Health." *Nature* 515. doi:10.1038/nature13959

Tucker, C. 2019. *A Planet of Three Billion.* Atlas Observatory Press.

UNEP [United Nations Environment Program], n.d. "Worldwide Food Waste." Accessed December 13, 2021. https://www.unep.org/thinkeatsave/get-informed/worldwide-food-waste

Union of Concerned Scientists. 2022. "1992 World Scientists' Warning to Humanity." Updated February 4, 2022. https://www.ucsusa.org/resources/1992-world-scientists-warning-humanity

USGS [United States Geological Survey]. 2018. "Where Is the Earth's Water?" USGS, June 6, 2018. https://www.usgs.gov/special-topics/water-science-school/science/where-earths-water

———. n.d. "Areas of Land Subsidence in California." Accessed April 30, 2022. https://ca.water.usgs.gov/land_subsidence/california-subsidence-areas.html

Walker, W., S. Gorelik, A. Baccini, J. Aragon-Osejo, et al. 2020. "The Role of Forest Conservation, Degradation, and Disturbance in the Carbon Dynamics of Amazon Indigenous Territories and Protected Areas." *PNAS* 117 (6): 3015–25. https://www.pnas.org/content/117/6/3015

Wong, W. 2015. "Touch Screens: Why a New Transparent Conducting Material Is Sorely Needed." The Conversation, November 18, 2015. https://theconversation.com/touch-screens-why-a-new-transparent-conducting-material-is-sorely-needed-34703

World Commission on Environment and Development. 1987. *Our Common Future*. Oxford: Oxford University Press.

Worm, B., E. Barbier, N. Beaumont, J. Duffy, et al. 2006. "Impacts of Biodiversity Loss on Ocean Ecosystem Services." *Science* 314 (5800): 787–90. doi:10.1126/science.1132294

WWF [World Wide Fund for Nature]. 2020. "Overshoot." https://wwf.panda.org/discover/knowledge_hub/all_publications/living_planet_report_timeline/lpr_2012/demands_on_our_planet/overshoot/

———. 2021. "Soil Erosion and Degradation: Overview." https://www.worldwildlife.org/threats/soil-erosion-and-degradation

Fertility measures and patterns

After reading this chapter, a student should be able to:

1   discuss reasons for high fertility, using the example of Niger;

2   explain how fertility typically declines associated with processes of mod-
    ernization, using the example of Thailand;

3   discuss the reasons for and challenges of very low fertility, using the exam-
    ple of South Korea;

4   explain some of the approaches designed to raise fertility.

Globally, women had on average 2.3 children each in 2020, but this average masks wide geographical variations from a high of 7.1 in Niger to a low of 0.8 in South Korea (figure 6.1). In this chapter we describe basic concepts of fertility and explore how and why fertility rates vary so much. Despite high fertility and huge population growth across much of the world in the 20th century, one of the most significant stories of the past 20 years has been dramatic fertility decline, with many countries approaching their lowest fertility rate on record. A growing list of countries, including in the Global South, now have below replacement level fertility. As we noted in chapter 4, the big questions on the horizon are how quickly Africa might follow this path to lower fertility and, conversely, whether family-friendly policies will encourage women to have more children once again in countries with very low fertility. The answers to these questions will have significant ramifications for environmental and economic futures.

Influences on fertility

Fertility describes how many babies an individual woman bears in her lifetime and provides a fascinating intersection between biological, social, and environmental influences. A woman's fertility is typically a lot lower than her **fecundity**, which refers to

DOI: 10.4324/9781003143253-6

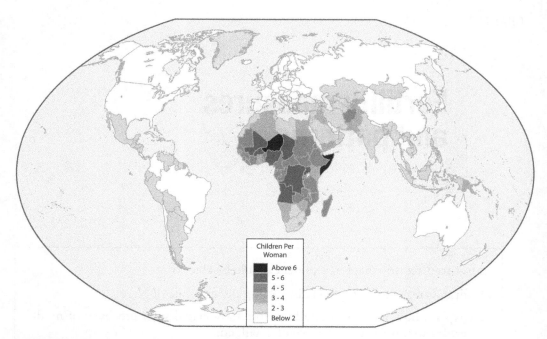

FIGURE 6.1 Total fertility rates by country, 2015–20
Data source: UNDESA (2019).

the potential maximum number of children she is physiologically able to bear. Fecundity is determined by biological factors that influence a woman's physical capacity to bear children, with things like age, genetics, and nutrition influencing the likelihood that a woman will conceive and/or bring a pregnancy to full term. Environmental influences, particularly pollutants, can also influence conception and gestation, potentially leading to either enhanced fertility or greater likelihood of infertility or miscarriage. Box 6.1 discusses some of these influences on fecundity and infertility.

Fertility, by contrast, refers to the *actual* number of children that women bear in a particular society and is heavily influenced by a wide variety of social factors, including cultural patterns (e.g., at what age men and women typically marry), economic factors (particularly the economic costs and benefits of having children), and political factors (e.g., whether governments have family-friendly policies in place; figure 6.2). Fertility shows particularly clear patterns related to how empowered women are in a society (often closely linked to female education), religious and other cultural expectations of childbearing, as well as access to effective methods of contraception. Many of these factors are closely tied to economic development, with high fertility typically associated with low-income countries as well as poorer communities within countries. Women with limited education, strong religious beliefs, or subject to **patriarchal** structures of society are particularly likely to have large families. Broadly speaking, fertility has declined over the past 100 years as couples are increasingly making conscious efforts to limit family size, and effective forms of contraception offer them the opportunity to do so reliably.

Fertility fluctuates in all societies as social conditions change. In Europe, for instance, we can recognize a "birth dearth" associated with the World Wars and then a subsequent rebound in fertility in the post-war years. In the United States, a

BOX 6.1 BIOLOGICAL INFLUENCES ON FECUNDITY AND INFERTILITY

Although women today typically give birth to fewer than four children, most women are biologically capable of having far more than that. This theoretical biological maximum is what we refer to as fecundity. Perhaps the most famous case of extreme fertility is Mrs Vassilyeva from Russia, who was recorded in the 1700s to have had 27 pregnancies, from which she delivered 69 children, including 16 pairs of twins, 7 sets of triplets, and 4 sets of quadruplets (Guinness World Records 2021)! For men, the maximum number of children is far higher, with many powerful men historically fathering large numbers of children with multiple wives and concubines. Perhaps most famous, the Mongolian ruler Genghis Khan is widely reported to have fathered hundreds of children prior to his death in 1227. Indeed, his genetic legacy is so far-reaching that millions of men today still show genetic markers that can be traced to his lineage (Callaway 2015).

By convention, most statistical measures assume that the childbearing population consists of women aged 15 to 49 years. However, fecundity is arguably increasing as the age at which puberty begins has become younger. Over the past century, the average age that girls start to menstruate has declined from about 16 or 17 at the turn of the 20th century to 12 or 13 today (Hernandez 2018), although this may now be approaching a biological limit (Pierce and Hardy 2012). Similarly, breast development has been found to be occurring earlier in populations all over the world (Eckert-Lind et al. 2020). Similar changes may be occurring in boys, but the data are less clear because onset of puberty is harder to measure in males (Pierce and Hardy 2012). The main explanation for these changes is improved diet, allowing the reproductive system to mature earlier. More worryingly, researchers are exploring the possibility that exposure to hormone-mimicking chemicals in the environment from everyday products like certain types of plastic may be playing a part in early onset of puberty (Hernandez 2018).

Traditionally, women's fertility has been affected by periodic food shortages and infections. Today, more stable conditions are allowing many women to be able to conceive throughout their childbearing years. Fertility treatments are also extending the potential for women to have children into their 40s. Despite these fertility improvements, **infertility** is a growing concern, with increasing numbers of women and couples seeking infertility treatment in many parts of the world. Whether this represents changing socio-cultural patterns or a physiological decline in fecundity is debated. On one hand, the fact that many couples are delaying childbearing until later in life is undoubtedly leading to more need for infertility treatment. Women in their 30s are only half as fertile as those in their 20s and miscarriage rates increase with age, with women over 35 at particular risk of infertility problems (ACOG 2014). Male fertility also declines with age, although more gradually. On the other hand, there are also signs that we may be seeing a genuine decline in some populations' biological ability to have children. Men appear to be particularly affected. Many reproductive disorders have increased in human populations over the past 50 years, including genital malformations, poor semen quality, and low sperm count (Bergman et al. 2012). In one meta-analysis, sperm count in men in Western countries was found to have declined by more than 50 percent between 1973 and 2011 (Levine et al. 2017). Corresponding reproductive problems in animals have been traced to exposure to endocrine-disrupting chemicals, suggesting that chemicals in the environment could be at least partly responsible for these declines (Bergman et al. 2012).

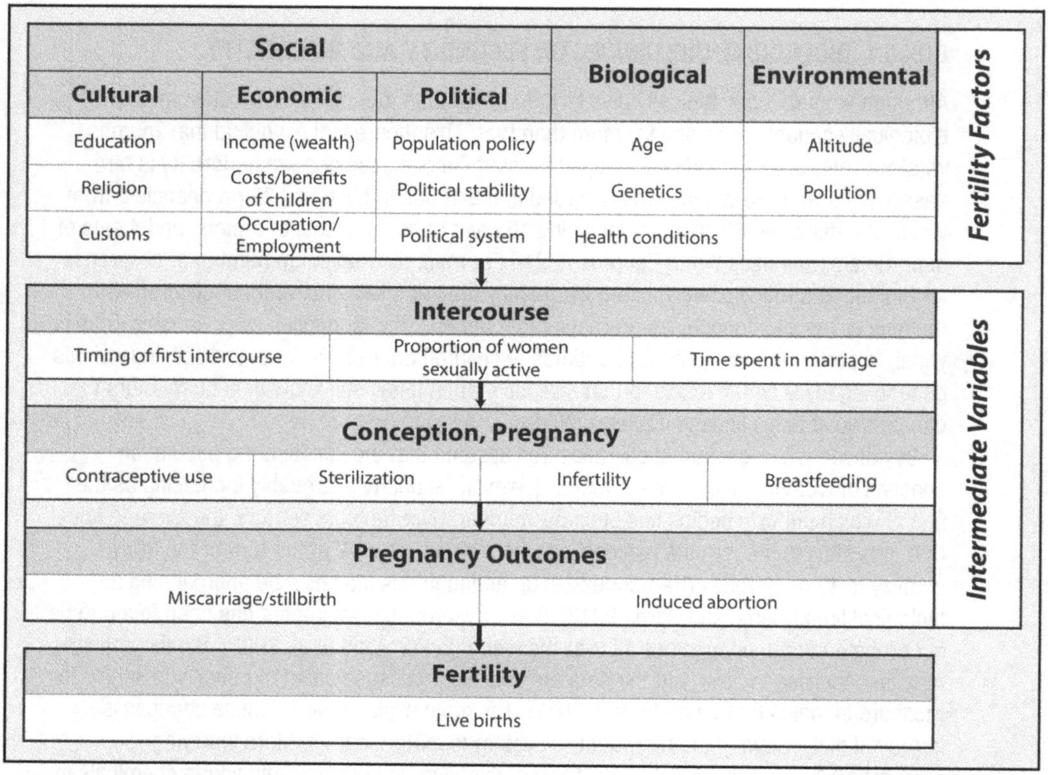

Social			Biological	Environmental	Fertility Factors
Cultural	**Economic**	**Political**			
Education	Income (wealth)	Population policy	Age	Altitude	
Religion	Costs/benefits of children	Political stability	Genetics	Pollution	
Customs	Occupation/ Employment	Political system	Health conditions		

Intercourse			Intermediate Variables
Timing of first intercourse	Proportion of women sexually active	Time spent in marriage	

Conception, Pregnancy			
Contraceptive use	Sterilization	Infertility	Breastfeeding

Pregnancy Outcomes	
Miscarriage/stillbirth	Induced abortion

Fertility
Live births

FIGURE 6.2 Influences on fertility
Inspired by: Davis and Blake (1956).

similar fertility decline has been associated with the challenging conditions of the Great Depression in the 1930s. By contrast, favorable economic and political circumstances in the post-war period, as well as delayed childbearing from during the wars, were associated with the United States' "baby boom" of 1946–64 (figure 6.3). This phenomenon was short-lived, however, with fertility rates soon dropping again to mirror longer-term trends.

Measures of fertility

The **crude birth rate** involves a simple count of the number of babies born in a particular population, divided by the midyear population of that same community, to provide an average for that population. The figure is then multiplied by 1,000 to provide a more meaningful whole number. In this way, we remove the influence of population size, so that we can fairly compare birth rates between large and small populations. For example, if we have 20,000 births in a population of 1 million people in a particular year, the crude birth rate would be 20 births per 1,000:

$$20,000 / 1,000,000 = 0.02$$

$$0.02 \times 1,000 = 20 \text{ births per } 1,000 \text{ people.}$$

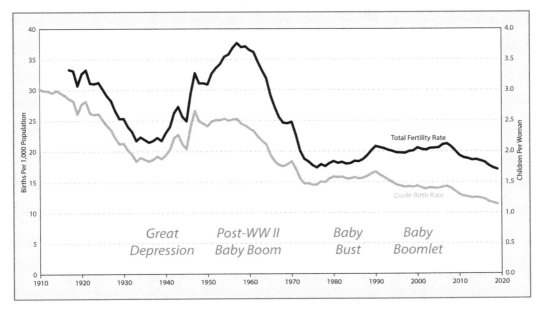

FIGURE 6.3 Fertility fluctuations in the United States, 1910–2020
Data source: CDC (n.d.).

In 2020, the global crude birth rate was approximately 19 per 1,000 but ranged widely from just 6 in Monaco and South Korea to 48 in Niger. Though we tend to assume that a very low crude birth rate indicates a society where most women are choosing to have small families, a low crude birth rate could also indicate a society where only a small proportion of its citizens are women of childbearing age. To provide an extreme example, we would likely see a very low crude birth rate in a military community of unmarried men. In this case, however, we would have to question the value of quoting a crude birth rate at all! Of greater significance, the crude birth rate in many rich countries today is decreasing not only through couples choosing to have small families but also through the process of population aging, whereby more and more of the population is beyond childbearing age.

To avert the potentially confounding influence of age and sex structure of the population on the data, we often see the **general fertility rate** or **total fertility rate** stated rather than the crude birth rate. These measures consider fertility in relation to the *number of women of childbearing age in the population*, rather than the size of the whole population, thus avoiding the problems noted in the previous paragraph. The general fertility rate is the total number of live births in a particular population divided by the number of women of childbearing age in the population multiplied by 1,000. The total fertility rate is a little more complicated to calculate and is designed to estimate the average number of children that a woman in a particular population would be expected to have, based on average childbearing rates in a given year. These measures typically provide a better insight into how childbearing varies among communities or over time than the crude birth rate. For example, in 2020, the total fertility rate for Europe as a whole was 1.5 babies per

TABLE 6.1 Total fertility rate by world region, 2020

Continent	Total fertility rate
Africa	4.4
World	*2.3*
Oceania	2.3
Asia	2.0
Latin America/Caribbean	2.0
Northern America	1.7
Europe	1.5

Data source: PRB (2020).

woman. If we compare that to Africa's total fertility rate of 4.4 babies per woman, we can legitimately ask what is leading women in Africa to have, on average, almost three times as many babies as European women (table 6.1). We could hypothesize that generally lower levels of education, more rural agricultural lifestyles, and more patriarchal societies in Africa might all be contributory factors, while recognizing that there are huge variations within both Africa and Europe.

Where counts of births are not available, the **child–woman ratio** can be used to estimate fertility, although it is really a measure of population structure rather than fertility patterns. This measure looks at the ratio of children under 5 to women of reproductive age and gives an indication of the youthfulness of the country (UNFPA, n.d.). In countries with very high fertility, the child–woman ratio may be as high as just below one (almost as many children as women), whereas affluent countries typically have ratios of only about three children to every ten women.

The idea of **completed fertility**, reflecting the number of children a woman has in her whole lifetime, is another useful measure. It is particularly important for understanding whether a drop in population fertility recorded at a particular point in time reflects women having fewer children overall or simply postponing having children until later in life, creating only a temporary decline in births. The completed fertility rate usually looks at how many children women in their 40s have (on the assumption that their childbearing is now complete) as a retrospective estimate of fertility patterns over the previous 20 or so years (Livingston 2019).

We can also quote **age-specific fertility rates** if we are interested in the fertility of specific age **cohorts** of women. For instance, we might want to compare the fertility rates of those in their 20s with those in their 40s. This can reveal intriguing patterns—in the United States, for instance, age-specific fertility rates over recent years reveal an overall decline in adolescent fertility, as well as a decline in births to women in their 20s (figure 6.4). However, births to those in their 30s and 40s have been increasing since the 1980s as women delay childbearing and fertility treatments have improved (box 6.2). Indeed, the US Census Bureau recently began collecting fertility data on women up to age 50 because the over-40s were becoming newly significant in birth statistics (Livingston 2019; Livingston and Cohn 2013).

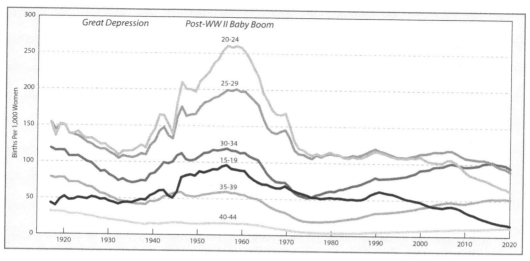

FIGURE 6.4 Age-specific fertility rates in the United States, 1915–2020
Data source: CDC (n.d.).

BOX 6.2 ASSISTED REPRODUCTIVE TECHNOLOGIES (ART)

Fertility treatments (or assisted reproductive technologies) have advanced rapidly since the first "test-tube baby" was born by in vitro fertilization (IVF) in England in 1978. Today, 1 in 60 babies in the United States are born by IVF; in Denmark, Israel, and Japan that figure is around 1 in 25 (*The Economist* 2019a). A wide variety of fertility interventions have been developed, ranging from drugs that help to stimulate ovulation, to the use of donated eggs and sperm, to in vitro fertilization (where egg and sperm are brought together outside the body and resulting embryos then implanted directly into the mother's uterus), to gestational surrogacy (where a surrogate undertakes an entire pregnancy to bring a baby to term for the intended parents). Traditionally, these procedures have been used to address biological aspects of infertility such as low sperm count, poor egg quality, or repeated miscarriage. Increasingly, fertility treatments are now also being used by couples in homosexual partnerships and single women, who are unable to have children owing to what is sometimes referred to as **social infertility**.

From a social justice perspective, many liberal and secular commentators argue that it is important to offer fertility treatments to groups experiencing social infertility. By contrast, social conservatives and religious commentators often consider the birth of children to be the exclusive right of male–female partnerships (sometimes only those in married unions), raising considerable controversies. Globally, this has resulted in a patchwork of rules and regulations. In many Islamic countries, fertility treatments are only available to married couples, and the use of donor eggs and sperm is sometimes controversial because it has been interpreted by some Islamic scholars as a form of adultery (Cha 2018). By contrast, some socially liberal countries allow single women and gay couples to receive fertility treatments. Mexico has recently emerged as a center for fertility treatments for homosexual couples from the Global North owing to its relatively low-cost treatments and liberal policies around who can receive them (Schurr 2017).

Fertility treatments raise further ethical questions beyond simple access. Many scholars have noted that significant power differentials exist between those who have the money and resources to pay for fertility treatments and those who provide the reproductive labor such as egg donation or surrogacy to assist them. Though some countries require that certain reproductive services be performed altruistically, others allow payments for them, leading to further inequities both for the reproductive laborers and in terms of access to services. Gestational surrogacy is particularly contentious in this respect. Many people are uncomfortable with women being paid to carry a baby, seeing it as commoditizing human life. In response, some countries such as the UK only allow gestational surrogacy to be performed altruistically. However, in such cases, gestational surrogates are arguably being asked to perform 9 months of unpaid labor and there is typically huge unmet demand for surrogates, pushing many couples to look overseas for services. The high cost of fertility treatment also encourages couples to look abroad for cheaper options, with a strong transnational market developing in reproductive services, generating further concerns related to lack of regulation of the industry. For some time, India and Southeast Asia were at the center of the global surrogacy industry, with critics exposing the restrictive conditions that many surrogate mothers had to tolerate. In India, for example, many women acting as surrogates had to live in surrogate houses where their diet, movements, and activities were all closely regulated (Bhattarcharjee 2019). In response to concerns around the industry, many countries, including India, have now largely shut their borders to transnational surrogacy. In response, other countries, including several in Eastern Europe, have become increasingly active in providing these services.

Reproductive technologies are frequently operating at the boundaries of science, forcing us also to consider thorny ethical questions. When does life begin (at conception, at birth, or somewhere in between)? Who has the "right" to become a parent (everyone, no one, only those in a formal union)? If an individual has donated eggs or sperm, do they have the right to anonymity or do their biological children have the right to know them as their biological parents? These ethical questions have significant ramifications when we try to formulate policy. For instance, how we deal with unused embryos at the end of a series of IVF cycles will depend on whether they are considered to be groups of cells or potential babies. Similarly, whether we offer fertility treatment to all couples, only to married heterosexual couples, or to single women, or also to single men, involves value judgments. Society needs to grapple with these questions to ensure that fertility policy reflects broader societal values, rather than just the values of a policymaking elite.

High fertility and fertility decline

One of the main assumptions of the **demographic transition model** is that fertility declines as a country develops economically. If we compare countries' fertility rates to their Human Development Index (a UN metric that estimates relative "development" using measures of health, education, and economy), we find a strong positive correlation between human development and declining fertility. Although this relationship may break down at very high levels of human development (Myrskylä, Kohler, and Billari 2009), in general, improvements in measures such as education, health provision, and the economic stability of a country seem to lead to a predictable decline in fertility. Here we explore two countries to consider in more detail why poor countries have traditionally had high birth rates (as exemplified by Niger), as well as some of the factors that can lead to a transition to low fertility as a country develops economically (as illustrated by Thailand).

Niger: a rapidly growing population

In 2020, Niger had both the highest total fertility rate (7 births per woman) and the highest birth rate (48 births per 1,000 population) in the world (PRB 2020). With a persistently high infant mortality rate (69 infant deaths per 1,000 live births) and short life expectancy (59 years; PRB 2020), Niger is one of a handful of countries that can still potentially be placed in stage 2 of the demographic transition model (high birth rates and declining mortality), although declines in fertility in recent years suggest that even Niger is at the beginning of stage 3. Regardless, Niger is currently in a stage of rapid population growth, with one of the highest rates of natural increase in the world (3.8 percent per year; PRB 2020). Some projections suggest that Niger may end up as the second most populous nation in Africa behind Nigeria (Potts et al. 2011). So, why does Niger have such high fertility?

Niger is a primarily agricultural country where children can be economically productive from a young age. Many children are already doing jobs in the fields as young as age 5 or 6, so children rapidly become an economic asset to the family—as a Nigerien proverb states, "A child comes with two hands and only one mouth" (*The Economist* 2019b, 57). A further economic incentive to have children is the tradition of family-based support for the elderly, which means that many parents are reliant on their children in old age. The social environment also strongly endorses large family size. Much of Niger's population is Muslim and adheres to the idea that children are blessings, with large family size indicating that your family is favored. Indeed, not having children carries considerable stigma in traditional Nigerien society. Early marriage is common, with women marrying and having their first birth in their mid- to late teens on average (Potts et al. 2011). Conservative expectations of women as childbearers persist in many villages, and access to education is limited, particularly for girls, so successfully raising children is one of the few ways for women to gain status (*The Economist* 2019b). In 2015, literacy rates in Niger were just 19 percent overall (27 percent for adult men and 11 percent for adult women), making it an outlier even for the Sahel region, where literacy rates generally range between 30 and 60 percent (Roser and Ortiz-Ospina 2016).

How quickly this situation will change is open to question. Although contraceptives have been provided free in Niger since 2002, contraceptive use remains limited (Potts et al. 2011). In 2020, only 11 percent of married women of childbearing age were using contraception (PRB 2020), and Niger's strongly pronatalist culture puts it in the unusual position of still having a desired family size so high that it has been argued that there is little unmet demand for contraception (Potts et al. 2011). As recently as 2012, less than 1 percent of Nigeriens reported wanting to limit their family to 2 children, with married men reporting that they wanted, on average, 12 children and women 9 (INS 2013). Many argue that Niger's extremely marginalized economic position (ranking last in the United Nations Human Development Index) is critical in this regard, with little change likely in fertility patterns until basic needs are met, particularly with respect to food security and education. Education and female empowerment are frequently listed as two key factors that can decrease fertility rates and so Niger's very conservative patriarchal society is pivotal in explaining high birth rates.

We could question whether it is appropriate for outsiders to promote contraception within a society that appears satisfied with high birth rates. Alternatively, we could counter that women are significantly disadvantaged in Niger, owing to high rates of childbearing and the attendant health risks and heavy domestic labor that involves, and that many women might choose smaller family sizes if provided with more education and alternative opportunities. From an economic standpoint, the rapidity of population growth in Niger makes raising people out of poverty challenging, with little likelihood that expansions in education can keep pace with growing numbers of children (Potts et al. 2011). From a sustainability standpoint, too, the situation is concerning. Niger is situated in the Sahel region of Africa (along the southern border of the Sahara Desert) and ranges from desert in the north to seasonal grasslands and shrublands in the south, with high susceptibility to over-grazing and topsoil loss. The Sahel has already seen its population more than triple since the 1960s, and many fear that the region has reached its ecological limits. To add to these concerns, a likely climate change scenario for the region is that summer rainfall will become increasingly erratic, leading to more frequent droughts. Some scholars predict that this could lead to communities in the region slipping into more persistent poverty, with food aid potentially becoming a necessity soon (Potts et al. 2011).

Fertility transition in Thailand

Thailand represents a country that illustrates how rapidly things can change when education and aspirations for women begin to increase. From an average of six births per woman in the early 1960s (comparable to Niger), marital fertility in Thailand fell by about 40 percent in just the 1970s (figure 6.5). Meanwhile contraceptive prevalence increased from under 15 percent in 1969 to nearly 60 percent in 1981, with more than half of recently married women indicating a preference for just two children by the early 1980s (Knodel, Havanon, and Pramualratana 1984).

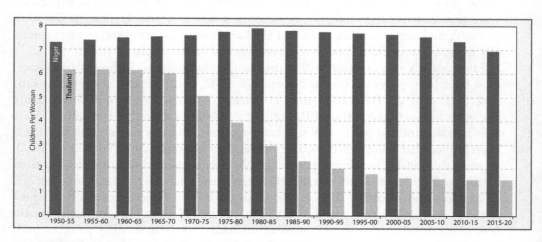

FIGURE 6.5 Total fertility rates in Niger and Thailand, 1950–2020
Data source: UNDESA (2019).

   A qualitative analysis of interviews with Thais at the time of Thailand's fertility transition provides some clues to changes in expectations and aspirations driving this fertility decline (Knodel, Havanon, and Pramualratana 1984). Industrialization and urbanization are often seen as contributing to declining fertility rates as children shift from being an economic asset to an economic burden. Considering the costs of raising children from an economic perspective appears to have become common among young Thais as early as the 1980s, with many parents reporting that raising children had become expensive, particularly as the economy became increasingly monetized and consumer aspirations rose. The standard of living in Thailand rose steadily over this period and the cost of having children was seen as increasing as parents' aspirations for their children grew to include consumer goods, modern medicine, and education. Education, in particular, became a key aspiration for many Thai parents. Education is costly both directly, through fees associated with schooling, as well as indirectly, through taking children out of economic production. Whereas in the past only sons might have been educated, increasing gender equity led many parents to want to educate daughters too. Farming began to be perceived as arduous and education the ticket out of an agricultural lifestyle. Shortages of agricultural land were also significant. Whereas in previous eras new land would have been cleared from the forest, increasing population densities meant that subdividing land among children became the only option and so ensuring that children had alternative sources of income became important. Though some concerns persisted about the need for children to support parents in old age, a shift occurred even here, with respondents reporting that a few children earning good salaries could be more important than relying on many children living in poverty (Knodel, Havanon, and Pramualratana 1984).
   A relatively stable government, which invested significantly in education and healthcare over this period, provided a supportive environment for these societal changes to occur, and a growing economy fueled rising consumer aspirations. Culturally, it is also significant that Thailand's largely Buddhist population had fewer concerns around contraception compared with populations in Catholic or Islamic contexts. Nonetheless, cultural barriers related to family planning still had to be surmounted. The leadership of a charismatic champion of reproductive health, Mechai Viravaidya, should not be underestimated in this respect. Known as "Mr Condom," Viravaidya became well known for his humorful approaches to breaking taboos around birth control, including a condom-themed restaurant chain and school-based education programs. Viravaidya is credited with bringing family planning messages to rural areas and under-served communities, as well as helping to stem the spread of HIV/AIDS in Thailand in the 1990s (WHO 2010).
   By the mid–1990s, the Thai fertility transition appeared to be reaching its conclusion, with most parents reporting a desire for just two children. Although concerns began to be voiced at this time that fertility might soon dip below the idealized two-child family, the desire to have a child of each sex initially appeared to prevent the birth rate from slipping below two children per woman (Knodel et al. 1996). However, by 2003, Thailand's fertility rate had dipped below replacement level to 1.9 births per woman and concerns were raised about the potential implications of Thailand's aging population structure (Prachuabmoh and Mithranon 2003). Today, Thailand's fertility statistics are comparable to countries of the Global North (despite its much more modest income per capita), with a total fertility rate

of just 1.5 babies per woman and 76 percent of women using modern methods of contraception (PRB 2020).

Low fertility

The Thai experience illustrates how quickly fertility can decline—the speed of this transition in many countries, as well as how low fertility rates have dipped, has surprised even population scholars. Traditionally, low fertility has often been associated with populations in crisis, where problems such as poor nutrition, disease, or social upheaval led to fertility reductions. This still occurs in certain circumstances today: for example, the HIV/AIDS crisis in sub-Saharan Africa may have lowered fertility through increased mortality in reproductive cohorts, reduced fecundity in those infected with the virus, and altered reproductive behavior associated with avoiding the virus (Benton and Newell 2013). Today, though, low fertility is more commonly associated with affluent countries and high living standards, with sub-replacement level fertility characterizing many countries of Europe and East Asia (figure 6.6). In these contexts, low fertility is consistent with the conditions found at stage 5 of the demographic transition model (low death rates and even lower birth rates). Here, having fewer children is often associated with couples wanting to focus more time and resources on their children in the hopes that this will lead to optimal life outcomes (Mace 2008). Increasingly, couples are also delaying child-bearing, particularly those with higher levels of education seeking to achieve career stability and other personal aspirations before starting a family (Billari, Liefbroer, and Philipov 2007). Fertility postponement clearly limits the total number of children a woman can have—both owing to the time required for individual pregnancies as well as the fact that older women are less fertile than younger women and so often take longer to get pregnant or even find themselves unable to conceive.

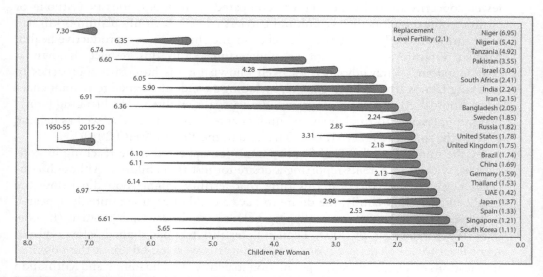

FIGURE 6.6 Fertility transition in select countries relative to replacement level, 1950–2020
Data source: UNDESA (2019); drawn after a similar graph at BBC (2019).

Fertility postponement across a whole cohort of women is therefore inevitably associated with an overall drop in fertility, although fertility can rise again slightly once this cohort of women decides to finally have their children. For example, if women traditionally had their first baby at 20 but a new generation of women is choosing to start their families at 30, there will be a period of reduced fertility until this cohort begins having children, even though they may still have the same number of children as the previous generation. Nonetheless, the longer generation turnover times that fertility postponement implies will also limit overall population size—if people in a society have their children at 30 on average, there are likely to be no more than three or four generations alive at once, whereas a society where the first child is typically born at fifteen could potentially see five generations alive at the same time.

The phenomenon of below replacement level fertility was first widely recorded in the affluent countries of Europe, where it has been referred to as the **second demographic transition** (Lesthaeghe and van de Kaa 1986). This transition includes sub–replacement level fertility, the decoupling of marriage and procreation, and greater acceptance of a variety of non-traditional living arrangements (Lesthaeghe and Surkyn 2008). Whereas the first demographic transition has been interpreted as strengthening the family as an institution, the second transition has been characterized as leading to a weakening of traditional expectations of family, including aspects such as higher divorce rates, increased cohabitation outside marriage, and greater acceptance of abortion and contraception (van de Kaa 2002). These changes have been linked to broader societal changes in Europe that occurred progressively in the post-war period that emphasized individual rights over communalism. This attitudinal shift paved the way for greater emphasis to be placed on the fulfillment of adult couples in society rather than the traditional emphasis on stable family structures as a mechanism for successful childrearing (Zaidi and Morgan 2017). Van de Kaa (2002, 7) summed up this attitudinal change:

> What appears to be a crucial element from a demographic point of view in the changes noted, is that man–woman relations are increasingly seen as a means of reciprocal emotional enrichment to which the birth of children may, or may not, be considered to be contributing. ... Marriage as an institution providing economic security and as an essentially permanent arrangement aimed at reproduction and enabling the rearing of children is no longer universally felt to be necessary.

These attitudinal changes were accompanied and propelled by other social and economic changes—particularly increasing female empowerment combined with rapidly rising costs of rearing a "successful" child and rising economic instability for many middle-class families in a new era of rapid **globalization**. By the early 21st century, aspirations for middle-class children in many affluent countries were rising from simply finishing high school to going to university and achieving a professional career. Our globalizing economy in many ways propelled this transition, as manufacturing jobs moved overseas to newly industrializing countries, leaving fewer well-paid blue-collar jobs in the Global North. For an increasing proportion of families in affluent countries, maintaining their position in the middle classes now meant looking toward professional jobs for their children, requiring significant training. This need for extended education means that many children are not economically productive until well into their 20s, and costs associated with educating children can be high.

At the same time, the declining value of wages in real terms and increasing aspirations in consumer societies meant that maintaining a family's standard of living began to require two incomes to support the household, pushing more women into the workforce and further increasing the pressure for smaller family size. Increasing female empowerment encouraged many women to look to career fulfillment as equally as important as, or even more important than, raising children, leading to an increase in the average age at marriage and first birth and an increasing proportion of women choosing to forgo motherhood entirely.

Europe's plummeting fertility may now have stabilized, and some affluent countries such as the UK and the Nordic countries have seen modest rises in fertility in recent years back toward replacement level—although a "high" birth rate for a European country is still only about 1.8 births per woman (PRB 2020). This uptick may be related to improvements in gender equity and **family-friendly policies** designed to reduce the burden of childrearing (Myrskylä, Kohler, and Billari 2009; Buchanan and Rotkirch 2013). Many European and East Asian governments nonetheless continue to fear the economic consequences of their aging populations and are attempting to actively promote childbearing through further ramping up family-friendly policies. This includes financial support for having children (either regular family support payments or one-time birth bonuses), as well as broader support for reproductive roles such as paid parental leave, subsidized childcare, and efforts to provide more flexible roles in the workplace, especially the option of part-time jobs. The extent to which these policies actually influence the birth rate is debatable, but some evidence suggests that family-friendly policies can lead to slight increases in fertility (Buchanan and Rotkirch 2013). In broadest terms, the European countries that offer the strongest support for women to be able to perform both productive (paid employment) and reproductive (childrearing) roles in society, particularly France and the Nordic countries, seem to have achieved slightly higher birth rates than neighboring countries (Gauthier 2013). Looking at specific policies, research suggests that financial incentives can contribute to a modest increase in the birth rate, although this may have more impact on the timing of births than overall number—in other words, parents already considering having children may choose to have them when a birth bonus is on offer (Thévenon and Gauthier 2011). Regarding leave entitlements and other caregiver support benefits, studies point to a slight positive impact on the birth rate, although evidence is again mixed (Duvander, Lappegård, and Andersson 2010; Gauthier 2013). Family-friendly policies seem to have a greater effect on women of lower socio-economic status and more limited education, perhaps because the financial stimulus is more significant for poorer women or because their attachment to the labor market is weaker (Gauthier 2013).

Overall, it is perhaps surprising that family-friendly policies do not have a more significant effect on the birth rate. However, as Gauthier (2013) pointed out, these policies assume that there is a **fertility gap**—a difference between how many children a couple wants and how many they actually have—caused by societal barriers to childbearing, and yet this gap may be overestimated. Furthermore, the barriers to childbearing that do exist would have to be issues that could be tackled by policy, and yet in one European study people asked about the major barriers to having more children reported concerns over their children's future as their top consideration—a rather intangible barrier to address (Fokkema and Esveldt 2006).

Although there remains significant debate over whether family-friendly policies are worth the cost and which policies might be most influential, there is certainly an argument to be made to assist parents with the economic and emotional costs of childrearing if raising children is viewed as benefiting society. Indeed, returning to the evidence that very high levels of human development correlate with an uptick in fertility, we could argue that improving factors like gender equity, labor market flexibility, and social security might promote higher fertility as a side effect of happier, more hopeful citizens, regardless of explicit fertility policies (Myrskylä, Kohler, and Billari 2009). As Gauthier (2013, 283) explained,

> Perhaps what policies can do is to act more broadly on the overall context in which young families live, for example, by creating greater happiness with work and with life in general, generating more confidence in the future, and helping create greater gender equality including in paid and unpaid work. Once these conditions are in place, we will perhaps observe a significant effect of policies on fertility.

The idea that "happy" citizens may have more children is also supported by the fact that we can find examples of situations of political, economic, and social turmoil that seem to have resulted in birth rates declining to critically low levels. This is perhaps best illustrated by the very low fertility rates of Russia and the ex-Soviet republics after the collapse of the Soviet Union. Although many ex-Soviet countries have seen slight rebounds recently, Eastern Europe still has some of the lowest fertility rates in the world (PRB 2020). People have traditionally assumed that these low fertility rates were a direct result of lacking the economic wherewithal to support children, but recent research suggests that social upheaval associated with the collapse of the Soviet Union may have been more influential than economic challenges per se (box 6.3).

Although many countries of the Global North now have sub–replacement level fertility, most of these same countries have continued to see population growth in recent years owing to immigration. Immigration not only brings new people to a country, directly increasing the population, but also tends to introduce people of childbearing age, thereby raising fertility. Furthermore, many immigrants move from high-fertility countries, bringing childbearing expectations from their home culture. For one generation at least, many immigrant communities therefore have higher fertility rates than their host communities (Sobotka 2008). In Scotland, for instance, the number of births to mothers born outside Scotland increased from 13 percent in 1977 to 24 percent in 2009, with non–Scots-born mothers responsible for 55 percent of the modest increase in the Scottish birth rate in the early 2000s (Scottish Government 2010). Similarly, in the United States, children of color now outnumber white children in almost one-third of states, reflecting in part the growth of recent immigrant populations (although not all children of color are recent immigrants, of course). Part of this is due to low fertility rates among white women but also the fact that populations of color tend to be younger—only 41 percent of white women are of childbearing age, compared with 57 percent of women

BOX 6.3 FERTILITY DECLINE IN RUSSIA

The collapse of the Soviet Union in the late 1980s and early 1990s provides a particularly dramatic example of political and economic turmoil leading to fertility decline. Russia's birth rate in 1989 was 2.0 births per woman; by 1999 it had plummeted to below 1.2 (Kumo 2010). Although many other countries experienced low fertility in the 1990s, this was often associated with a rise in the average age of childbearing as women entered the workforce in large numbers. Russia was unusual in that many women continued to have their first child at a young age (typically in their early 20s) but then stopped at just one child. In this respect, longstanding cultural preferences for early motherhood appear to have collided with the economic challenges of an economy that was spiraling downwards (Perelli-Harris and Isupova 2013). This led to the so-called Russian demographic cross when birth rates fell below death rates in 1992 (figure 6.7).

Scholars have suggested several contributory factors to Russia's fertility decline: falling incomes and rising unemployment made affording the costs of raising a child harder, the collapse of some of Russia's state-run systems—particularly heavily subsidized childcare—placed the burden of childrearing increasingly on parents rather than on the state, and uncertainties over the future may have pushed some couples to delay childbearing (Kumo 2010). The relative role of these different factors is still debated. The role of economic challenges may have been overstated, however, because relatively few sub-national studies have found a clear connection between fertility and income or fear of job loss (Kumo 2010; Perelli-Harris and Isupova 2013). In addition, even today, now that some of these economic upheavals have stabilized, fertility remains below replacement level in every country of Eastern Europe (PRB 2020).

Instead, it may have been broader upheavals to family functioning and social norms that proved most significant in lowering Russia's fertility rate. Stress and fear of loss of control contributed to a rise in alcohol use, accidents, and suicide, with associated increases in

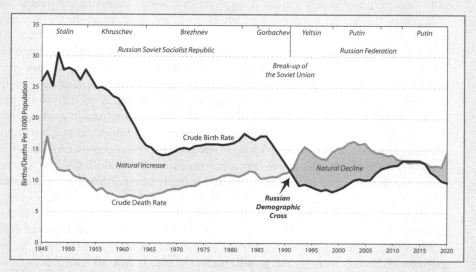

FIGURE 6.7 Fertility and mortality rates in Russia, 1946–2020
Data source: Federal State Statistics Service (n.d.).

mortality and morbidity (Gavrilova et al. 2001). Some evidence suggests that families where a parent, particularly the father, experienced a loss of status associated with their employment were less likely to have a second child (Perelli-Harris and Isupova 2013). Thus, parents may have chosen to have fewer children owing to stressful circumstances rather than a literal lack of the economic resources to support children (Perelli-Harris 2006). These societal upheavals also caused changes in family structure with implications for the birth rate: marriage rates declined, and divorce and the proportion of births outside marriage increased. Processes of modernization and Westernization that were occurring in Russia at the time, as Russians were suddenly exposed to new ideas from the outside world, likely also spurred on these broader trends weakening family structure (Perelli-Harris and Isupova 2013).

By the mid-2000s, President Putin considered fertility decline to be "the most acute problem facing our nation today" (Perelli-Harris and Isupova 2013, 148) and a presidential order was issued in 2007 to halt population decline by 2025 (Kumo 2010). The Russian government responded with "one of the most ambitious pro-natalist policies the world has ever seen" (Perelli-Harris and Isupova 2013, 142), including generous child allowances, subsidized daycare, and payments to help cover the costs of childcare. Additionally, the state provided a lump sum of "maternity capital" to mothers of second and higher-order children that could be spent on a list of specific purposes, including children's education or improving the family home. However, qualitative research on the impacts of these policies suggest that many women were skeptical that the plan was sufficient to meet their actual needs related to having more children, particularly because the maternity capital was received as a lump sum when the child turned 3, providing little assistance in the early years and no recurring financial help. The number of births did increase slightly following these incentives, however—the fertility rate had risen to 1.54 births per woman by 2009—but has still not rebounded sufficiently to reach replacement level—by 2020, the fertility rate was still only 1.6 (Perelli-Harris and Isupova 2013; PRB 2020). The maternity capital program ceased in 2016.

More recently, other trends are complicating matters. Border closures and deaths associated with COVID-19 led to an estimated population decline of about half a million people in 2020—the first decline recorded in 15 years. A continued decrease of perhaps 1.2 million people is expected from 2020 to 2024 (Khurshudyan 2020; Al Jazeera 2021). Reports suggest that official figures may be a gross underestimate of Russian coronavirus losses (Cocco and Ivanova 2022). A rising number of Russian women now appear content to remain childless, in stark contrast to previous generations where social pressures saw most women having at least one child. Also, Russia's major cities, which have experienced the greatest economic recovery, now have lower than average fertility rates compared with the rest of Russia, suggesting that economic stagnation can no longer be held responsible for patterns of low fertility. Today, it appears that the pressures of balancing two careers with parenthood may be responsible for persistently low fertility, in a pattern very similar to the rest of Europe (Perelli-Harris and Isupova 2013). The United Nations predicts that Russia will continue to see population decline to the tune of perhaps 20 million people by 2100 (UNDESA 2019). Things may only worsen in the wake of Russia's invasion of Ukraine, if economic turmoil and military losses once again suppress birth rates in Russia, particularly as highly skilled workers and wealthy entrepreneurs leave the country in large numbers (Cocco and Ivanova 2022).

TABLE 6.2 Lowest fertility countries and territories, 2020

Country/region	Total fertility rate
South Korea	0.8
Hong Kong, Puerto Rico	0.9
Singapore, Malta	1.1
Ukraine, Spain, Italy, Bosnia and Herzegovina, *Macau*	1.2
San Marino, Andorra, Moldova, Bermuda, North Macedonia, Cyprus, Greece, *Japan*	1.3
United Arab Emirates, Finland, Poland, Belarus, Portugal, Canada, St. Lucia, Mauritius, Austria	1.4
Switzerland, Liechtenstein, Serbia, Lithuania, Croatia, Norway, Thailand, Russia, Germany	1.5

East Asian countries and territories (listed in italics) provide some of the lowest fertility rates in the world. Many of the remaining countries with extremely low fertility are in Eastern and Southern Europe.
Data source: World Bank (n.d.).

of color (Sáenz and Poston 2020). Over time, the fertility behavior of immigrants usually begins to mirror that of the resident population, however, so this effect is often short-lived (Sobotka 2008).

East Asia has not generally seen the high levels of immigration experienced by affluent Western countries, helping to explain why East Asian countries today have some of the lowest fertility rates in the world (table 6.2). Although East Asia has followed many of the same trends seen in Europe, including increased labor force participation for women, financial pressures to raise "successful" children, and changing aspirations related to career and family, East Asia's fertility trajectory has been distinct in other ways. In particular, cost of living—especially education and housing—is exceptionally high in many East Asian countries (*The Economist* 2022). East Asian communities have also tended to be slower to reject traditional patriarchal domestic expectations, and a growing number of East Asian women appear to be rejecting relationships, marriage, and children altogether, citing economic pressures and gendered expectations as reasons to avoid settling down in traditional family structures (Jones 2013; BBC 2019). These social changes have elicited a strong reaction from the more conservative establishment: in South Korea, women who choose to stay single, often continuing to live with their parents, are sometimes referred to as "gold misses" as a critique of their supposedly selfish and individualistic lifestyles; in Japan, the term "parasite singles" has been used in a similar way (Jones 2013). To explore these place-specific factors further, we turn to an in-depth look at South Korea.

South Korea

In 2020, South Korea had the lowest fertility rate in the world (0.8 babies per woman, down from more than 5 in the 1950s; figure 6.8). Here, the economic pressures of raising children have run to an extreme, particular with respect to education (Jones 2013). Parents' aspirations for rearing a "successful" child may include getting children into specific schools and then particular universities, often involving paying for additional tutoring in costly *hagwon* or cramming schools (Hektor 2020). Cost of

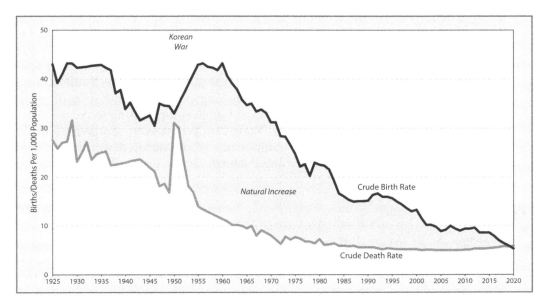

FIGURE 6.8 Birth and death rates in South Korea, 1925–2020
Data source: Statistics Korea (n.d.).

living, particularly housing, is also very high in many South Korean cities and so affording a residence large enough to accommodate several children is challenging.

Aspirations have risen not just for children but also for women. As in most industrialized countries, women have joined the paid workforce in South Korea in large numbers, and many women today delay having children while pursuing their career. South Korean society remains quite strongly gendered, however, with many domestic and caring responsibilities left largely to women, setting up significant tensions. Until recently, women were expected to leave work on starting a family, and to this day there is often only limited flexibility for women to combine career mobility and promotion with having a family. Evidence suggests that affluent countries where women have entered the labor market without significant changes to family values and policies are now experiencing some of the lowest levels of fertility (Buchanan and Rotkirch 2013). Factors such as these have resulted in many women choosing to marry later once their careers are established or even to forgo marriage altogether—in 2015, just 23 percent of South Korean women aged 25 to 29 were married, compared with 90 percent in 1970 (Quick and D'Efilippo 2019). Given that very few children are born outside marriage in East Asia—perhaps 2 percent in South Korea—low marriage rates have a particularly significant impact on the birth rate (*The Economist* 2022).

South Korea's declining birth rate is leading to a decrease in the working-age population while the elderly population continues to grow, making South Korea one of the most rapidly aging countries in the world, signaling an impending economic crisis. Today, South Korea has one of the longest life expectancies in the world at 83 years—a huge increase from just 42 years in the early 1950s—and life expectancy is predicted to continue to rise (Quick and D'Efilippo 2019; PRB 2020). One recent study forecasts that South Korean women may be the first demographic group in the world to have an average life expectancy that breaks the 90-year barrier—a life expectancy considered unattainable for any population group by many demographers until recently (Kontis et al. 2017). By the mid-2060s, 40 percent of South Koreans may be over 65, up from 15 percent today, and population size may begin to contract as soon as 2024 (BBC 2019). The South Korean government has responded by recently establishing explicitly family-friendly policies such as childcare leave, parental leave, and support for housing to try to relieve some of the burden of raising children (Kang 2013). However, competitive workplace environments that have not traditionally accommodated working parents remain a significant barrier to increasing the birth rate.

Conclusion

Fertility shows paradoxical trends at present, with some countries struggling to slow population growth even as others are trying to raise birth rates. Some countries such as Thailand and China have been so successful at curbing birth rates that they are now faced with trying to raise them again. The potential implications of these trends are significant and are picked up in several other chapters. We have already noted some of the challenges of aging populations in chapter 3. Migration—a potential solution to regional differences in fertility—brings other challenges related to cultural mixing, dislocation of communities, and concerns around nationalism, as we explore in chapters 10 and 11. Finally, as we turn to in the next chapter, directing future fertility patterns involves policy decisions that bring with them a plethora of social justice concerns.

Discussion questions

1   Should fertility treatment be considered a human right, along the same lines as access to basic healthcare? Should countries that provide universal healthcare pay for fertility treatment for their citizens? How should "social infertility" be involved in this discussion?

2   Why is it proving so challenging to raise birth rates in affluent countries? Think about the aspirations and circumstances of couples and women in making fertility decisions. How might a greater understanding of women's aspirations help guide future efforts to raise birth rates?

3   The "second demographic transition" suggests that increasing emphasis is being placed on the fulfillment of adults in society, resulting in a weakening of the institution of the family. What are some of the pros and cons of this transition?

4   In some European countries, both fathers and mothers are eligible for parental leave. Iceland recently instituted a "use-it-or-lose-it" policy, which requires that fathers take some of this leave or the days are lost. What could be some of the consequences of this policy?

5   Given our current sustainability crisis, is it appropriate for governments to try to increase birth rates?

Suggested readings

The Economist. 2019. "Thanks to Education, Global Fertility Could Fall Faster than Expected." February 19, 2019, 56–8. https://www.economist.com/international/2019/02/02/thanks-to-education-global-fertility-could-fall-faster-than-expected

Mace, R. 2008. "Reproducing in Cities." *Science* 319 (5864): 764–6. https://www.science.org/doi/10.1126/science.1153960

Population Reference Bureau [PRB]. 2019. "US fertility drops to historic low in 2019." PRB, September 17, 2019. https://www.prb.org/u-s-fertility-drops-to-historic-low-in-2019/

Quick, M., and V. D'Efilippo. 2019. "South Korea's Population Paradox." BBC, October 14, 2019. https://www.bbc.com/worklife/article/20191010-south-koreas-population-paradox

Glossary

age-specific fertility rate: number of births in a population of a particular age divided by the number of women of childbearing age in that population, multiplied by 1,000

child–woman ratio: the ratio of children under 5 to women of reproductive age

cohort: a demographic group that shares certain experiences; often defined by age

completed fertility: the number of children a woman has in her whole lifetime

crude birth rate: the number of live births in a particular population divided by the midyear population of that same population multiplied by 1,000

demographic transition model: model that describes changes in birth and death rates as countries develop economically

family-friendly policies: policies designed to support families, such as low-cost childcare and parental leave

fecundity: the biological capability of an individual or community to produce offspring

fertility: the actual number of offspring being produced by an individual or community

fertility gap: the difference between the number of children that people would like to have and the number of children they actually have

general fertility rate: the total number of live births in a particular population divided by the number of women of childbearing age in the population multiplied by 1,000

globalization: the growing interconnectedness of the global economy and associated cultural and economic shifts

infertility: inability to produce offspring; *see also* **social infertility**

patriarchy: social system dominated by men, marginalizing women

second demographic transition: shift to below-replacement fertility associated with social changes such as declining emphasis on marriage and the disconnection of sexual activity from procreation

social infertility: inability to have children owing to lifestyle factors such as living in a same-sex relationship or being single

total fertility rate: the average number of children that a woman in a particular population would be expected to have in a given year, based on average childbearing rates

Works cited

ACOG [American College of Obstetricians and Gynecologists]. 2014. "Female Age-Related Fertility Decline." Committee Opinion 589, March 2014. https://www.acog.org/clinical/clinical-guidance/committee-opinion/articles/2014/03/female-age-related-fertility-decline

Al Jazeera. 2021. "Russia: Population Shrinks for First Time in 15 years." Al Jazeera, January 29, 2021. https://www.aljazeera.com/news/2021/1/29/russia-population-shrinks-for-first-time-in-15-years

BBC. 2019. "South Korea's Population Paradox." BBC Worklife, October 14, 2019. https://www.bbc.com/worklife/article/20191010-south-koreas-population-paradox

Benton, L., and M.-L. Newell. 2013. "Childbearing and the Impact of HIV: The South African Experience." In *Fertility Rates and Population Decline*, edited by A. Buchanan and A. Rotkirch, 166–84. London: Palgrave Macmillan.

Bergman, Å, J. Heindel, S. Jobling, K. Kidd, and R. Zoeller. 2012. *State of the Science of Endocrine-Disrupting Chemicals*. WHO and UNEP.

Bhattacharjee, D. 2019. "'It Is a Jail Which Does Not Let Us Be ...' Negotiating Spaces of Commercial Surrogacy by Reproductive Labourers in India." In *Reproductive Geographies: Bodies, Places, and Politics*, edited by M. England, M. Fannin, and H. Hazen, 106–21. London: Routledge.

Billari, F., A. Liefbroer, and D. Philipov. 2007. "The Postponement of Childbearing in Europe: Driving Forces and Implications." *Vienna Yearbook of Population Research*, 1–17. doi:10.1553/populationyearbook2006s1

Buchanan, A., and A. Rotkirch. 2013. "No Time for Children? The Key Questions." In *Fertility Rates and Population Decline*, edited by A. Buchanan and A. Rotkirch, 3–21. London: Palgrave Macmillan.

Callaway, E. 2015. "Genghis Khan's Genetic Legacy Has Competition." *Nature News*, January 23, 2015. https://www.nature.com/news/genghis-khan-s-genetic-legacy-has-competition-1.16767

CDC [Centers for Disease Control and Prevention]. n.d. "National Vital Statistics System." National Center for Health Statistics. Accessed February 2, 2022. https://www.cdc.gov/nchs/nvss/births.htm

Cha, A. 2018. "Fertility Frontier: Gifts from God." *The Washington Post*, April 27, 2018. https://www.washingtonpost.com/graphics/2018/national/how-religion-is-coming-to-terms-with-modern-fertility-methods/

Cocco, F., and P. Ivanova. 2022. "Ukraine War Threatens to Deepen Russia's Demographic Crisis." *The Financial Times*, April 3, 2022. https://www.ft.com/content/8c576a9c-ba65-4fb1-967a-fc4fa5457c62

Davis, K., and J. Blake. 1956. "Social Structure and Fertility: An Analytic Framework." *Economic Development and Cultural Change* 4 (3): 211–35. https://doi.org/10.1086/449714.

Duvander, A.-Z., T. Lappegård, and G. Andersson. 2010. "Family Policy and Fertility: A Comparative Study of the Impact of Fathers' and Mothers' Use of Parental Leave and Continued Childbearing in Norway and Sweden." *Journal of European Social Policy* 20 (1): 45–57. https://doi.org/10.1177/0958928709352541

Eckert-Lind, C., A. Busch, J. Petersen, F. Biro, et al. 2020. "Worldwide Secular Trends in Age at Pubertal Onset Assessed by Breast Development in Girls." *JAMA Pediatrics* 174 (4): e195881. doi:10.1001/jamapediatrics.2019.5881

The Economist. 2019a. "Seed Capital: The Fertility Business Is Booming." August 10, 2019. https://www.economist.com/business/2019/08/08/the-fertility-business-is-booming

———. 2019b. "Thanks to Education, Global Fertility Could Fall Faster than Expected." February 19, 2019, 56–8. https://www.economist.com/international/2019/02/02/thanks-to-education-global-fertility-could-fall-faster-than-expected

———. 2022. "Asia's Advanced Economies Now Have Lower Birth Rate than Japan." May 19, 2022. https://www.economist.com/asia/2022/05/19/asias-advanced-economies-now-have-lower-birth-rates-than-japan

Federal State Statistics Service. n.d. "Demographic Yearbook of Russia." Accessed March 16, 2022. https://eng.rosstat.gov.ru/

Fokkema, T., and I. Esveldt. 2006. "Work Package 7: Child-friendly Policies." Bundesinstitut für Bevölkerungsforschung. https://www.bib.bund.de/EN/Research/Ageing/Projects/Archive/DIALOG/Reports-Papers/dialog_ps_no7.pdf?__blob=publicationFile&v=3

Gauthier, A. 2013. "Family Policies and Fertility—Do Policies Make a Difference?" In *Fertility Rates and Population Decline*, edited by A. Buchanan and A. Rotkirch, 269–87. London: Palgrave Macmillan.

Gavrilova, N., G. Evdokushkina, V. Semyonova, and G. Year. 2001. "Economic Crises, Stress, and Mortality in Russia." The Population Association of America Annual Meeting, 28–31 March 2001, Washington, DC. https://citeseerx.ist.psu.edu/viewdoc/download?doi=10.1.1.2.7026&rep=rep1&type=pdf

Guinness World Records. 2021. "Most Prolific Mother Ever." https://www.guinnessworldrecords.com/world-records/most-prolific-mother-ever

Hektor, A. 2020. "The Korean Sampo Generation: The Lowest Birthrate in the World." https://sweden-science-innovation.blog/seoul/the-korean-sampo-generation-the-lowest-birth-rate-in-the-world/

Hernandez, D. 2018. "The Decreasing Age of Puberty." *VitalRecord*. News from Texas A&M Health, January 10, 2018. https://vitalrecord.tamhsc.edu/decreasing-age-puberty/

INS [Institut National de la Statistique, Niger]. 2013. *Enquête Démographique et de Santé et à Indicateurs Multiples (EDSN-MICS IV) 2012*. Calverton, MD: ICF International. https://dhsprogram.com/pubs/pdf/FR277/FR277.pdf

Jones, G. 2013. "The Growth of the One-Child Family and Other Changes in the Low Fertility Countries of Asia." In *Fertility Rates and Population Decline*, edited by A. Buchanan and A. Rotkirch, 44–61. London: Palgrave Macmillan.

Kang, Y.-H. 2013. "Is Family-Friendly Policy (FFP) Working in the Private Sector of South Korea?" *SpringerPlus* 2: 561. http://www.springerplus.com/content/2/1/561

Khurshudyan, I. 2020. "In Siberian Coal Country, Signs of Russia's Shrinking Population Are Everywhere. It 'Haunts' Putin." *The Washington Post*, December 1, 2020. https://www.washingtonpost.com/world/2020/12/01/russia-population-decline-putin/

Knodel, J., N. Havanon, and A. Pramualratana. 1984. "Fertility Transition in Thailand: A Qualitative Analysis." *Population and Development Review* 10 (2): 297–328. doi.org/10.2307/1973084

Knodel, J., V. Prachuabmoh Ruffolo, P. Ratanalangkarn, and K. Wongboonsin. 1996. "Reproductive Preferences and Fertility Trends in Post-transition Thailand." *Studies in Family Planning* 27 (6): 307–18. https://doi.org/10.2307/2138026

Kontis, V., J. Bennett, C. Mathers, G. Li, et al. 2017. "Future Life Expectancy in 35 Industrialised Countries: Projections with a Bayesian Model Ensemble." *The Lancet* 389: 1323–35. doi:10.1016/S0140-6736(16)32381-9

Kumo, K. 2010. "Explaining Fertility Trends in Russia." VOX[EU]CEPR, June 2, 2010. https://voxeu.org/article/explaining-fertility-trends-russia

Lesthaeghe, R., and J. Surkyn. 2008. "When History Moves On: The Foundations and Diffusion of a Second Demographic Transition." In *Ideational Perspectives on International Family Change*, edited by R. Jayakody, A. Thornton, and W. Axinn, 81–118. New York: Routledge.

Lesthaeghe, R., and D. van de Kaa. 1986. "Twee Demografische Transities?" In *Bevolking: Groei en Krimp*, edited by R. Lesthaeghe and D. van de Kaa, 9–24. Deventer, Netherlands: Mens en Maatschappij, Van Loghum Slaterus.

Levine, H., N. Jørgensen, A. Martino-Andrade, J. Mendiola, et al. 2017. "Temporal Trends in Sperm Count: A Systematic Review and Meta-regression Analysis." *Human Reproduction Update* 23 (6): 646–59. https://doi.org/10.1093/humupd/dmx022

Livingston, G. 2019. "Is U.S. Fertility at an All-time Low? Two of Three Measures Point to Yes." Pew Research Center, May 22, 2019. https://www.pewresearch.org/fact-tank/2019/05/22/u-s-fertility-rate-explained/

Livingston, G., and D. Cohn. 2013. "Birth Rates Hit Record Low for Those Under 25, Still on the Rise for Those 40+." Pew Research Center, July 3, 2013. https://www.pewresearch.org/fact-tank/2013/07/03/birth-rates-hit-record-low-for-those-under-25-still-on-the-rise-for-those-40/

Mace, R. 2008. "Reproducing in Cities." *Science* 319 (5864): 764–6. doi:10.1126/science.1153960

Myrskylä, M., H.-P. Kohler, and F. Billari. 2009. "Advances in Development Reverses Fertility Declines." *Nature* 460: 741–3. doi:10.1038/nature08230

Perelli-Harris, B. 2006. "The Influence of Informal Work and Subjective Wellbeing on Childbearing in Post-Soviet Russia." *Population and Development Review* 32: 729–53.

Perelli-Harris B., and O. Isupova. 2013. "Crisis and Control: Russia's Dramatic Fertility Decline and Efforts to Increase It. In *Fertility Rates and Population Decline*, edited by A. Buchanan and A. Rotkirch, 141–56. London: Palgrave Macmillan.

Pierce, M., and R. Hardy. 2012. "Commentary: The Decreasing Age of Puberty—As Much a Psychosocial as Biological Problem? *International Journal of Epidemiology* 41 (1): 300–2. doi:10.1093/ije/dyr227

Potts, M., V. Gidi, M. Campbell, and S. Zureick. 2011. "Niger: Too Little, Too Late." *International Perspectives on Sexual and Reproductive Health* 37 (2): 95–101. https://www.jstor.org/stable/41229000

Prachuabmoh, V., and P. Mithranon. 2003. "Below-Replacement Fertility in Thailand and Its Policy Implications." *Journal of Population Research* 20: 35–50. https://doi.org/10.1007/BF03031794

PRB [Population Reference Bureau]. 2020. *World Population Datasheet 2020.* Washington, DC: PRB. https://www.prb.org/wp-content/uploads/2020/07/letter-booklet-2020-world-population.pdf

Quick, M., and V. D'Efilippo. 2019. "South Korea's Population Paradox." BBC, October 14, 2019. https://www.bbc.com/worklife/article/20191010-south-koreas-population-paradox

Roser, M., and E. Ortiz-Ospina. 2016. "Literacy." OurWorldInData https://ourworldindata.org/literacy

Sáenz, R., and D. Poston. 2020. "Children of Color Already Make Up the Majority of Kids in Many US States." The Conversation, January 9, 2020. https://theconversation.com/children-of-color-already-make-up-the-majority-of-kids-in-many-us-states-128499

Schurr, C. 2017. "From Biopolitics to Bioeconomies: The ART of (Re-)producing White Futures in Mexico's Surrogacy Market." *Environment and Planning D: Society and Space* 35 (2): 241–62.

Scottish Government. 2010. "Demographic Change in Scotland." November 26, 2010. https://www.gov.scot/publications/demographic-change-scotland/

Sobotka, T. 2008. "The Rising Importance of Migrants for Childbearing in Europe." *Demographic Research* 19 (9): 225–48.

Statistics Korea. n.d. "Domestic Status." Accessed April 14, 2022. http://kostat.go.kr/portal/eng/resources/3/1/index.static

Thévenon, O., and A. Gauthier. 2011. "Family Policies in Developed Countries: A 'Fertility Booster' with Side-Effects." *Community, Work and Family* 14 (2): 197–216. https://doi.org/10.1080/13668803.2011.571400

UNDESA [United Nations Department of Economic and Social Affairs]. 2019. "World Population Prospects 2019." https://population.un.org/wpp/

UNFPA [United Nations Population Fund]. n.d. "Child/Woman Ratio." Accessed November 5, 2021. https://papp.iussp.org/sessions/papp101_s04/PAPP101_s04_050_010.html

van de Kaa, D. 2002. "The Idea of a Second Demographic Transition in Industrialized Countries." *Japanese Journal of Population.* https://www.researchgate.net/publication/253714045_The_Idea_of_a_Second_Demographic_Transition_in_Industrialized_Countries

WHO [World Health Organization]. 2010. "Thailand's New Condom Crusade." *Bulletin of the World Health Organization* 88 (6): 401–80. https://www.who.int/bulletin/volumes/88/6/10-010610/en/

World Bank. n.d. "Fertility Rate, Total (Births per Woman)." Accessed September 21, 2021. https://data.worldbank.org/indicator/SP.DYN.TFRT.IN?most_recent_value_desc=true&year_high_desc=true

Zaidi, B., and S. Morgan. 2017. "The Second Demographic Transition Theory: A Review and Appraisal." *Annual Review of Sociology* 43: 472–92. doi:10.1146/annurev-soc-060116-053442

7

Ethical issues in fertility

After reading this chapter, a student should be able to:

1 discuss some of the key controversies associated with population issues;

2 consider the problems associated with pro-natal policies;

3 explain the potentially coercive nature of policies designed to limit fertility.

Today, many people have access to a range of options to control their fertility, and the idea that women and couples have the freedom to control the number and spacing of their children is considered a human right. Nonetheless, fertility policies open up a variety of ethical questions related to how governments and policymakers try to influence individual decisions over family, and it is essential that we understand some of the injustices of the past as we think about fertility policies into the future. In this chapter we expand our discussion of fertility by focusing on the social justice and ethical concerns that surround fertility, contraception, and population policy.

Contraception and abortion

People have made efforts to control their fertility for thousands of years. Scholars have often assumed that this initially occurred largely through abortion and infanticide, given limited understandings of the process of conception in the ancient world, but there is some circumstantial evidence that people may have attempted to reduce their fertility prior to conception as well. Pessaries may have been used to try to physically block sperm from reaching the egg, using substances such as honey, gums, and even crocodile dung, and herbal potions or douches may have been administered to prevent conception or cause miscarriage. How effective these methods were remains contested, but demographic evidence from censuses, as well as evidence of pregnancy from pelvic bones of women from this time, suggests that many women in antiquity may have attempted to limit their family size (Riddle and Estes 1992; Riddle, Estes, and Russell 1994).

DOI: 10.4324/9781003143253-7

By the Middle Ages, much of this pharmacological knowledge seems to have been lost and religious proscriptions on fertility began to increase, although people were still making efforts to control their fertility (Riddle and Estes 1992). Many of these methods were relatively crude, including sheaths made from animal intestines or linen (a precursor to the modern-day condom) and unsafe "back-street" abortions that posed a significant risk to the mother's life.

The development of effective contraceptive technologies in the modern era allowed women and couples to control their fertility more effectively and safely. This began with the development of the diaphragm in the 1870s, along with more effective condoms and intra-uterine devices, culminating in the introduction of the contraceptive pill in the 1950s. Many people embraced these technologies as a new era of **family planning** where open access to effective contraception would enhance individuals' **reproductive freedom**. Indeed, reliable forms of contraception have allowed the decoupling of sexual activity from reproduction, and modern condoms have provided an effective means to protect people from sexually transmitted diseases, allowing individuals to take greater control over their reproductive health.

Nonetheless, well into the 21st century there is still a strong unmet demand for contraception, with over 200 million women in the Global South estimated to have an unmet need for contraception and perhaps 40 percent of pregnancies worldwide unwanted or mistimed, particularly in sub-Saharan Africa, the Arabian Peninsula, and Latin America. Of these unintended pregnancies, about half result in abortion, 13 percent in miscarriage, and 38 percent in unintended birth (Sedgh, Singh, and Hussain 2014). Preventing unwanted pregnancies is a key aspect of maternal health, given strong links between unsafe abortion and maternal mortality and morbidity. Pregnancy brings particular risks for adolescent girls, including physical challenges associated with small body size, as well as psychological and social impacts such as increasing the risk that girls will drop out of school (UNESCO 2017). Furthermore, stillbirths and newborn deaths are 50 percent more common among babies born to adolescents than among babies of women in their 20s. About 95 percent of adolescent births occur in the Global South, where complications from pregnancy and birth are a leading cause of death among teenage girls, indicating barriers to meeting adolescent health targets in poorer countries (WHO 2014). Although not all adolescent births are unintended, we can often question a teenager's ability to advocate for herself, even when pregnancy occurs within marriage—gender-based violence is often implicated in situations where early marriage involves a huge power disparity between a young girl and much older man, for example (UNESCO 2017). In high-income contexts, too, evidence suggests that unintended pregnancies have negative impacts. In the United States, unintended pregnancy has been linked to lower incidence and duration of breastfeeding, mental health challenges for mothers, and poorer educational and behavioral health outcomes for children (Logan et al. 2007). If unmet demands for contraception were met worldwide, 21 million unplanned births and 26 million abortions could be averted annually (Sedgh, Singh, and Hussain 2014).

For reasons such as these, liberal, feminist, and secular commentators have often actively promoted family planning as essential to maternal and infant health, as well as women's rights and social justice goals. Many policymakers argue that increasing reproductive freedom has huge benefits for society: limiting the number of unintended pregnancies, reducing the burden of domestic responsibilities for women, and enhancing reproductive freedom. From this perspective, an individual's right to control their own fertility using contraception (and more controversially also abortion) is placed ahead of ethical concerns with the technology itself. Also significant in this approach is a frequent emphasis on adolescent health, reflecting the perspective that adolescents have the right to control their own reproductive health, regardless of whether they are married or their sexual preferences. For many social conservatives, this degree of sexual permissiveness is a step too far, with overarching moral concerns around contraception, and particularly abortion, seen as outweighing any potential benefits. Here, a significant moral distinction is often drawn between **contraception** (which prevents the union of egg and sperm, preventing conception) and **abortion** (which terminates a pregnancy after conception has occurred), with public opinion generally considering contraception less morally concerning than abortion. Conservative Christian and Muslim groups have been particularly vocal in their concerns that family planning is a threat to traditional family values (box 7.1).

Family planning campaigns

Despite widespread support for family planning among liberals, many scholars emphasize that contraception can be used in both empowering and disempowering ways. In general, where family planning has been used as a tool to meet demographic targets, policies have been critiqued as disempowering or even coercive, whereas efforts to promote reproductive choice are usually considered empowering. As such, family planning campaigns require careful planning and oversight. One key aspect of ensuring that family planning is an empowering experience is to provide choice among contraceptive methods, so that each individual can select the right method for their own unique circumstances and goals. A wide variety of contraceptive methods exist, which can be weighed up in practical terms, particularly variations in cost, convenience, and effectiveness (table 7.1). For instance, many methods that require the user's regular participation are less effective than those that are implanted or applied for months or even years, owing to the possibility of user error; on the other hand, they offer higher levels of flexibility because they are totally in the user's control. With some methods, there are also additional health impacts to take into consideration—in particular, condoms can prevent the spread of sexually transmitted infections (STIs), and hormonal methods of contraception disrupt hormone levels, which can carry small health risks (such as a slightly elevated risk of stroke or heart attack) or benefits (such as reducing premenstrual symptoms).

Beyond these practical considerations, we can also look at contraceptives from a social justice perspective, revealing further ways in which different methods can empower or disempower the user. Particularly critical in this respect is whether the method is permanent or reversible. Many commentators have noticed a preference for permanent methods of contraception—sterilization—in programs where the primary concern is to reduce population size. Although sterilization is a cost-effective way to reduce fertility at the population level, from an individual perspective it is less often an optimal contraceptive choice, given that many people use contraception

BOX 7.1 CATHOLIC AND ISLAMIC ATTITUDES TOWARD CONTRACEPTION AND ABORTION

Given the powerful role of religion in influencing norms around children and the family, it is not surprising that many religious groups have taken a stance on contraception and abortion. In general, Christian and Islamic traditions have expressed the greatest opposition to contraceptive technologies, or even the very notion of interfering with conception at all.

The Catholic Church has been most vociferous in its condemnation of artificial birth control, arguing that children are gifts from God and that technological means of preventing conception are therefore immoral. Despite this condemnation from the institutional Catholic Church, contraceptives are widely used by many Catholics. The sway of Catholic beliefs on formal reproductive policy is strong, however, because Vatican City (the seat of the Pope) is recognized by the United Nations as an independent state with its own representation at UN meetings, meaning that Roman Catholicism has traditionally had a much stronger voice than other faiths at international population conferences (Goldberg 2009).

When it comes to abortion, some Catholic communities are vehement in their opposition, with Catholic doctrine used to argue that abortion should be banned on the grounds that it can be equated to ending a human life. Today, some of the most restrictive laws on abortion can be found in Catholic countries such as Malta and Poland. In 2021, Poland's conservative government, which has strong ties to the Catholic Church, voted to ban abortion except in cases of rape, incest, or where the mother's life is in danger—among the most restrictive abortion laws enacted today (BBC News 2021). Several countries in Latin America also have very restrictive regulations on abortion associated with their Catholic ties. Ireland, by contrast, despite its longstanding reputation of reflecting Catholic doctrine in restrictive abortion laws, has recently relaxed its stance. This patchwork of regulations leads to significant cross-border flows of women seeking abortions in countries or states with less restrictive policies, as well as transnational efforts to supply women in countries with restrictive abortion policies with abortion pills from overseas.

In Islamic belief systems, the situation is also complex. Although Islam is strongly in support of the family and children, the Koran does not specifically mention contraception and most schools of Islamic law do not prohibit it. Many Islamic governments have actively supported family planning campaigns as consistent with Islamic beliefs. However, some conservative Muslim scholars have interpreted certain passages from the Koran as banning contraception, and the idea that Islam does not allow contraception has spread in some Islamic communities (Atighetchi 1994). Broadly speaking, abortion is tolerated in many Islamic communities but not encouraged, particularly at later stages of pregnancy. Access to abortion varies widely in Muslim-influenced countries, ranging from being legal unconditionally in some countries (e.g., Turkey) to being restricted to circumstances where the mother's life is in danger in others (e.g., in Egypt and Indonesia). In general, Islamic scholars promote conservative approaches to family that encourage children within marriage.

simply to delay childbearing or at least to keep their options open. Many family planning programs of the mid- to late 20th century have thus been found wanting for their heavy focus on sterilization. Sterilization programs have also been critiqued from a gender perspective. Typically, women have often been targeted by sterilization programs on the grounds that they have the greatest incentive to agree to the

TABLE 7.1 Comparison of contraceptive methods

Contraceptive type	Description	Estimated effectiveness	Permanency	Expert assistance	Examples
Barrier Methods					
Surgical sterilization	Surgical techniques to prevent the normal release of eggs or sperm	Highly effective (>99% reliable)	Permanent; hard to reverse, especially for women	Requires surgeon and healthcare facility	Tubal ligation (women), vasectomy (men)
Barrier methods, implanted	Prevents sperm from reaching egg; some also have a hormonal component	Highly effective (>99% reliable)	Some implants may remain in place for more than 10 years; can be removed at any time	Must be inserted and removed by healthcare professional	Intra-uterine device (IUD)
Barrier methods, individual use	Barrier inserted/applied before sex to prevent sperm reaching egg	Somewhat effective (71–88% reliable)	Must be used prior to each sexual encounter	Use and removal controlled by user	Condom, cap, diaphragm
Hormonal Methods					
Contraceptive implants	Device is implanted under the skin; releases hormones to prevent pregnancy	Highly effective (>99% reliable)	Effective for up to 5 years but can be removed at any time	Must be inserted and removed by healthcare professional	e.g., Norplant
Birth control injections	Regular injections deliver hormones to prevent pregnancy	Moderately effective (94% reliable)	Typically need to be repeated every 3 months	Must be administered by healthcare professional	e.g., Depo-Provera
Birth control patches/rings	Release hormones to prevent pregnancy over the course of a week or month	Moderately effective (91% reliable)	Must be replaced weekly to monthly	Applied and removed by the user	e.g., Xulane
Birth control pill	Oral pill taken daily	Moderately effective (91% reliable)	Must be taken daily	Controlled by user	Combination pill; progestin-only "mini pills"
"Natural" Methods					
Withdrawal (a.k.a. "coitus interruptus")	Withdrawal before ejaculation	Somewhat effective (78% reliable)	Method must be used during every sexual encounter	Controlled by male partner	
Fertility awareness methods (a.k.a. "rhythm method")	Women track their fertile period and abstain from sex when they believe they may be fertile	Somewhat effective (76–88% reliable)	Must be tracked daily	Controlled by user	Tracking temperature, vaginal discharge, or menstrual cycle can all help chart ovulation

Information compiled from: NHS (2022) and Planned Parenthood (2021).

procedure, given their greater burden associated with birth and childrearing. Feminist critics note that this pattern has developed despite the fact that the surgery is more dangerous (and more expensive) for women than for men. Puerto Rico's efforts to lower fertility rates in its family planning campaigns of the late 20th century, for instance, were heavily premised on sterilizing women, leading to social equity concerns. By 1974, more than one-third of Puerto Rican women had been sterilized, at an average age of just 26, giving Puerto Rico one of the highest sterilization rates in the world. Many Puerto Rican women have reported that they were misinformed about the procedure, believing in many cases that the procedure could be reversed (although reversal is possible, it is often unsuccessful). One study found that more than 20 percent of Puerto Rican respondents expressed regret at having been sterilized (Boring, Rochat, and Becerra 1988). To provide another example, research suggests that access to sterilization has been used as a political tool in Northeast Brazil, with politicians helping to arrange sterilizations for low-income constituents and the number of sterilizations performed rising in election years (Caetano and Potter 2004). In Brazil, Dalsgaard (2004) suggested that a culture of sterilization has arisen owing to limited access to alternative family planning options, forcing us to question how empowering sterilization can be in a situation where it is the only option open to many low-income women, and even then perhaps only through political wrangling.

A second key power issue is whether the user is in control of the method or whether they need the assistance or agreement of another person for the method to be effective. This can operate in two ways. First, some methods can be used by women without the participation of their partner. Although ideally a couple makes choices over contraception together, in strongly patriarchal societies women may have little decision-making power. In such cases, many feminist scholars consider that it is important to offer women the option to control their own fertility without their partner's knowledge. Even with a method that can be used discreetly such as the contraceptive pill, access may still be a challenge, however. In patriarchal societies, women may be expected to seek permission from a senior or male family member before seeking healthcare or be accompanied when seeing a doctor, creating barriers to access. In response, some family planning programs have explicitly acknowledged girls' and women's right to access contraceptives independently and have attempted to provide education and outreach to vulnerable populations. Power imbalances are also relevant where minors may want to seek contraception without a parent's consent. In some healthcare systems such as the UK's National Health Service, teenagers are reassured that they can seek contraception without their parents' knowledge (NHS 2018). Despite these efforts, access to contraception remains a challenge for many women and controversies persist around the appropriateness of providing contraception to certain groups, particularly adolescents.

Further issues are raised as we think about whether a woman needs medical assistance to use the method. In cases where access to healthcare is poor, there may be practical barriers to obtaining regular assistance from a health professional. Of greater interest to our power argument here, however, are circumstances where a woman's wishes are over-ruled by healthcare professionals or policymakers in the belief of a greater good. For instance, some young women in the UK have argued that they have not been granted sterilization on the grounds that they will change their minds in the future, even though men of the same age are offered vasectomies with little question. Though we may believe that it is only reasonable for doctors to consider the possibility of future regret among their patients, these concerns can come across as paternalistic in such cases (McQueen 2017).

The issue of healthcare involvement moves beyond paternalism toward outright coercion if we look at the 1990s case of Norplant—a long-acting reversible contraceptive (LARC) that is implanted under the skin by a healthcare provider and then releases small doses of hormones to prevent pregnancy for up to 5 years. Norplant represented a technological leap forward as a highly effective, reversible method of contraception that was both convenient and cost-effective. However, it was the way in which it was used that raised concerns. Norplant soon became popular in the United States with some politicians who saw the new technology as a cost-effective way to prevent pregnancies among teens and women on a low income (also often racial/ethnic minorities), for whom high fertility was deemed undesirable. Many states made efforts to try to ensure that Norplant was easily available to low-income communities through subsidized implantation of the device. This in itself is not problematic. However, legislation was also proposed that would have offered significant financial incentives for women on welfare to use the device, or even threatened them with losing their benefits if they refused. Similarly, proposals were put forward to offer women shorter jail sentences if they accepted the implant. Although most of this legislation was eventually struck down, critics have highlighted the coercive nature of these proposals (Roberts 1997). Some women also reported that they found it hard to get their doctor to remove the device before the 5 years of its efficacy was up, if they decided they wanted to have a child or even because of side effects. In such cases, government funding was often used to insert the device, but women found it harder to get the cost of removing the device covered. Critics have suggested that this demonstrates that policies around Norplant were designed specifically to reduce births in particular populations, rather than to improve reproductive choice for women (Roberts 1997). By the early 2000s, Norplant had been taken off the market in the United States and UK owing to concerns about side effects and the possibility of coercion, although its use continued in many parts of the Global South until 2008. New types of LARCs have since been developed, providing a potentially powerful additional reproductive option for women, even as concerns over the possibility of coercion persist.

Access to abortion raises further challenges. Given how contentious abortion is, a wide variety of laws restrict access to abortion (figure 7.1). Some countries allow abortion only in situations of rape or incest, where the mother's life is in danger, or where the fetus has significant physical problems (figure 7.1, inset box). At the other end of the spectrum, abortion is acceptable in some countries to the point where it is provided on request, in some cases even becoming the primary form of family planning. In many parts of Eastern Europe during the communist era, for instance, abortion was promoted as convenient and with relatively few side effects compared with hormonal contraception (Popov and David 1999; Perelli-Harris and Isupova 2013). In most countries, abortion policy falls somewhere between these extremes, with many societies endorsing access to abortion as an important part of reproductive health services but considering that forms of contraception that operate prior to conception are preferable over terminating pregnancies. In such cases, abortions are often available with only limited administrative barriers but usually as part of a wide range of family planning options.

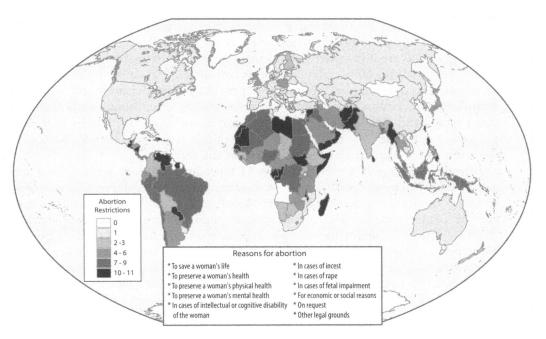

FIGURE 7.1 Number of abortion restrictions by country, 2017
The shading on the map illustrates the number of restrictions on abortion by country—the darker the shade, the more restrictions. The inset box lists reasons for abortion that are often accepted to access legal abortion—different countries accept different combinations of these reasons to support access to abortion.
Data source: UNDESA (2017b).

New technologies are now blurring the boundaries between contraception and traditional surgical abortions, leading to further debates. The development of a combination of safe and effective drugs to terminate pregnancy (mifepristone and misoprostol)—sometimes called the "abortion pill"—has provided the option of a drug-based termination during approximately the first 10 weeks of pregnancy. Although far removed from traditional surgical abortions in their operation, opponents emphasize that these drugs still cause the artificial rejection of an embryo that has the potential to develop into a baby, which many religious conservatives consider unconscionable. For those with fewer moral concerns around abortion, the pill is a breakthrough because it allows an unintended pregnancy to be avoided with minimal risk to the mother. The "morning after pill" or "emergency contraception" also blurs the boundaries between contraception and abortion. It can be taken up to 5 days after unprotected sex to prevent pregnancy, before a pregnancy has even been confirmed, avoiding delay and some emotional heartache for many women. For those strongly opposed to abortion, however, this technology is also rejected because it may still cause the rejection of a fertilized embryo. In this regard, even some forms of contraception meet with moral censure for their possible abortifacient (abortion-causing) properties. At this extreme, intra-uterine devices are sometimes spurned because it is possible that they can prevent the implantation of a fertilized egg, although their usual mode of action is to prevent conception.

Resolving these differing viewpoints is challenging because they come from deep-seated philosophical beliefs about birth, marriage, the autonomy of women, and when life begins. Conservative attitudes have proved remarkably resistant to change, even in the light of empirical scientific evidence, because religious frameworks rely on alternative belief systems that may not be swayed by scientific arguments. For instance, family planning programs have often come under fire from social conservatives on the grounds that providing sex education and access to reproductive services will encourage promiscuousness, even though empirical evidence refutes this idea (Dreweke 2019). Indeed, many Western countries have seen a significant decline in sexual activity among adolescents since the 1990s, despite the widening of reproductive health education and services.

Broadly speaking, however, the story of reproductive health services over the past 50 years has been a general increase in acceptability of and access to contraception in communities around the world. Globally, 84 percent of governments now provide family planning services directly through government facilities, with a further 9 percent supporting private provision of services (UNDESA 2017a). Nonetheless, governmental attitudes toward fertility vary widely, with some countries attempting to raise fertility and others to lower it. Although countries with higher fertility rates tend to have policies to lower fertility, whereas those with lower fertility tend to want to raise or maintain fertility, wide variations in policy exist even at very similar fertility rates (figure 7.2).

Large inequalities in service provision also persist at sub-national scales, with rural women often significantly under-served compared with urban women. There is therefore still significant scope to improve access to family planning—worldwide, 12 percent of women of childbearing age who are married (or in a similar union) report that they wish to stop or delay childbearing but are not using any method of contraception; this figure rises to 17 percent if we consider those not using modern forms of contraception (UNDESA 2017a). In 24 countries in sub-Saharan Africa, more than 50 percent of women of childbearing age with a need for contraception are not using modern methods (PRB 2020). Even in countries that have made huge strides in providing contraception, the situation is often precarious. In Indonesia,

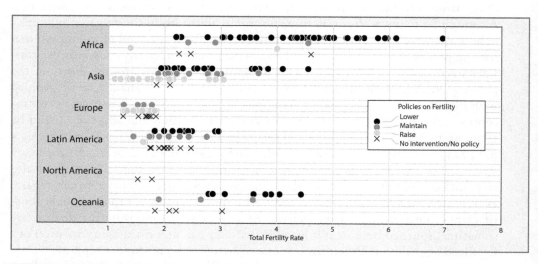

FIGURE 7.2 Government policies on fertility in relation to total fertility rates by country
Data source: UNDESA (2015b).

for instance, the National Population and Family Planning Agency reported in 2020 that about 10 million married couples had stopped using contraception during the COVID-19 pandemic because of closure of clinics or fears of contracting the virus at health facilities (as cited in Paddock and Sijabat 2020). In Iran, access to contraception has waxed and waned with different political administrations—in the early 2000s, Iran was lauded for having one of the most successful and least coercive family planning campaigns in the low-income world; today, the situation has changed dramatically in response to the government's reinstatement of pro-natal ideals (box 7.2).

BOX 7.2 IRAN AND THE POLITICIZATION OF FERTILITY

Iran provides a prime example of the politicization of fertility, with Iranian population policies swinging back and forth between pro-natal and promoting family planning (figure 7.3). Iran's first official population policy was developed in 1967. This policy primarily targeted urban populations and was focused around providing access to the contraceptive pill. At the same time, women's rights related to marriage and custody of children began to improve (Hoodfar and Assadpour 2000). Although limited in scope, these gender-based reforms were strongly criticized by the Islamic religious establishment, led by Ayatollah Khomeini. Historically, contraception had been permitted under Islamic law, but religious hardliners began to condemn family planning as an imperialist plot aimed at weakening Muslim populations. With the dawn of Iran's Islamic revolution and the transition of Iran into an Islamic Republic in 1979, Iran's family planning program collapsed and access to contraceptives became erratic, generating a black market in contraceptives (Hoodfar and Assadpour 2000).

Under the early years of Iran's development as an Islamic Republic in the early 1980s, marriage and procreation were celebrated as promoting the Muslim family. Age of marriage was reduced and polygynous (multiple wives) and temporary marriages were legalized and even encouraged by the state (Hoodfar and Assadpour 2000). Population growth was

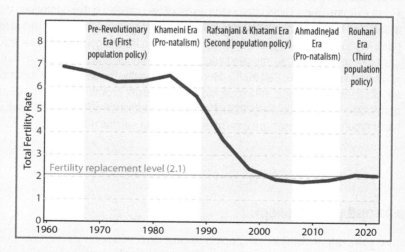

FIGURE 7.3 Fertility rates and population policy in Iran
Data source: Macrotrends (n.d.).

promoted as a way to defend the Islamic Republic, with fertility peaking at about 6.5 births per woman in 1983 during the Iran–Iraq War. By the end of the war, however, astute political leaders began to recognize the challenges of a rapidly growing population, and during the presidency of Hashemi Rafsanjani (1989–97), Iran embarked on a reproductive health program that provided free contraception and counseling to married couples via voluntary family planning services (Cincotta and Sadjadpour 2017). Services were provided via a network of state-subsidized clinics, even in remote rural areas, staffed mainly by female technicians (Roudi 2012). Fertility declined to about two births per woman, and this campaign has been credited as being one of the most successful family planning campaigns in the Global South (Hoodfar and Assadpour 2000; Cincotta and Sadjadpour 2017), with Iran's fertility reduction occurring faster than China's and without resorting to coercion (Roudi-Fahimi 2002).

Today, Iran is benefiting from a **demographic dividend** associated with the success of this campaign. With a median population age of 31 years, the government currently has an ideal opportunity to maximize returns from its large working-age population, as well as to plan for the future by reforming its healthcare and pensions policies to accommodate population aging (Cincotta and Sadjadpour 2017). This situation is temporary, however. With persistently low birth rates and rising life expectancies, an increasing proportion of Iran's population is moving into elderly cohorts, with United Nations projections suggesting that favorable economic conditions will last only until the 2040s. At this point, one-fifth of the Iranian population is likely to be over 65 and Iran's demographic makeup may look similar to Germany's today (Cincotta and Sadjadpour 2017).

Meanwhile, Iran's policies have once again been reversed, calling Iran's demographic future into question yet again. With the election of Mahmoud Ahmadinejad in 2005 and the support of religious hardliner supreme leader Ayatollah Ali Khamenei, religious conservative views were reasserted as part of a populist movement. Political leaders reframed the birth rate as a national security issue and suggested that Iran's population should double. Foreign enemies have been blamed for engineering Iran's population decline and family planning characterized as an idea from the secular world that threatens to undermine Islamic ideals. These extreme views culminated in the arrest of two academic demographers, who were accused of manipulating statistics to hide Iran's supposed "population crisis" of low birth rates (Etehad and Mostaghim 2019). Since 2010, the Iranian government has introduced cash payments for births, improved access to infertility treatments, and instituted parental leave policies (AP 2010; Safi 2020). More controversially, the government has reduced access to family planning, with state hospitals no longer providing free contraceptives or performing vasectomies (Safi 2020). Despite these efforts, Iran's birth rate continues to hover around the replacement level of two babies per woman (PRB 2020).

Eugenics and pro-natalism

As the example of Iran shows, governments are often very interested in influencing population size and composition, with fertility policy seen as a primary mechanism by which this can be achieved. Early modern fertility policies were often **pro-natalist**, concerned with promoting economic and military strength as well as settling new territory. With early understandings of genetics in the late 1800s, the problematic idea of using fertility policy to "improve" the population also emerged. Although the idea of selective breeding of humans dates back at least to the Ancient Greeks, it was new revelations about Darwinian evolution that really propelled the idea, with the British scholar Sir Francis Galton coining the term **eugenics** in 1883. Eugenics—literally "good growing"—is the idea that humankind should attempt to improve the human race by selectively encouraging children among those deemed "fittest," while actively discouraging or even preventing births among those deemed "unfit." The ultimate goal was to try to remove certain diseases, disabilities, or even characteristics from the human gene pool. Those targeted as "unfit" have included groups such as people with epilepsy or mental illness, individuals defined as criminals, those living in poverty, and even people from particular ethnic groups. Today, eugenics is widely discredited and the idea that certain groups should be prevented from having children is considered to contravene human rights.

Although the term eugenics originated in Britain, it was in the United States that it was enthusiastically embraced by prominent citizens. Changing racial patterns associated with the immigration of new Americans began to be seen by "old stock" white Americans as a threat, fomenting concerns that white populations would soon be outnumbered by the rapid reproduction of more recently immigrated populations of color (Roberts 1997). By the early 1900s, eugenic societies were becoming established across the United States and a series of pseudo-scientific eugenics conferences were held. These were accompanied by eugenic policies, including selective immigration, restrictions on marriage, and state-mandated sterilizations. As early as 1896, Connecticut made it illegal for people with epilepsy or the "feeble-minded" to marry; by 1913 at least 24 US states had laws restricting marriage for those considered to be imbeciles, drunkards, paupers, or criminals, among other categories (Roberts 1997). It was the United States' experimentation with policies of forced sterilization that marked an even darker turn in US fertility policy, however. Starting in 1907, Indiana became the first among many states to enact an involuntary sterilization law that allowed state institutions to sterilize people from "undesirable" groups, including those considered to be imbeciles or criminals and people with epilepsy (Roberts 1997). By the 1930s, following the Great Depression, African American communities began to be targeted, with rising numbers of black women involuntarily sterilized in institutions in the 1930s and '40s. Racist concerns were voiced that black communities were reproducing too rapidly and were too poor or dissolute to effectively care for their children (resulting in increasing numbers on welfare) and even that intermarriage would lead to "mongrelization" of the races (Roberts 1997).

By the 1940s, attention shifted to Nazi Germany, where eugenic policies became even more horrific, as loss of fertility began to turn into loss of life for certain populations, including Jews, as well as those with disabilities or mental limitations. In reaction to the Nazis' inexcusable policies, eugenics finally began to be widely discredited. The legacy of previous prejudices did not disappear immediately, however. In the United States, for instance, although forced sterilizations became less common for overtly eugenic reasons, the sterilization of targeted populations may have persisted well into the 1970s, with claims that doctors continued to perform unnecessary hysterectomies on certain groups of women, particularly African Americans, as part of routine healthcare (Roberts 1997).

In the modern era, such inhumanities have diminished and preventing births in a particular group is officially considered **genocide** by the United Nations (UN, n.d.b). However, critical scholars point to persistent injustices, now mostly perpetuated through **pro-natal policies** that offer incentives to favored groups to have children rather than efforts to deter births. In Singapore in the 1980s, for instance, the government offered incentives to encourage births among highly educated women, while concurrently promoting sterilization among less educated citizens. Explicitly, this was supposed to address human capital concerns, but commentators note that it also had the impact of promoting births among Singaporeans of Chinese origin, while discouraging births among those from Indian and Malay minorities, owing to educational inequalities among the three groups (Heng and Devan 1992; King 2002). In response to criticism, the policies were repealed after only 2 years, to be quickly refocused toward greater support for all mothers and subsequently all citizens (King 2002). To provide another example, in France in the 1990s the right-wing National Front party of Jean-Marie LePen argued for reserving pro-natal family benefits for "the French"—explicitly excluding minority populations—although such exclusions were never enacted by French lawmakers and did not gain widespread support (King 1998).

Beyond concerns over the fairness of pro-natal policies to different sub-populations, they can also be applied in ways that are outright coercive. In extreme cases, women may lose access to contraception altogether, with subsequent loss of control over their own fertility and detrimental consequences for maternal and child health. This is evidenced particularly clearly in the case of Romania.

Pro-natalism in Romania

When dictator Nicolae Ceaușescu came to power in Romania in 1965, he rapidly enacted pro-natalist policies, believing that a large number of people entering the workforce could stimulate Romania's economy. The state generated pro-natalist propaganda and provided financial incentives for births, combined with stringent restrictions on abortion (Kivu 1993). As in much of the Soviet bloc at that time, abortion had been the primary method of contraception in Romania, with liberal abortion policies over the previous decade. After the decree was enacted in 1966, only women over the age of 40 (subsequently 45) who already had at least four (subsequently five) children in their care were allowed to obtain abortions through official channels. The importation of contraceptives was also banned, leaving them available only on the black market (Kligman 1992). Those who remained childless after the age of 25 were subject to heavy taxes, and divorces were hard to obtain (Bachman 1989). By the 1980s, the policies had become even more extreme, with women subjected to regular gynecological exams to ensure that pregnancies were brought to term and childless couples interviewed about their sexual activities (Bachman 1989). Doctors were pressured to ensure that demographic targets were met and could be fined or imprisoned for providing an abortion (Kligman 1992).

For many women, economic constraints argued strongly against large families at this time. Basic goods, including food, were frequently in short supply and rationing was common. Financial incentives to support children were meager and far from compensated the cost of raising children, particularly because many women had entered the workforce and yet were also expected to continue to bear most of

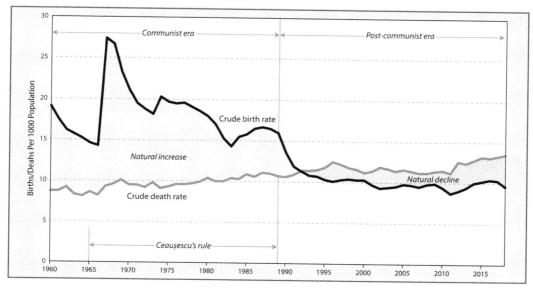

FIGURE 7.4 Fertility and mortality rates in Romania, 1960–2018
Data source: National Institute of Statistics, Romania (2021).

the responsibility for childcare and housework. Although abortions declined significantly in the early years of the program, by the 1980s illicit abortions had become the primary means of birth control for Romanian women (Kligman 1992). Infant and maternal mortality rose sharply owing to the lack of access to safe reproductive health services (figure 7.4). In 1967, after restrictions had been in place on abortions for just 1 year, the number of abortion-related maternal deaths had tripled, and by 1987 more than five out of every six maternal deaths were from complications from abortions (Kligman 1998).

After the overthrow of Ceaușescu in 1989, further implications of the policies became clear to the rest of the world as the dire conditions in Romania's overflowing orphanages were revealed. With little opportunity to limit their fertility and few economic resources to support their children, many women turned to Romania's state-run institutions to care for the children they could not afford, with perhaps as many as 300,000 children living in institutions in 1989 (Gaines 2006). Many of the children who were given up for adoption had disabilities and the overstretched orphanages were unable to provide a standard of care sufficient for healthy children, let alone for thousands of children with special physical and neurological needs. With significant input of funds from the European Union and reform from the Romanian government, the situation has greatly improved in the intervening years (Rogers 2009), with the institutionalized population dropping by 90 percent by 2006 (Gaines 2006).

Access to contraception and abortion in the post-communist era has shown a complex story. Although family planning rules were liberalized after the fall of Ceaușescu, in the absence of sound family planning education, a large proportion of women continued to rely heavily on abortion into the 2000s, when Romania had the highest number of abortions and highest death rate from abortion in Europe (Leidig 2005). Reproductive health education has improved for younger generations and Romanians have been moving toward greater use of contraception. However, recent evidence suggests that women may now be having trouble obtaining abortions once again. Although Romanian law decrees that abortions should be offered up to the 14th week of pregnancy, women have begun reporting that some doctors are refusing to perform the procedure, this time owing to pressure from groups close to the Orthodox and Catholic Church (Vdovîi and Bird 2019). As the case of Romania illustrates, attitudes toward family planning are constantly evolving.

Fertility reduction policies

We have already discussed several examples of pro-natal fertility policy, but the more typical trend since the 1950s has been policies aimed at fertility reduction. This modern era of fertility control began as mounting concerns were raised in the Global North about a "population explosion" in the Global South. Many scholars have noted the race-based assumptions that undergirded these concerns, with fears openly voiced by some policymakers that the growth of populations of color might lead to global political and economic instability and even the spread of communism. For others, the concern was increasing competition over resources or environmental degradation or the challenges of raising rapidly growing populations out of poverty. A number of population-based organizations were set up in this era. For instance, the US-based Population Council was set up in 1952, with its main aim to control population growth. The organization funded research into more effective methods of contraception and lobbied the US government to contribute further funding to the cause. The organization continues to develop and promote new, more effective forms of contraception—most recently long-acting reversible contraceptives (LARCs).

The United Nations has also long been involved in population issues, acting as an important facilitator of dialogue among individual countries to try to generate consensus on population policy. Since the UN's creation in 1945, a number of notable events have occurred (table 7.2). Early population conferences in Rome (1954) and Belgrade (1965) were academic in nature and attended by population experts (rather than policymakers), mostly encouraging the sharing of information on population issues. By the late 1960s, the UN had begun to focus on human rights, which quickly became intertwined with population issues. Following the International Conference on Human Rights in 1968, a UN resolution was developed, including the key passage: "[...] couples have a basic human right to decide freely and responsibly on the number and spacing of their children and a right to adequate education and information in this respect" (UN 1968). This was a seminal moment in population policy because it placed fertility policy squarely within the framework of human rights.

UN involvement with population issues has expanded the significance of human rights as a fundamental principle for fertility policy ever since. The creation of the UN Population Fund in 1969 provided an organization whose mission was to guide the UN's population programs, based on this declaration of human rights. Known now as the UNFPA, the organization today "calls for the realization of reproductive rights for all and supports access to a wide range of sexual and reproductive health services—including voluntary family planning, maternal health care and

TABLE 7.2 United Nations world population conferences

Year	Event	Significance
1954	World Population Conference, Rome	Focused on sharing information on population and improving understanding of the global demographic situation; promoted the creation of regional training centers to address population issues
1965	World Population Conference, Belgrade	Focused on analysis of fertility patterns for development planning; most attendees were population experts
1974	World Population Conference, Bucharest	Sharp divide between the "incrementalist" position of many Western states that saw population growth as impeding economic development and the "redistribution" position of other countries led by Argentina and Algeria that argued that population problems were a result, not the cause, of economic under-development and could be addressed with better redistribution of resources
		The *World Population Plan of Action* was the output from the conference, which emphasized the need for economic development in addition to fertility planning, as well as interconnections between the two
1984	International Conference on Population, Mexico City	Expanded and endorsed the *World Population Plan of Action*, emphasizing family planning as a key aspect of economic development. Many affluent countries expressed their willingness to increase support for family planning
		Human rights presented as critical to population planning
1994	International Conference on Population and Development, Cairo	Reaffirmed connections between population and development
		Emphasized the need to respect universally recognized human rights of the individual rather than just demographic targets; introduced concept of sexual and reproductive health
		Emphasized the significance of empowering women

Sources: UN (n.d.a, n.d.c).

comprehensive sexuality education" (UNFPA 2018). This shift in focus from demographic targets (sometimes denoted "population control" by critics) to broader issues of reproductive health occurred progressively through the 1970s, 1980s, and 1990s. The 1974 World Population Conference in Bucharest marked a newly developing divide between (mostly affluent) countries that saw population growth as a problem that impeded the development of poorer countries and an increasingly vocal group of countries from the Global South that argued that rapid population growth was not so much the *problem* as the *result* of global inequalities. The idea that "development is the best contraceptive" was popular among representatives of poorer countries at this conference. This represented a clear divide between a neo-Malthusian emphasis on population size and a Marxist perspective focusing on inequality as the root of the problem. This debate persisted into the subsequent International Conference on Population and Development in Mexico City (1984) and came to a head in the 1994 Cairo conference, which finally fully asserted the need for human rights to be prioritized over demographic targets and reoriented population discussions around the broader notion of reproductive and sexual health, with gender as an integral aspect of the discussion. This Cairo conference is often seen as marking a pivotal change in global attitudes toward population policy, directing them firmly away from strict demographic targets and toward efforts to enhance reproductive freedom and gender equity.

These latter two conferences (Mexico City and Cairo) also marked the rise of a new era of social conservatism in population policy, however. Early conferences had viewed family planning as largely value neutral—simply technologies that could be applied to achieve fertility reduction. By the Mexico City and Cairo conferences, however, this rise of secular and liberal approaches to fertility was being viewed with alarm by religious conservatives as having the potential to undermine traditional family values (Goldberg 2009). Although contraceptive technologies had long met with resistance in many traditional societies in the Global South, this had usually been interpreted as reflecting backwardness and parochialism in these communities, with contraceptive technologies persistently pushed as part of broader moves toward modernization. It was only when a backlash began to build among religious conservatives in more affluent countries that these concerns began to be taken seriously. It was in the United States that this trend was most influential, given the huge economic and political power that the United States held over global population policy. Traditionally a major proponent of population programs, the United States began to shift its stance with the rise of social conservatism during the Reagan administration. This came to a head in 1984 at the Mexico City conference when the United States declared that it would no longer fund any organizations that "perform or actively promote abortion as a method of family planning," even if US money was not involved in that arm of their programming. This "Mexico City Policy" is referred to as the "Global Gag Rule" by its opponents because it prevents organizations from providing any information on abortion or from lobbying governments to legalize abortion if they want to receive US funding (KFF 2021). In the United States, the Mexico City Policy has repeatedly been repealed by Democratic administrations and reinstated by Republican administrations ever since, until the policy was significantly expanded under the Republican Trump presidency, arguably compromising a variety of other public health programs including those targeting HIV/AIDS, maternal and child health, malaria, and nutrition. The Mexico City Policy was once again repealed in 2021 by the Democratic Biden administration (KFF 2021). Given the large amount of funding that the United States has traditionally provided to global health programs, this vacillation in public health policy has been especially damaging, with many organizations seeing their funding repeatedly cut and reinstated at short notice, even if abortion-related activities are only a tiny portion of the services they provide.

The Cairo conference represented a particularly significant transformation in global population policy. The Catholic Church joined forces with several conservative Muslim organizations to express their opposition to traditional population programs, which they saw as representing an inexorable rise of secularism, liberalism, and feminism. Goldberg (2009) reflected that, in many ways, the Cairo conference saw delegates attempting to renegotiate traditional norms in a newly globalized world, with women's role in this new world order under particular scrutiny. Feminist lobbyists had recognized the need to be at the negotiating table as well and demanded that women's rights not get lost in the reshuffle, as they pressed for future policies to focus on reproductive choice rather than demographic targets. In an ironic twist, some members of the feminist lobby teamed up with Catholic and Muslim delegates to attempt to turn public opinion away from traditional population policies focused on demographic targets. Despite the very different perspectives of the feminist and religious conservative lobbies, the Cairo conference has been interpreted as a success for them because traditional population programs were increasingly rejected. The final document that emerged from Cairo was a mixed bag of progressive and more conservative viewpoints. It required that population programs turn away from demographic targets but also affirmed the need for greater empowerment of women and noted that adolescents should have access to sex education and reproductive health services (Goldberg 2009). In the end, the "UN documents spoke both of ensuring equal rights for men and women and of respecting traditional cultures, never acknowledging that these two goals could be in direct conflict" (Goldberg 2009, 120).

Overall, global reproductive policy has therefore seen a significant shift over the past 50 years. In the early years, fertility scholars debated the most effective policies and best technologies to apply to achieve demographic targets, specifically lowering birth rates. Many of these early, supposedly value-free programs have since been criticized for being overly concerned with demographic targets at the expense of individual rights. We turn to the case of China in the next section as an example of just such a demographically driven population program. The UN's Declaration of Human Rights in the 1960s marked an initial turning point as the right to make fertility choices was declared a human right. It took many more years before that idea was resolutely put into practice, however, with the Cairo conference seen as a pivotal point where traditional policies of "population control" were finally rejected and emphasis was placed more firmly on policies encouraging reproductive empowerment through family planning. These newer programs typically include the following aspects: (1) access to a variety of contraceptive options (including both permanent and temporary contraception), (2) a focus on adolescent health and the need for high-quality sex education, and (3) gender-based policies that recognize that women often bear the brunt of fertility decisions but may be disempowered to make their own fertility choices. These policies are liberal, progressive, and feminist in nature and have generated a backlash from some religious and conservative communities. This is where we stand today, with controversy over current reproductive policy drawn along lines between liberals and conservatives, rather than the rich world and the poor world as in the past.

China's population policies

China has enacted some of the strictest population policies in the world, giving its story particular significance to discussions of population control. As the most populous nation in the world, China's policies are also far-reaching, influencing the fertility decisions of almost one-fifth of the global population.

Ironically, China began its fertility policies with pro-natal approaches. When the Communist party came to power in 1949, they began their rule with the assumption that poverty was a result of maldistribution of resources rather than overall population size and that people were a great source of power to a country. As such, they established policies outlawing abortion and contraception and offering financial incentives for large families (Xie 2000). By the mid-1950s, this approach had changed, however, as rapid population growth began to be seen as a potential barrier to China's economic development. The challenges of rapid population growth were further reinforced by famines that occurred during China's "Great Leap Forward" from 1958 to 1961. Although these famines were arguably caused by

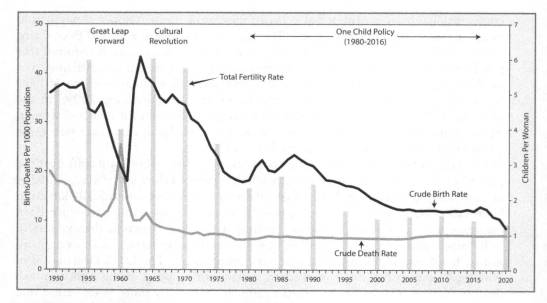

FIGURE 7.5 Fertility and mortality rates in China, 1949–2020
Data source: National Bureau of Statistics of China (n.d.).

problematic social and economic policies, it is hard to avoid the conclusion that a rapidly growing population is also a contributor to food shortages. As such, family planning and abortion were once again made easily available to Chinese couples. Figure 7.5 illustrates how birth and death rates have fluctuated since the beginning of China's Communist regime.

It was not long before explicit fertility limitation was added to Chinese policies. China's third family planning campaign of 1971 became the model for China's new approach. This campaign focused on late marriage and childbearing, birth spacing, and fertility limitation. Couples were encouraged to delay marriage until their early 20s in rural areas and late 20s in urban areas, with couples allowed just two children in cities and three children in the countryside; births were to be spaced by at least 3 years (Attané 2002). By 1980, the government had tightened restrictions to just one child per couple in recognition that fertility targets were not being met, marking the beginning of China's "one-child policy."

China's efficient public health system ensured good access to contraceptives and abortions for women throughout this period. China's strong communalist (prioritizing the group over individuals) culture is a further key to how the Chinese government encouraged its citizens to comply with these restrictions, with low fertility promoted as patriotic. Nonetheless, more draconian measures were also used to ensure compliance. Penalties for "excess" births sometimes equated to a 5 to 10 percent reduction of a family's total income for up to 16 years, plus parents were expected to pay all medical and educational expenses for "excess children." By contrast, "only children" received considerable government assistance, including priority access to nurseries, schools, clinics, jobs, and even extra grain rations. Policies were enforced through community members and family planning workers being encouraged to report possible infractions (Hartmann 1987). The Chinese government maintains that the policies themselves were not coercive, but lower-ranking officials who were expected to meet birth quotas in their local area are widely believed to have coerced women in a number of ways—some critics suggest that this local-scale coercion is exactly how the policies were designed to work. Reports suggest that women with unsanctioned pregnancies may have been threatened with dismissal from work, loss of property, and even sometimes forced sterilizations and abortions. In one extreme case from Guangdong in 1981, pregnant women without birth permits were reported to have been sent for forced abortions in handcuffs, and peasant farmers who had violated birth restrictions were denied water and electricity supplies (Scharping 2003). It is hard to know how frequently extreme penalties like this occurred, but the policies have been widely criticized for contravening Chinese couples' human right to determine the number and spacing of their children, as well as feminist concerns over how women are treated with respect to factors such as control over their own reproductive health and inviolability of the body (Scharping 2003).

Popular resistance to the policies within China mounted over time. One study suggests that fertility in 1988 may have been at least 30 percent higher than authorized fertility across a significant number of Chinese provinces, particularly in rural areas. Affluent families bribed local officials, and others reported siblings born up to several years apart as twins to avoid sanctions (Huang, Lei, and Zhao 2016). More recently, fertility treatments were used to try to initiate multiple births (twins and triplets), which were not penalized.

Popular resistance forced a mild relaxation of policies in 1984. The new policies allowed rural couples to have a second child if the first was a girl (in some places, all rural couples). Restrictions were also relaxed for ethnic minorities, who were allowed two or more births per couple (Yi 2007). Although regulations varied from province to province—with more densely populated eastern provinces typically maintaining stricter requirements—overall this relaxation of the "one-child" policy allowed 1.47 children per couple (Baochang et al. 2007). In 2013, the policy was again relaxed, allowing couples to have two children, provided that both parents were only children (UNDESA 2015a). Despite an enthusiastic response to the removal of restrictions, 6 months after the relaxation just 2.5 percent of the 11 million couples eligible to have a second birth had applied for permission (UNDESA 2015a). In response to government concerns about the rapid aging of China's population, in 2016, all couples were allowed to have two children (Feng, Baochang, and Cai 2016). Finally, in May 2021, the Chinese government announced that all couples could have three children in response to census data indicating a sharp decline in birth rates and unexpectedly few couples having two children in response to the earlier relaxation of fertility policy (Stanway and Munroe 2021).

The success of these policies is still widely debated. The policies were successful in terms of reducing the birth rate; some estimates suggest the avoidance of 250 million births by the year 2000 (Xie 2000), equivalent to the population of a large country. Indeed, the Chinese government has often emphasized the scale of the problem it was facing in justifying the strictness of its policies. In the early 2000s, China's population was still growing by nearly 15 million people per year and officials estimated that this rate could have been twice that figure in the absence of their strict population policies (Scharping 2003). However, the program has also had several unintended consequences. China is now one of the world's most rapidly aging populations, with almost 14 percent of China's population over 65 years in

2010, forecast to rise to 25 percent by 2030 (UNDESA 2015a). The rapidity of this change poses particular challenges through generating a 4–2–1 constellation of four grandparents to two parents to one child, posing significant burdens of care on individuals at the bottom of that configuration. A further consequence has been a distorted sex ratio, with approximately 22 million more boys than girls born in China between 1980 and 2000, probably owing mostly to sex-selective abortions. The Chinese government has responded by attempting to raise the status of girls in society, along with a ban on sex-selective abortions in 1989 and a ban on ultrasounds to determine the sex of the fetus in 2002 (Ebenstein and Sharygin 2009). These measures, along with rapid socio-economic development in the intervening years, appear to have been somewhat successful, with China's sex ratio dropping from a high of around 120 boys to 100 girls in 2008 to 110 to 100 in 2019 (Tang 2021).

Many commentators point to the rapid fertility decline seen in other countries of East Asia and suggest that China may have seen a significant contraction in fertility over this period even without such strict policies (figure 7.6). Indeed, most of China's fertility decline occurred in the 1970s, in the early years of China's fertility limitation policies but before the most severe "one-child" policy was instituted, dropping from 5.8 births per woman in 1970 to 2.7 in 1979 (UNDESA 2015a). Whether fertility will now rebound is open to question, but many scholars have expressed doubts that fertility will increase significantly in upcoming years, owing to increasing urbanization, rising education levels, growing consumer aspirations, challenges related to the cost of living, and persistent inequalities in domestic

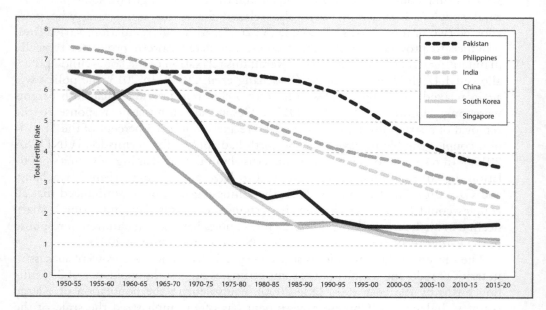

FIGURE 7.6 Comparison of fertility decline in several Asian countries, 1950–2020
China's fertility decline closely mirrors that of South Korea and Singapore, even though China was the only one of these nations to institute strict population policies. As such, some commentators have suggested that China may have seen a significant decline in its fertility rate even in the absence of its population policies. On the other hand, much slower fertility declines have been seen in other Asian countries such as the Philippines and India, which provide an alternative model for what might have happened in China in the absence of its population policies.
Data source: UNDESA (2019).

gender roles. Years of fertility limitation have now apparently normalized small family size, with rural couples increasingly mirroring urban low-fertility patterns (Guo, Gietel-Basten, and Gu 2019). In one recent fertility survey from 2017, women expressed a desire for, on average, 1.96 children, although they reported intentions to actually have just 1.76 children, indicating that women still perceive constraints on their fertility (Zhuang, Jiang, and Li 2020). Results from China's most recent census, completed in November 2020, were delayed, with leaks suggesting that this may have been because fertility decline had been more dramatic than expected. Chinese planners had been working on the assumption of a birth rate of 1.8, but some Chinese scholars as well as the World Bank suggest that the rate is more likely between 1.6 and 1.7, with some sources suggesting as low as 1.5 babies per woman (*The Economist* 2021). China is now moving toward family-friendly policies to try to stimulate the birth rate, including proposals for maternity and paternity leave, child support payments, and tax allowances for children (Gietel-Basten 2016). China has shown little enthusiasm for encouraging immigration, leading some commentators to suggest that China's demographic decline could lead it to slip in its competition for global power with the United States, which is projected to continue to grow over the next few years through immigration (*The Economist* 2021).

Conclusion

Governments have long tried to influence the size of their populations through policy, generating explicitly political approaches to population issues, as well as corresponding social justice concerns. Population policies of the modern era have historically been divided between the rich world wanting to limit fertility for reasons related to resource scarcity and global political stability and the poor world seeking to avoid unnecessary meddling in their demographic futures. More recently, however, the lines of engagement have increasingly shifted to disagreements between liberal, secular, and feminist scholars/policymakers seeking reproductive freedom and religious and social conservatives promoting a return to family values.

Fertility issues therefore continue to sit precariously on a moral knife-edge. With tight connections to religious and moral belief systems, fertility is inevitably contentious when considering issues such as abortion and family planning. At the same time, the significance of fertility decisions to the economy is likely to keep fertility at the top of the political agenda. Whether governments have the right to try to influence their citizens' fertility decisions remains contested, and yet governments have a strong incentive to consider it within their domain of responsibility, particularly as issues of resource scarcity and aging populations leave governments with conflicting goals with respect to population.

Discussion questions

1   Discuss how contraception can be used in both empowering and disempowering ways.

2   Should governments try to manipulate the size of their population, or should fertility decisions be totally within the control of individuals?

3   What are some of the different lenses through which different groups view fertility issues? How might we achieve consensus on controversial population issues, given these very different lenses through which different groups see the issues at hand?

4   Why do you think so many Chinese chose not to have an additional child once restrictions were lifted? What might be some of the barriers that Chinese couples face beyond government policies?

Suggested readings

Dreweke, J. 2019. "Promiscuity Propaganda: Access to Information and Services Does Not Lead to Increases in Sexual Activity." *Guttmacher Policy Review* 22: 29–36. https://www. guttmacher.org/sites/default/files/article_files/gpr2202919.pdf

Gietel-Basten, S. 2016. "How China Is Rolling Out the Red Carpet for Couples Who Have More than One Child." *The Conversation*, April 7, 2016. https://theconversation.com/ how-china-is-rolling-out-the-red-carpet-for-couples-who-have-more-than-one-child-57299.

King, L. 2002. "Demographic Trends, Pronatalism, and Nationalist Ideologies in the Late Twentieth Century." *Ethnic and Racial Studies* 25 (3): 367–89.

Sedgh, G., S. Singh, and R. Hussain. 2014. "Intended and Unintended Pregnancies Worldwide in 2012 and Recent Trends." *Studies in Family Planning* 45 (3): 301–14. https://onlinelibrary. wiley.com/doi/epdf/10.1111/j.1728-4465.2014.00393.x

Glossary

abortion: termination of a pregnancy

contraception: technologies used to prevent conception

demographic dividend: the temporary economic boost an economy receives when a population has a large proportion of workers owing to recent declines in the birth rate

eugenics: the premise of trying to improve a human population through selective breeding; the idea is now widely discredited owing to its discriminatory implications

family planning: the practice of controlling family size and spacing, especially using contraception

genocide: targeted killing of a particular group of people; the UN definition also includes deliberate attempts to prevent births in a particular group of people

pro-natal policies: policies designed to increase the size of a population

reproductive freedom: empowering men and women to make informed choices related to their own fertility

Works cited

AP [Associated Press]. 2010. "Iran's Leader Introduces Plan to Encourage Population Growth by Paying Families." *New York Times*, July 27, 2010. https://www.nytimes.com/2010/ 07/28/world/middleeast/28iran.html

Atighetchi, D. 1994. "The Position of Islamic Tradition on Contraception." *Medical Law* 13 (7–8): 717–25.

Attané, I. 2002. "China's Family Planning Policy: An Overview of Its Past and Future." *Studies in Family Planning* 33 (1): 103–13.

Bachman, R. 1989. *Romania: A Country Study*. Washington, DC: GPO for the Library of Congress. http://countrystudies.us/romania/37.htm

Baochang, G., W. Feng, G. Zhigang, and Z. Erli. 2007. "China's Local and National Fertility Policies at the End of the Twentieth Century." *Population and Development Review* 33 (1): 129–48. doi:10.1111/j.1728-4457.2007.00161.x

BBC News. 2021. "Poland Enforces Controversial Near-Total Abortion Ban." BBC, January 31, 2021. https://www.bbc.com/news/world-europe-55838210

Boring, C., R. Rochat, and J. Becerra. 1988. "Sterilization Regret among Puerto Rican Women." *Fertility and Sterility* 49 (6): 973–81.

Caetano, A., and J. Potter. 2004. "Politics and Female Sterilization in Northeast Brazil." *Population and Development Review* 30 (1): 79–108. https://www.jstor.org/stable/3401499

Cincotta, R., and K. Sadjadpour. 2017. "Iran in Transition: The Implications of the Islamic Republic's Changing Demographics." Carnegie Endowment for International Peace, December 18, 2017. https://carnegieendowment.org/2017/12/18/iran-in-transition-implications-of-islamic-republic-s-changing-demographics-pub-75042

Dalsgaard, A. 2004. *Matters of Life and Longing: Female Sterilization in Northeast Brazil.* Copenhagen: Museum Tusculanum Press and University of Copenhagen.

Dreweke, J. 2019. "Promiscuity Propaganda: Access to Information and Services Does Not Lead to Increases in Sexual Activity." *Guttmacher Policy Review* 22: 29–36. https://www.guttmacher.org/sites/default/files/article_files/gpr2202919.pdf

Ebenstein, A., and E. Sharygin. 2009. "The Consequences of the 'Missing Girls' of China." *The World Bank Economic Review* 23 (3): 399–425. doi:10.1093/wber/lhp012

The Economist. 2021. "Is China's Population Shrinking?" May 1, 2021. https://www.economist.com/china/2021/04/29/is-chinas-population-shrinking

Etehad, M., and R. Mostaghim. 2019. "Iran Arrests Demographers, the Latest Targets amid an Escalative Crackdown on Academics and Activists." *Los Angeles Times*, January 7, 2019. https://www.latimes.com/world/la-fg-iran-demographer-arrests-20190107-story.html

Feng, W., G. Baochang, and Y. Cai. 2016. "The End of China's One-Child Policy." *Studies in Family Planning* 47 (1): 83–6. https://doi.org/10.1111/j.1728-4465.2016.00052.x

Gaines, S. 2006. "The Mission Continues." *The Guardian*, December 27, 2006. https://www.theguardian.com/society/2006/dec/27/voluntarysector

Gietel-Basten, S. 2016. "How China Is Rolling Out the Red Carpet for Couples Who Have More than One Child." The Conversation, April 7, 2016. https://theconversation.com/how-china-is-rolling-out-the-red-carpet-for-couples-who-have-more-than-one-child-57299

Goldberg, M. 2009. "Cairo and Beijing." In *The Means of Reproduction*, 103–20. New York: Penguin.

Guo, Z., S. Gietel-Basten, and B. Gu. 2019. "The Lowest Fertility Rates in the World? Evidence from the 2015 Chinese 1% Sample Census." *China Population and Development Studies* 2: 245–58. https://link.springer.com/article/10.1007/s42379-018-0012-1#ref-CR14

Hartmann, B. 1987. *Reproductive Rights and Wrongs: The Global Politics of Population Control and Contraceptive Choice*. New York: HarperCollins.

Heng, G., and J. Devan. 1992. "State Fatherhood: The Politics of Nationalism, Sexuality, and Race in Singapore." In *Nationalisms and Sexualities*, edited by A. Parker, M. Russo, D. Sommer, and P. Yaeger, 343–64. New York: Routledge.

Hoodfar, H., and S. Assadpour. 2000. "The Politics of Population Policy in the Islamic Republic of Iran." *Studies in Family Planning* 31 (1): 19–34. https://www.jstor.org/stable/172166

Huang, W., X. Lei, and Y. Zhao. 2016. "One-Child Policy and the Rise of Man-Made Twins." *The Review of Economics and Statistics* 98 (3): 467–76.

KFF [Kaiser Family Foundation]. 2021. "The Mexico City Policy: An Explainer." Global Health Policy, January 28, 2021. https://www.kff.org/global-health-policy/fact-sheet/mexico-city-policy-explainer/

King, L. 1998. "'France Needs Children': Pronatalism, Nationalism and Women's Equity." *The Sociological Quarterly* 39 (1): 33–52.

———. 2002. "Demographic Trends, Pronatalism, and Nationalist Ideologies in the Late Twentieth Century." *Ethnic and Racial Studies* 25 (3): 367–89.

Kivu, M. 1993. "Une Rétrospective: La Politique Démographique en Roumanie 1945–1989." In *Annales de Démographie Historique* 1993: 107–26. http://www.jstor.org/stable/44385618

Kligman, G. 1992. "The Politics of Reproduction in Ceaușescu's Romania: A Case Study in Political Culture." *East European Politics and Societies* 6 (3): 364–418.

———. 1998. *The Politics of Duplicity: Controlling Reproduction in Ceaușescu's Romania.* Berkeley, University of California Press.

Leidig, M. 2005. "Romania Still Faces High Abortion Rate 16 Years after Fall of Ceaușescu." *British Medical Journal* 331 (7524): 1043. doi:10.1136/bmj.331.7524.1043-a

Logan, C., E. Holcombe, J. Manlove, and S. Ryan. 2007. *The Consequences of Unintended Childbearing: A White Paper.* Washington, DC: Child Trends.

Macrotrends. n.d. "Iran Fertility Rate 1950–2022." Accessed February 6, 2022. https://www.macrotrends.net/countries/IRN/iran/fertility-rate

McQueen, P. 2017. "Autonomy, Age, and Sterilization Requests." *Journal of Medical Ethics* 43: 310–13. http://dx.doi.org/10.1136/medethics-2016-103551

National Bureau of Statistics of China. n.d. "China Statistical Yearbook." Accessed March 23, 2020. http://www.stats.gov.cn/english/Statisticaldata/AnnualData/

National Institute of Statistics, Romania. 2021. "Romanian Statistical Yearbook 2019." https://insse.ro/cms/en/content/statistical-yearbooks-romania

NHS [National Health Service, UK]. 2018. "If I Use a Sexual Health Service Will They Tell My Parents?" https://www.nhs.uk/live-well/sexual-health/confidentiality-at-sexual-health-services/

———. 2022. "The Different Types of Contraception." NHS Inform. https://www.nhsinform.scot/healthy-living/contraception/getting-started/the-different-types-of-contraception

Paddock, R., and D. Sijabat. 2020. "Indonesia's New Coronavirus Concern: A Pandemic Baby Boom." *New York Times*, October 2, 2020. https://www.nytimes.com/2020/06/10/world/asia/indonesia-coronvirus-baby-boom.html

Perelli-Harris, B., and O. Isupova. 2013. "Crisis and Control: Russia's Dramatic Fertility Decline and Efforts to Increase It." In *Fertility Rates and Population Decline: No Time for Children*, edited by A. Buchanan and A. Rotkirch, 141–56. Basingstoke: Palgrave Macmillan.

Planned Parenthood. 2021. "Birth Control." https://www.plannedparenthood.org/learn/birth-control

Popov, A., and H. David. 1999. "Russian Federation and USSR Successor States." In *From Abortion to Contraception*, edited by H. David, 223–78. Westport, CT: Greenwood Press.

PRB [Population Reference Bureau]. 2020. *World Population Datasheet 2020.* Washington, DC: PRB. https://www.prb.org/wp-content/uploads/2020/07/letter-booklet-2020-world-population.pdf

Riddle, J., and J. Estes. 1992. "Contraceptives in Ancient and Medieval Times." *American Scientist*, May–June 1992. https://www.jstor.org/stable/29774642?seq=2#metadata_info_tab_contents

Riddle, J., J. Estes, and J. Russell. 1994. "Ever Since Eve …: Birth Control in the Ancient World." *Archaeology* March–April: 29–34.

Roberts, D. 1997. *Killing the Black Body.* New York: Pantheon Books.

Rogers, C. 2009. "What Became of Romania's Neglected Orphans?" BBC News, December 22, 2009. http://news.bbc.co.uk/2/hi/europe/8425001.stm

Roudi, F. 2012. "Iran Is Reversing Its Population Policy." *Viewpoints*, no. 7. Woodrow Wilson International Center for Scholars. https://www.wilsoncenter.org/sites/default/files/media/documents/publication/iran_is_reversing_its_population_policy.pdf

Roudi-Fahimi, F. 2002. "Iran's Family Planning Program: Responding to a Nation's Needs." Population Reference Bureau. http://www.igwg.org/pdf/IransFamPlanProg_Eng.pdf

Safi, M. 2020. "Iran Ends Provision by State of Contraceptives and Vasectomies." *The Guardian*, June 15, 2020. https://www.theguardian.com/world/2020/jun/15/iran-bans-vasectomies-and-contraceptives-to-improve-birth-rate

Scharping, T. 2003. *Birth Control in China 1949–2000: Population Policy and Demographic Development*. New York: Routledge.

Sedgh, G., S. Singh, and R. Hussain. 2014. "Intended and Unintended Pregnancies Worldwide in 2012 and Recent Trends." *Studies in Family Planning* 45 (3): 301–14. https://onlinelibrary.wiley.com/doi/epdf/10.1111/j.1728-4465.2014.00393.x

Stanway, D., and T. Munroe. 2021. "Three-Child Policy: China Lifts Cap on Births in Major Policy Shift." Reuters, June 1, 2021. https://www.reuters.com/world/china/china-says-each-couple-can-have-three-children-change-policy-2021-05-31/

Tang, M. 2021. "Addressing Skewed Sex Ratio at Birth in China: Practices and Challenges." *China Population and Development Studies* 4: 319–26. https://doi.org/10.1007/s42379-020-00075-1

UN [United Nations]. 1968. Resolution XVIII: Human Rights Aspects of Family Planning, Final Act of the International Conference on Human Rights. U.N. Doc. A/CONF. 32/41, p.4. https://undocs.org/pdf?symbol=en/A/CONF.32/41

———. n.d.a. "Conferences: Population." Accessed March 5, 2022. https://www.un.org/en/conferences/population

———. n.d.b. "Genocide." Accessed June 12, 2021. https://www.un.org/en/genocideprevention/genocide.shtml

———. n.d.c. "Outcomes on Population." Accessed June 26, 2021. https://www.un.org/en/development/devagenda/population.shtml

UNDESA [United Nations Department of Economic and Social Affairs]. 2015a. "Below-Replacement Fertility in China: Policy Response Is Long Overdue." East-West Center, Policy Brief, no. 5. https://www.un.org/en/development/desa/population/events/pdf/expert/24/Policy_Briefs/PB_China.pdf

———. 2015b. "World Population Policies 2015: Fertility, Family Planning, and Reproductive Health." https://www.un.org/development/desa/pd/data/world-population-policies

———. 2017a. "Government Policies to Raise or Lower the Fertility Level." Population Facts, December 2017, No. 2017/10. https://www.un.org/en/development/desa/population/publications/pdf/popfacts/PopFacts_2017-10.pdf

———. 2017b. "World Population Policies 2017: Abortion Laws and Policies." https://www.un.org/en/development/desa/population/theme/policy/wpp2017.asp

———. 2019. "World Population Prospects 2019." https://population.un.org/wpp/

UNESCO [United Nations Educational, Scientific and Cultural Organization]. 2017. "Early and Unintended Pregnancy: Recommendations for the Education Sector." Document ED-2017/WS/27. https://unesdoc.unesco.org/ark:/48223/pf0000248418

UNFPA [United Nations Population Fund]. 2018. "About Us." https://www.unfpa.org/about-us

Vdovîi, L., and M. Bird. 2019. "Over 30 Percent of Hospitals in Romania Are Refusing Legal Abortions." The Black Sea, November 29, 2019. https://theblacksea.eu/stories/quarter-hospitals-romania-are-refusing-legal-abortions/

WHO [World Health Organization]. 2014. "Adolescent Pregnancy." https://apps.who.int/iris/bitstream/handle/10665/112320/WHO_RHR_14.08_eng.pdf

Xie, Z. 2000. "Population Policy and the Family Planning Programme." In *The Changing Population of China*, edited by X. Peng and Z. Guo, 51–63, Oxford, UK: Blackwell.

Yi, Z. 2007. "Options for Fertility Policy Transition in China." *Population and Development Review* 33 (2): 215–46. doi:10.1111/j.1728-4457.2007.00168.x

Zhuang, Y., Y. Jiang, and B. Li. 2020. "Fertility Intention and Related Factors in China: Findings from the 2017 National Fertility Survey." *China Population and Development Studies* 4: 114–26. https://doi.org/10.1007/s42379-020-00053-7

8

Mortality and morbidity

After reading this chapter, a student should be able to:

1 understand the use, importance, and limitations of major health measures such as life expectancy and mortality rates;

2 explain how patterns in health have changed over time, particularly with respect to the epidemiologic transition.

Everyone dies eventually. At what age and with how many years of ill health differ significantly from person to person and community to community, however. Whereas doctors treat individual patients, other health specialists focus on population-scale patterns in death (**mortality**) and ill health (**morbidity**). The crucial role of environmental factors in influencing population-scale patterns in health makes it a key area of geographic study, with an entire sub-field—geography of health—devoted to exploring the influence of physical and social environments on health. The significance of the natural and built environments to health means that sustainability issues are interwoven with health in fundamental ways—pollution, landscape change, and interactions with animals all have potentially profound influences on health, as well as being key environmental issues. With respect to the social environment, health has important points of intersection with factors such as race, gender, and socio-economic status, raising critical social justice questions. As such, health topics provide a powerful link between populations, sustainability, and social justice. In this chapter, we think about key measures of population health and how we can model and analyze health patterns.

PART I: HEALTH INDICATORS

Health indicators like mortality rates and life expectancy summarize health outcomes across entire populations. Many health indicators are closely correlated with other population characteristics such as socio-economic status. A familiarity with

DOI: 10.4324/9781003143253-8

key health indicators is critical for understanding health patterns at the scale of populations, so we begin by considering some of these important measures.

Life expectancy

Life expectancy is defined as the average number of years that an individual from a particular population is expected to live. This is usually calculated as the life expectancy of a baby born in a particular year, assuming that conditions do not change over their lifetime. In reality, we know that life expectancy *is* likely to change as a child grows up—indeed, life expectancy has been increasing in most countries for many years. In the UK, for instance, a baby born in the 1840s had a life expectancy of about 40 years; by 2016, this had doubled to nearly 80 (Ruggeri 2018). Geographers are particularly interested in spatial variations in life expectancy, which can reveal stark inequalities. In 2020, average global life expectancy was 73 years, but in many parts of the high-income world life expectancies are now in the 80s, whereas life expectancy remains in the 50s in the poorest countries of Africa where inadequate sanitation and limited diet still pose significant challenges (PRB 2020; figure 8.1). Furthermore, even in countries where people generally enjoy good health, there are always sub-populations with poorer health statistics. If we consider Australia, for instance, indigenous Australians had life expectancies about 8 years shorter than those of non-indigenous populations in 2015–17 (AIHW 2021).

We expect women to live a few years longer than men owing to physiological differences, and so life expectancies may be further broken down by sex. Globally,

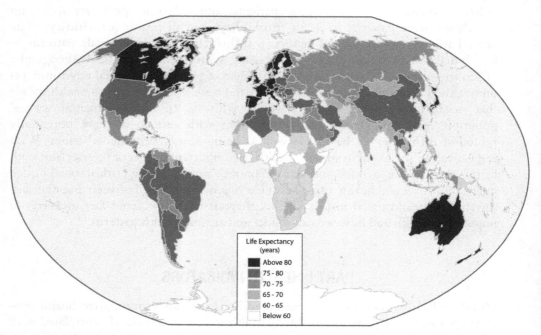

FIGURE 8.1 Life expectancy by country, 2015–20
Data source: UNDESA (2019).

males currently have a life expectancy of 70 and females have a life expectancy of 75 years (PRB 2020). It is important to note that variations in life expectancy between men and women are related to not only physiological differences between the sexes but also gendered differences in lifestyles and how men and women are treated in different societies. Traditionally, many communities have had a shorter female life expectancy than male life expectancy, associated with high rates of maternal mortality that overwhelmed women's physiological advantages in old age. Additionally, we must look to **patriarchal** structures of society that leave women with poorer access to healthcare and nutrition, as well as disadvantaged in society more broadly. Many patriarchal practices still compromise women's health, ranging from the physical burden of strenuous activities that many women and girls are asked to perform such as collecting firewood and water, to girls being less likely to be taken to a doctor or offered nutritious foods than their brothers. Nonetheless, in 2020, life expectancy was greater for women than for men in all countries, reflecting improvements in gender equity in many countries (figure 8.2). Indeed, we must now consider whether we may be seeing a crisis in men's health in some places, with several countries, including Russia, Kazakhstan, Latvia, Belarus, Syria, and Myanmar, now reporting an advantage in female life expectancy of almost a decade. This forces us to consider what has happened to compromise men's health in these communities. In Russia and some of the ex-Soviet republics, challenging economic circumstances appear to be taking a particular toll on men's health, with alcoholism, suicide, and heart disease (associated with stress and poor diet) causing

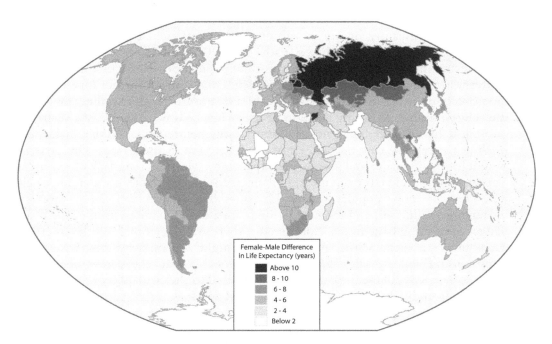

FIGURE 8.2 Female–male disparities in life expectancy by country, 2015–20
The map shows the number of extra years women are expected to live beyond men by country, with female life expectancy exceeding male in every country in 2015–20.
Data source: UNDESA (2019).

early deaths for many men. In other cases, such as Myanmar and Syria, conflict may be lowering male life expectancies.

We must also recognize that life expectancy is an average and that relatively few people die at the average age predicted for a particular population. For populations with high infant mortality, life expectancy can be particularly misleading because the large number of infant deaths pulls down the average age of death, even though many people may live to old age if they survive childhood. Pliny, a famous author from Ancient Rome, wrote about a whole list of illustrious Roman citizens who lived well into their 80s and even to over 100, and yet average life expectancy was perhaps only 30 to 35 years at the time owing to high child mortality. As such, life expectancy "hasn't increased so much because we're living far longer than we used to as a species. It's increased because more of us, as individuals, are making it that far" (Ruggeri 2018).

Subtleties such as this become apparent when considering the data provided in **life tables** (table 8.1). Life tables represent statistical calculations of the likelihood of an individual of a particular age dying before their next birthday. The idea of generating detailed mortality calculations in this way dates back to the 1600s when a man named John Graunt began tabulating survivorship from one age group to the next in the population of London. Today, life tables are used by demographers in statistical studies of population and by actuaries whose job it is to calculate insurance risks and premiums. From a life table, we could see, for instance, that a person who makes it to the age of 15 in a population with high child mortality might then have good odds of leading a relatively long life—a fact that could easily be obscured by looking only at life expectancy in that same community.

Cases such as these emphasize another important aspect of life expectancy—it can differ significantly among sub-groups within a particular population. In such cases, breaking down mortality statistics by sub-population—using factors like socio-economic status, race, or urban vs rural populations—may reveal important patterns as well as highlight social justice issues that warrant further investigation. Socio-economic status is always a key consideration in health studies, owing to the huge number of health-related factors influenced by affluence. As such, we must usually account for socio-economic status before other subtler influences such as patterns by race become apparent. There are many reasons why poverty might lead to ill health and shorter life expectancy. Most obvious, lack of resources can prevent individuals from meeting basic needs like food, clean water, and shelter. For very poor populations, health issues like vitamin deficiency diseases, water-borne illnesses, and infected wounds are a direct manifestation of the challenges of meeting basic needs. Additionally, there are many other, subtler intersections between social status and health that might explain why class has such a profound influence on health, even in the Global North where most people can meet basic needs. These disparities may be related to the availability of healthcare, how likely individuals are to utilize medical services, or disparities in who is exposed to risk factors for disease or subjected to dangerous work conditions (Höpflinger 2012). A growing body of research suggests that even factors such as discrimination and prejudice can have profound negative impacts on health over the long term, potentially leading to weaker immune systems and premature aging through mechanisms such as higher levels of stress hormones circulating in the body (box 8.1).

TABLE 8.1 Abridged life table for the United States, 2018

Age (years)	Probability of dying between ages x and x+n	Number surviving to age x	Number dying between ages x and x+n	Person-years lived between ages x and x+n	Total number of person-years lived above age x	Expectation of life at age x
	n qx	lx	n dx	n Lx	Tx	ex
0–1	0.005650	100,000	565	99,505	7,873,749	78.7
1–5	0.000965	99,435	96	397,512	7,774,244	78.2
5–10	0.000577	99,339	57	496,540	7,376,732	74.3
10–15	0.000745	99,282	74	496,262	6,880,192	69.3
15–20	0.002454	99,208	243	495,513	6,383,930	64.3
20–25	0.004500	98,964	445	493,776	5,888,417	59.5
25–30	0.005810	98,519	572	491,206	5,394,642	54.8
30–35	0.007070	97,946	692	488,056	4,903,436	50.1
35–40	0.008693	97,254	845	484,214	4,415,380	45.4
40–45	0.010792	96,408	1,040	479,565	3,931,166	40.8
45–50	0.015496	95,368	1,478	473,369	3,451,601	36.2
50–55	0.023617	93,890	2,217	464,296	2,978,232	31.7
55–60	0.036021	91,673	3,302	450,565	2,513,936	27.4
60–65	0.051414	88,371	4,543	431,009	2,063,371	23.3
65–70	0.071504	83,827	5,994	404,854	1,632,362	19.5
70–75	0.105404	77,833	8,204	369,767	1,227,508	15.8
75–80	0.164987	69,629	11,488	320,917	857,740	12.3
80–85	0.263857	58,141	15,341	253,950	536,823	9.2
85–90	0.417699	42,800	17,878	169,639	282,873	6.6
90–95	0.617625	24,923	15,393	83,692	113,234	4.5
95–100	0.798470	9,530	7,609	25,355	29,542	3.1
100+	1.000000	1,921	1,921	4,187	4,187	2.2

Life tables are designed to show the statistical likelihood of survival at particular ages. Here the second column shows the probability of dying between ages x and x+1. The final column shows how many additional years an average person of that age would be expected to live.

Source: Arias and Xu (2020).

Mortality

Mortality rates (death rates) provide another key measure of population health. A mortality rate represents the number of people dying from a particular condition, or from all causes, relative to overall population size. In 2020, the global death rate was seven deaths per 1,000 people (PRB 2020). To reiterate a very important point, remember that a rate controls for differences in population size and so it is possible to

BOX 8.1 HEALTH AND INEQUALITY

A social justice framework asks us to explore why health statistics often show clear inequalities by factors such as race and socio-economic status. As already noted, lack of resources provides obvious connections between poverty and poor health, which also has implications for sub-populations that are poorer on average such as certain racial minority groups. However, a considerable body of research suggests that the influence of social inequalities may go deeper than this, particularly given evidence that even affluent members of unequal societies often have poorer health than their counterparts in more equal societies (Pearce and Dorling 2009). Mounting evidence suggests that living in an unequal society leads to stresses that take a toll on everyone's health but pose a particular burden for those of lower socio-economic status or who suffer discrimination within that society.

A famous piece of research from the UK—the Whitehall Studies—explored health outcomes among civil servants working at different pay grades and ranks in the latter part of the 20th century (Marmot et al. 1991). This research found that mortality rates, as well as rates of a variety of chronic diseases such as diabetes and heart disease, formed a clear gradient, with those in the lowest status jobs experiencing shorter, unhealthier lives than those in mid-level positions, who in turn had worse health than those in the highest-level jobs. Men in the lowest occupational grade were found to have rates of coronary heart disease three times higher than their high-grade counterparts, for example (Marmot et al. 1991). There are several possible explanations for these findings. One hypothesis is that the poor are more likely to engage in risky behaviors such as smoking, which are then responsible for the negative health outcomes. However, this only accounted for about one-third of the health disparities identified (Marmot et al. 1997). Another possibility is that stress associated with low status may harm health through a physiological response in which the body pumps out a constant stream of stress hormones, keeping the body in a high-alert status that has detrimental health implications over the long term (Mayer and Sarin 2005). This idea has been explored particularly with respect to the stress of discrimination and marginalization associated with minority status and appears to be influential (Geronimus 2001), as we explore further in the next chapter.

fairly compare a rate from a country with a small population with one from a much more populous country. An additional factor that must be considered with respect to mortality rates, however, is that they are greatly influenced by the age structure of the population. In short, we expect higher death rates in older populations, even if those elderly populations are healthy. As such, we would expect the death rate in Nigeria to be suppressed somewhat by its young population; by contrast, a relatively high death rate in Japan could simply reflect its population being elderly rather than unhealthy. A death rate that has not been adjusted for age structure—a **crude death rate**—is thus a poor indicator of the overall "healthiness" of a particular population. Figure 8.3 helps illustrates this point. Although the countries with the highest crude death rates are among the poorest countries, such as Chad and the Central African Republic (CAR), we also see relatively high crude death rates among some affluent countries like Japan and Portugal. In Chad and the CAR, high death rates are associated with basic problems such as lack of sanitation, poor nutrition, and high rates of infectious disease. Indeed, these countries illustrate especially challenging

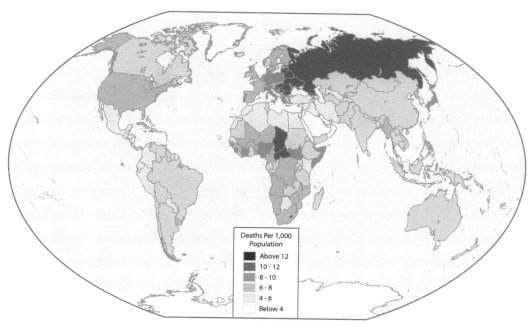

FIGURE 8.3 Crude death rates by country, 2015–20
Data source: UNDESA (2019).

situations because their young populations are suppressing their death rates and yet they still have high death rates by global standards. By contrast, Japan and Portugal have generally healthy populations; here high crude death rates reflect the advanced age of these countries' populations. Some countries with aging populations *and* persistent health challenges, such as Russia and Belarus, have especially high crude death rates. At the other end of the spectrum, low crude death rates are typically found in young populations that have made strides toward improving population health such as Peru.

To avoid the potentially confounding influence of age structure, mortality rates are usually **age-adjusted**. This is a mathematical process that calculates the death rates of different populations assuming that they all have the same age structure so that they can be fairly compared. The importance of age-adjusted rates becomes apparent if we consider a disease such as heart disease. Typically, people only begin to develop heart disease in later life, and so a young population will inevitably show low rates of heart disease. Even if a large proportion of the population are smokers who eat an unhealthy diet and get very little exercise, we would still not expect them to develop heart disease until perhaps their 50s. As such, low crude rates of heart disease could be indicative of a healthy population or a young population, making it difficult to draw useful conclusions from rates that have not been age-adjusted. Once we have age-adjusted the data, we have removed the influence of the population's age structure and so high rates of *age-adjusted* heart disease mortality would indicate a genuinely unhealthy population. We can show this using mortality rates from heart disease in the United States (figure 8.4). Crude death rates from heart disease are relatively high in parts of the Great Lakes region and Southern United States (map A), but it is impossible to tell whether these high rates

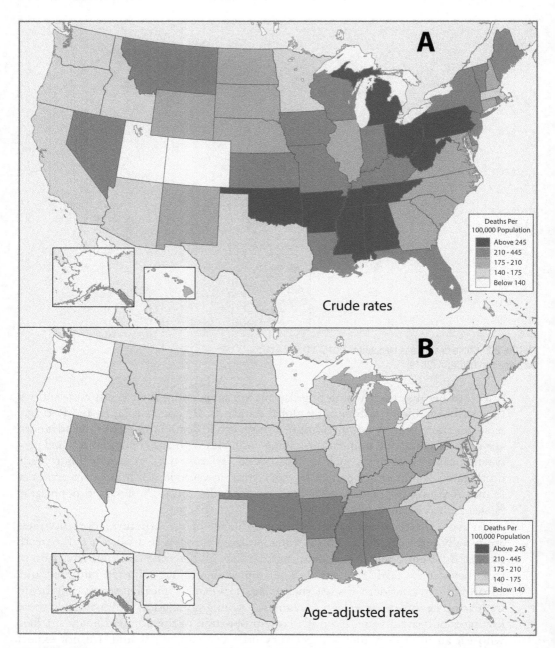

FIGURE 8.4 Comparison of crude and age-adjusted mortality rates due to heart diseases, United States, 2019
Data source: Xu et al. (2021).

are caused by older populations or unhealthy conditions. Once we age-adjust the data (map B), the influence of age structure is removed, and we get a much better indication of places with genuinely unhealthy conditions with respect to heart disease, most notably the Appalachian region and parts of the South. This region is so well known for its poor health statistics that it is sometimes called the "Stroke Belt"—poverty is thought to contribute to high rates of smoking, poor diet, and low rates of exercise in the region, explaining its poor health statistics. By contrast, in the Northeast and Florida, health looks better after age adjustment, suggesting that the older age structure of these states is at least partly responsible for their higher crude rates of heart disease. It is important to note at this point that not all causes of death affect the elderly the most. Many sexually transmitted diseases such as HIV/AIDS affect those in middle age groups most significantly, for instance, and age adjustment is equally critical in such cases.

Another option for dealing with populations with different age structures is to quote mortality rates for just a particular age group within the population. The **infant mortality rate**—the number of deaths in infants under age 1 year per 1,000 live births—is perhaps the most widely used of these statistics, but you may also see **child mortality rates** (deaths under age 5), **peri-natal mortality** (deaths in the period around birth), adolescent mortality, or mortality rates for those over 65. In such cases, we do not need to age-adjust the data because we are considering deaths only in a very specific age category, which will not be influenced by the population's overall age structure.

Infant mortality is often considered to be a particularly powerful indicator of population health. Infant survival has close connections to a wide range of human development indicators, including access to healthcare, levels of education (particularly female), quality of sanitation, and access to nutritious foods—a high rate of children dying in infancy is a clear reflection of a society failing to meet basic needs. At the global scale, infant mortality rates closely match levels of affluence. In sub-Saharan Africa, several countries have infant mortality rates over 65 babies per 1,000 live births (e.g., Sierra Leone, Chad, and Somalia; PRB 2020; figure 8.5).

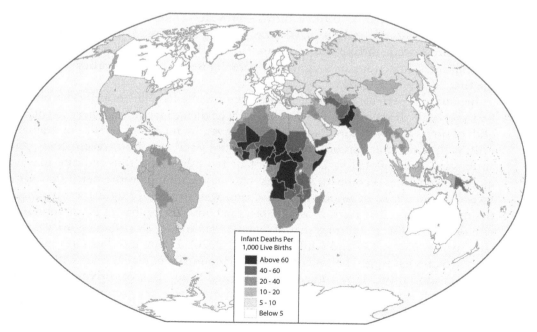

FIGURE 8.5 Infant mortality rates by country, 2015–20
Data source: UNDESA (2019).

These infant deaths are associated with factors like malnutrition, poor sanitation, and infectious disease. Many middle-income countries have made huge strides in tackling these problems with basic interventions such as improved sanitation, vaccinations, antibiotics, and pre-natal healthcare and have managed to dramatically cut their infant mortality rate. Much of Latin America now has infant mortality rates below 20 per 1,000, for instance. For high-income countries, infant mortality is typically under 5, with a few countries (e.g., Singapore, Iceland, and Sweden) achieving rates under 2 per 1,000 (PRB 2020). Even with the best healthcare, a small number of babies will die at birth or soon after from **congenital anomalies**, but most rich countries prevent most babies dying from infectious diseases or complications during childbirth through near-universal pre-natal care and skilled birth attendants. Even the risk of congenital anomalies can be reduced through limiting toxic environmental exposures and providing pre-natal care to advise on matters such as optimal diet. As a result, peri-natal mortality has dropped significantly over the past century in most affluent countries.

In poorer parts of the world, infancy remains a precarious time. In the pre-natal period, poor maternal diet, lack of pre-natal care, and environmental exposures lead to higher rates of low birth weight and premature birth, which are associated with poorer birth outcomes. Complications during the process of birth itself, such as breech presentation (where the baby is not positioned headfirst), can be life-threatening to mother and infant. These problems can be minimized if skilled birth attendants are available to identify complications early on and perform interventions—ranging from turning a breech baby in utero, to identifying high blood pressure in mothers, to offering cesarean sections—but access to pre-natal care and well-equipped birth attendants is often limited in poor communities. The peri-natal mortality rate—the number of infant deaths between 28 weeks gestation and 7 days after birth per 1,000 live births—provides a good indicator of problems associated with birth and pre-natal care. Further problems arise from infections, which often occur during the birth itself or in the first weeks of life. In poorer communities, infections are a significant cause of **neo-natal mortality**—deaths in the first 28 days after birth.

Beyond the peri-natal period, many more infants have historically died in the first year of life from infections—in especially challenging circumstances an infant's odds of survival may be less than 50–50. However, infant deaths can be cut dramatically with basic interventions such as sanitation, good nutrition, vaccinations, and antibiotics. Wide global variations in infant mortality rates thus indicate massive global inequalities—compare Nigeria's rate of 67 infant deaths per 1,000 with Sweden's mere 2. Even among rich countries, we can identify significant inequalities in infant mortality rates. Most notable, the United States has an infant mortality rate of 5.7, almost three times that of Sweden and higher than some poorer countries such as Cuba (4.0) and Chile (5.6; PRB 2020). Figure 8.6 illustrates how the United States has lagged behind other affluent countries in recent years with respect to infant mortality. Many argue that the United States' lack of **universal healthcare**, which leaves millions of citizens uninsured, is a key factor explaining the United States' higher infant mortality than comparable affluent countries. The fact that African Americans in the United States have infant mortality rates more than double those of non-Hispanic white women emphasizes how racial inequalities may also contribute to poor health indicators. African American women are more

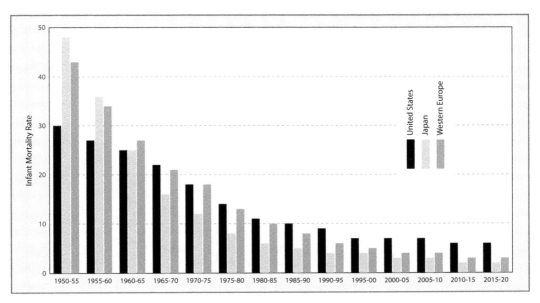

FIGURE 8.6 Infant mortality rates in the United States, Japan, and Western Europe, 1950–2020
Data source: UNDESA (2019).

than twice as likely as non-Hispanic white women in the United States to receive late or no pre-natal care, illustrating significant barriers to accessing healthcare (CDC 2020).

At the other extreme, some poorer countries have infant mortality rates almost comparable to those of affluent countries, including Cuba, China, and Costa Rica—these are countries that have prioritized universal access to **primary healthcare**, education, and social programs. If we plot national income against infant mortality rates on a scatterplot, we can begin to identify countries with better or worse health outcomes than expected for their income and think about the lessons we might learn from them. Figure 8.7 shows a negative relationship between national income (GDP) and infant mortality rates, although this relationship is non-linear. Among poor countries, we see significant declines in infant mortality rates for even modest increases in income, owing to the huge impact of basic interventions like sanitation and improved nutrition on infant mortality. At the other end of the graph, we see low infant mortality rates at a wide range of income levels in middle- and high-income countries, ranging from Costa Rica with its modest income to the United Arab Emirates with its huge wealth. We discuss how Costa Rica has achieved its very low infant mortality rate despite its limited income in box 8.2. If we imagine the best fit line that runs through the data points on this scatterplot, we can identify other **outliers** (data points that do not fit the overall pattern). Here, Equatorial Guinea, Turkmenistan, and Trinidad and Tobago stand out as having far worse infant mortality rates than would be expected for their income, and we can consider what improvements should be made in such settings. Equatorial Guinea, for instance, has unusually high wealth compared with other sub-Saharan African countries owing to oil revenues. However, this wealth remains concentrated in just a few hands and has done little to improve conditions for the

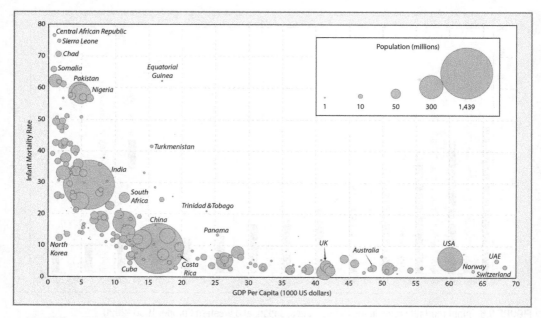

FIGURE 8.7 Infant mortality rate and gross domestic product (GDP) per capita, 2019
Data source: UNDESA (2019) and OurWorldinData (n.d.).

majority of Equatorial Guinea's citizens, many of whom continue to live in poverty, explaining why its infant mortality rate remains typical of other poor countries of sub-Saharan Africa.

Maternal mortality rates are another indicator of the overall strength of a country's social development. Maternal mortality rates show the number of women dying from complications of pregnancy or childbirth and are usually quoted as the number of women who die within 42 days of childbirth per 100,000 live births. Pregnancy and childbirth have traditionally been a vulnerable time for women, with large numbers of women throughout history dying from pre-existing conditions aggravated by pregnancy or from hemorrhage or infection associated with childbirth or abortion. Until recently, high maternal mortality rates significantly suppressed women's overall life expectancy in many countries. Today, most causes of maternal mortality are considered preventable given skilled health professionals with access to appropriate equipment. Many countries have made huge strides in tackling maternal mortality, with better pre-natal care picking up potential problems earlier on in pregnancy and trained birth attendants helping to avoid issues during the birth itself. Access to safe abortion services and reductions in adolescent pregnancy rates have also been very influential. More generally, better nutrition and healthcare for women and girls throughout their lives make women more resistant to infection and more able to withstand the rigors of childbirth (e.g., less likely to be anemic or stunted in growth). As such, maternal mortality rates have dropped significantly over the past 50 years. Between 2000 and 2017 alone, maternal mortality rates worldwide dropped by 38 percent, although we are still not on track to achieve the Sustainable Development Goal of reducing global maternal mortality to 70 deaths per 100,000 by 2030 (UNFPA 2019) and some countries lag significantly in their efforts to ensure safe births (box 8.3).

BOX 8.2 PRIMARY HEALTHCARE IN COSTA RICA

Costa Rica is frequently considered to be a role model for low- and middle-income countries in terms of improving health indicators. Despite an income of just $11,000 per capita (less than one-quarter that of the United States or one-third that of the UK), many of Costa Rica's health indicators today are on a par with countries in the affluent world. Costa Rica's infant mortality rate is just nine per 1,000 live births and life expectancy is 79 years. Eighty percent of women of childbearing age use contraception and the total fertility rate is just 1.9 births per woman (PRB 2020), with 99 percent of births attended by a skilled birth attendant (UNICEF, n.d.). Costa Rica also consistently performs well on high-priority health outcomes such as the proportion of children treated for diarrheal diseases and mortality rates among adults from chronic diseases (Pesec et al. 2017). How has Costa Rica achieved such strong health indicators, despite its modest income? The answer is relatively simple and yet hard-won: consistent investment in health and social programs that are accessible to all.

Since the 1940s, when health insurance covered just 3 percent of the population, Costa Rica has significantly expanded the reach of its social security system toward universal health coverage, reaching 88 percent of the population by 2005 (Rodriguez 2005). Today over 94 percent of Costa Ricans have access to government healthcare (Caja Costarricense de Seguro Social 2012), funded by a combination of employer, corporation, and government contributions, with private healthcare options running alongside public ones (Baker and Gallicchio 2020). Costa Rica has been able to direct funding to health and social programs thanks to a relatively stable political system that prioritized equity early on. Costa Rica was also notable for choosing against funding an army, diverting money from military to social programs (Pesec et al. 2017).

Despite the importance of directing funding toward health and social programs, Costa Rica actually spends less than the global average on healthcare, both in per capita terms and as a proportion of gross domestic product (Pesec et al. 2017). Costa Rica is thus revealed to be directing its limited health funding in directions where it has most impact. This has involved a heavy focus on **primary healthcare**—providing basic interventions to the majority in a community-focused manner. Costa Rica has also excelled in extending formal healthcare to remote rural populations. Since 1994, Costa Rica's healthcare delivery system has been based around multidisciplinary teams, including a physician, nurse, pharmacist, community health worker, and medical clerk. Every patient is assigned to a primary care team and then the team provides a cohesive unit to serve each patient's needs and follow up with patients who slip through the cracks. Healthcare is delivered via numerous clinics that reach well into the most rural parts of the country, and healthcare workers provide home visits to those who cannot get to clinics (Gawande 2021). This has proved to be a remarkably efficient way of delivering healthcare in an equitable manner (Pesec et al. 2017).

Of course, Costa Rica's system is not perfect—heavy focus on equitable provision of primary healthcare inevitably means that the system does not invest as heavily in complex treatments for complicated conditions, for which people may need to turn to the private healthcare system. Costa Rica's system has also been subject to significant stagnation during periods of economic hardship, such as during the 1980s when debt crippled many Latin American economies. Nonetheless, the strides made by Costa Rica in providing equitable, primary healthcare to the majority have been highlighted as a potential model for other countries of the Global South.

BOX 8.3 MATERNAL MORTALITY IN AFGHANISTAN

As recently as the year 2000, Afghanistan's maternal mortality rate was a staggering 1,450 deaths per 100,000 live births, with just 17 percent of births attended by a skilled birth attendant (WHO, n.d.). Since then, conditions have improved considerably, with the Afghan government prioritizing maternal mortality as a key public health goal since 2002 (Bartlett et al. 2017). Despite improvements, Afghanistan remained the only country outside of sub-Saharan Africa to be categorized as having a "very high" maternal mortality rate in 2017, with 638 maternal deaths per 100,000 live births (UNFPA 2019). The immediate causes of most of these deaths include "hemorrhage, obstructed labor, infection, high blood pressure, and unsafe abortion" but the wider causes relate to **social determinants of health** (Najafizada, Bourgeault, and Labonté 2017). In short, poverty, poor access to healthcare, and gender inequality—key social determinants of health—underlie these problematic statistics (UNFPA 2020).

One of the obvious challenges to health in Afghanistan is poor-quality and decaying infrastructure, a legacy of Afghanistan's political and economic instability. For many women, particularly in rural areas, a simple lack of providers and well-equipped facilities limits access to healthcare (Najafizada, Bourgeault, and Labonté 2017). Many women also lack transportation to get to a clinic. By 2012, nearly 80 percent of women in the capital Kabul were receiving antenatal care and delivering at healthcare facilities, but rural areas remained significantly under-served (Bartlett et al. 2017). In 2020, over 40 percent of births in Afghanistan were still not being overseen by a skilled professional (UNFPA 2020).

It is not just access to facilities but also utilization of those facilities that is problematic, with women not necessarily using facilities even when they are available (Najafizada, Bourgeault, and Labonté 2017). Attending facilities is especially challenging for rural women owing to poverty, illiteracy, and women's involvement in farmwork, which provides little time to attend to healthcare needs (Hadi et al. 2007). The idea that formal care is not important or necessary is another key reason that women do not seek antenatal or birth care at a health facility, even in urban areas (Bartlett et al. 2017). Information such as this suggests that improved education on the benefits of peri-natal care is essential. Indeed, studies in Afghanistan report that higher levels of education are associated with greater use of healthcare facilities, family planning, and birth attendants, as well as fewer adolescent pregnancies and longer spacings between births—all of which are recognized to improve birth outcomes. Educated women are also more likely to recognize obstetric emergencies and seek medical care more quickly (Najafizada, Bourgeault, and Labonté 2017). Illiteracy and limited education can therefore be seen as significant underlying causes of Afghanistan's poor maternal health outcomes.

Afghanistan's strongly patriarchal society is frequently cited as a further barrier to improving maternal health outcomes. Practices such as early marriage, limited autonomy for women around sexual activities, and the need to obtain a male relative's permission to access healthcare all constrain women's reproductive health (Najafizada, Bourgeault, and Labonté 2017). In 2020, just 19 percent of married women reported using contraception (PRB 2020). In one study on women's attitudes in Afghanistan, 93 percent of women reported needing permission from a male relative to seek medical care and three-quarters of women saw it as their duty to have sex with their husband even when they did not want to (Van Egmond et al. 2004). Although attitudes have been changing, low levels of female employment in Afghanistan tend to keep women within the domestic sphere, helping to maintain traditional family structures and attitudes (Najafizada, Bourgeault, and Labonté 2017).

To address these problems, the Afghan government had prioritized maternal health and was investing in healthcare facilities and personnel, prior to the reinstatement of the Taliban regime in 2021. Non-governmental efforts were also underway—for instance, the United Nations Population Fund had been supporting efforts to train and equip community midwives (UNFPA 2020). More work needed to be done to try to improve outcomes for rural women, however. Additionally, scholars recognized the need to continue to improve access to education for girls and to empower women in Afghanistan's patriarchal society (Najafizada, Bourgeault, and Labonté 2017). Unfortunately, this progress has been halted by the 2021 takeover of the Taliban, whose previous regime in Afghanistan was extremely oppressive, especially for women. Early reports of the reinstated Taliban regime suggest rapidly deteriorating maternal health services (Jung and Maroof 2021), with the UN Population Fund estimating that the number of Afghans with access to family planning may halve between 2021 and 2025, associated with a possible 51,000 additional maternal deaths and almost 5 million unintended pregnancies by 2025 (UNFPA 2021).

Today, maternal mortality is highest in the world's poorest countries; in 2017, 86 percent of all maternal deaths occurred in sub-Saharan Africa and South Asia, with Nigeria and India alone accounting for one-third of those deaths, given their large populations (UNFPA 2019; figure 8.8). A woman's lifetime risk of dying from

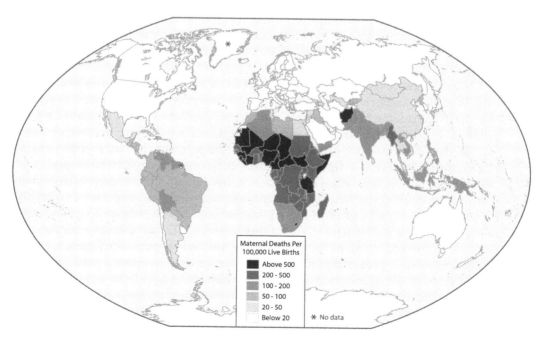

FIGURE 8.8 Maternal mortality rates by country, 2017
Data source: UNDESA (2019).

pregnancy-related complications is 1 in 38 in sub-Saharan Africa, 1 in 240 in South Asia, and 1 in 5,400 in high-income countries (UNICEF 2019). We see a clear correlation between maternal mortality and affluence at the global scale, but, as we saw with infant mortality, these patterns get more complicated when we look at variations within high-income countries. At this scale, individual policies become more significant, with countries that offer universal healthcare and strong social policies typically doing best in terms of preventing maternal deaths. In this respect, Australia and New Zealand excel, with a woman's lifetime risk of dying from maternal causes just 1 in 7,800 (UNFPA 2019).

High rates of premature death and disability in a population are not only unfortunate for the individuals affected but also take an economic toll on broader society owing to lost workdays and the cost of providing care. The economic impacts of the COVID-19 pandemic have been astronomical, for instance, in terms of lost productivity, and it is still unknown how long those suffering from "long COVID," who continue to experience symptoms long after infection, will be compromised in their ability to work. The HIV/AIDS pandemic of the late 20th century provides a further example of an epidemic that was devastating not only in terms of loss of human life but also through lost productivity. In the 1990s, life expectancy in several sub-Saharan African countries (e.g., Zambia and Tanzania) actually dropped from the 50s to the 40s as a result of high death rates from HIV/AIDS. As a primarily sexually transmitted infection, cases were focused in middle age groups, reducing the number of workers and caretakers in the population. In the worst affected areas of southern Africa, some villages became largely depopulated of working-age adults, leaving grandparents caring for small children and adolescents working to support siblings and elderly relatives. By the late 1990s, an effective therapy had been found, greatly reducing the death toll of the disease. Nonetheless, millions of people continue to live with HIV/AIDS, especially in central and southern Africa, and HIV caused 1.7 percent of deaths globally in 2017 and more than one-quarter of all deaths in Botswana and South Africa in 2019 (Roser and Ritchie 2019).

We can draw several important threads from this discussion of health indicators. First, it is important that we create and interpret health measures correctly. We must consider whether we are looking at absolute counts or rates and the potential impact of age structure on our data. Once we are sure that we are reading the data correctly, we can then interpret the patterns that we see. In general, health statistics reveal a tight correlation between health and affluence, with poorer populations typically experiencing worse health. Other marginalized populations—including minority groups and women—may experience a further toll on their health associated with discrimination and health inequalities. The good news is that these problematic connections between affluence, empowerment, and health can be addressed by policy. We see this when the correlation between health and wealth breaks down among higher income countries, with countries that offer good social programs and universal access to healthcare typically showing better health indicators than countries with greater inequalities. In poorer regions, too, sound policies can enable countries to achieve stronger health indicators than would be expected for their income, as illustrated by Costa Rica's investment in primary healthcare.

PART II: PATTERNS IN DEATH AND DISEASE OVER TIME

Although life spans have generally increased throughout human history, there have been periods when population health has declined. Additionally, the conditions causing the greatest toll on human health have changed over time, including a key shift from infectious to non-infectious causes of death known as the **epidemiologic transition**. In this section we consider how population-scale changes have interacted with a changing disease landscape, as well as some of the models used to explain these shifts.

Death rates play a key role in the demographic transition model, which posits that death rates decline as economic development proceeds. This leads to a period of population increase until birth rates also decline, restabilizing population size. Recent evidence suggests that death rates may then increase again slightly, associated with population aging. These are not the only patterns we can identify in mortality rates, however—clear patterns are also apparent in the *cause* of death over time.

Infectious disease, malnutrition, and injuries

In pre-industrial populations, infectious diseases, malnutrition, and injuries were all common, associated with harsh living conditions that kept average life expectancy short. In many cases, these factors were inter-related, with injuries leading to infections and malnutrition diminishing the effectiveness of the immune system. This latter connection is particularly significant: poor nutrition often leads to higher rates of infectious disease, as well as poorer outcomes from those infections, including higher mortality rates.

With the adoption of agriculture, a major shift in diet occurred from the diverse diet of the hunter gatherer to a diet focused on carbohydrate-rich grains. These starchy crops soon comprised the majority of agriculturalists' calorie intake. Though the carbohydrate provided sufficient energy, it was often at the expense of micro-nutrients (vitamins and minerals) and protein, leading to an explosion in nutrient deficiency diseases (table 8.2). Beriberi, caused by deficiency of thiamine (vitamin B-1), was common until recently in Asian communities subsisting largely on white rice. Pellagra, caused by niacin (vitamin B-3) deficiency, was common in the Southeast United States well into the 20th century among poor farmers subsisting largely on corn. To this day, vitamin A deficiency, associated with a diet lacking vegetables, remains the leading cause of preventable blindness in children worldwide.

In extreme cases, hunger can lead to famine where a whole population is at risk of starvation. Though we have traditionally attributed famines to environmental causes such as droughts, floods, and pests, today the fact that we have enough food to feed everyone at the global scale has led some commentators to argue that inequalities in economic systems as well as political mismanagement, rather than environmental problems, cause famines (Sen 1983). Ethiopia's Tigray region, for instance, experienced famine in 2021 associated with civil war in the region, alarming many observers that this might generate a replay of the devastating famines that swept the region in the 1980s during an earlier civil war. Although sufficient food could potentially be shipped in from outside the region, the region had

TABLE 8.2 Select nutrient deficiency diseases

Disease	Deficiency	Symptoms/outcomes	Common interventions
Kwashiorkor	Protein and other essential nutrients	Fluid retention and swelling; reduced immune function; stunted growth	Use of therapeutic foods
Vitamin A deficiency	Vitamin A	Vision loss, especially night vision; compromised immune function	Consumption of leafy greens; food fortification (esp. grains and milk)
Beriberi	Vitamin B-1 (thiamine)	Weakness; lack of energy; muscle pain; heart and circulatory failures	Food fortification or supplements
Pellagra	Vitamin B-3 (niacin)	Diarrhea; dermatitis; dementia	Food fortification or supplements
Scurvy	Vitamin C	Bleeding gums, poor wound healing	Consumption of citrus fruits
Rickets	Vitamin D	Malformed bones	Food fortification (esp. milk) or supplements
Goiter	Iodine	Swollen glands in neck; developmental limitations in children	Fortification of salt or supplements
Anemia	Iron	Weakness	Consumption of meat, leafy greens, or supplements

been isolated by enemy forces, and global apathy given the remote location and limited political and economic power of the population affected was held responsible for diminishing the humanitarian response (Gladstone 2021).

Today, famine is unusual, but **malnutrition** (poor nutrition) remains a significant problem. The World Health Organization has suggested that **undernutrition** (having insufficient food) still contributes to 45 percent of all child deaths worldwide (WHO 2021). The good news is that, in many regions, significant improvements have been made in recent decades as a result of increasing affluence, health education, and artificial fortification of foods with micro-nutrients. These factors have largely eradicated micro-nutrient deficiency diseases from the Global North, where they are now mostly restricted to populations with problems such as alcoholism, eating disorders, or medical conditions that prevent the absorption of nutrients. Ironically, however, we are now beginning to see a resurgence of micro-nutrient deficiencies as diets rich in processed foods provide abundant calories but insufficient micro-nutrients. Indeed, obesity is recognized as a form of malnutrition alongside undernutrition and micro-nutrient deficiencies. Furthermore, **food insecurity**, or uncertain access to food, remains a problem for marginalized populations in many regions, even in affluent countries (box 8.4).

Injuries, associated with unsafe living and working conditions, also persist to the present day as a significant source of morbidity and mortality. Accidents, particularly motor vehicle crashes, remain the leading cause of death of people (especially men) in their 20s in many affluent countries (Kent 2010). Higher rates of accidental deaths among males are associated with high rates of male employment in physically challenging industries such as forestry and fishing, as well as a greater propensity for risk-taking among young men, as evidenced by the higher insurance premiums charged for young male drivers than for young females. Although the absolute numbers of deaths in this age group may not be large, from a population

BOX 8.4 FOOD INSECURITY

We can recognize three general levels of diet. At the most basic level are diets that meet an individual's energy needs through providing sufficient calories. Here, people have enough energy to meet daily needs but may still experience micro-nutrient or protein deficiencies. The next stage is a diet that meets all energy and nutrient needs. Finally, "healthy diets" are defined as meeting the recommended intake of a diversity of desirable food groups. Though energy needs can be met relatively cheaply across the world, consuming a healthy diet may be up to five times more expensive and exceeds average expenditure on food in most countries of the Global South—in 2017, a healthy diet was estimated to be unaffordable for more than 3 billion people (FAO 2020). Poor-quality diets are a principal contributor to "the multiple burdens of malnutrition—stunting, wasting, micronutrient deficiencies, overweight and obesity" (FAO 2020). The healthcare costs associated with these problems are significant.

Food insecurity remains a problem (even in affluent countries), despite rising levels of obesity (even in poor countries), indicating huge inequalities in our food distribution system. Although the availability of staple foods like cereals and tubers has risen in low-income countries, the supply of fresh fruits and vegetables is only sufficient to meet recommended dietary needs in Asia and upper-middle-income countries (FAO 2020). Micro-nutrient deficiency diseases therefore persist in many poor populations. Availability of nutrient-rich meat and dairy is highest in high-income countries, with Asia also showing significant increases (FAO 2020). Though an expansion in access to animal protein is good news for poor populations, consuming meat and dairy to excess is having detrimental health implications in many affluent societies, as well as negative environmental implications.

Undernutrition represents an extreme form of food insecurity. For others, food insecurity may be associated more with uncertainties over food supply than absolute shortages, leading to significant psychological stress if not actual hunger (figure 8.9). In very poor countries, food insecurity often leads to an overall reduction in food consumption, whereas

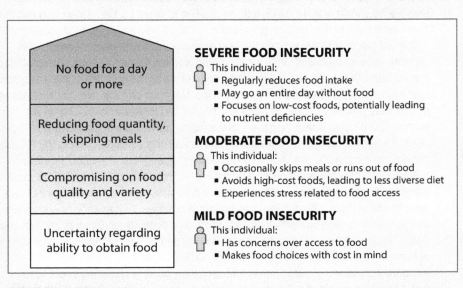

FIGURE 8.9 Food insecurity
Source: Drawn after a similar image at FAO (2022).

in middle-income countries it is associated more often with a shift away from expensive meat and dairy products to cheaper staple foods (FAO 2020). In rich countries, food insecurity is also rising as inequalities increase. In affluent regions, food insecurity can even lead to obesity, if lack of resources drives individuals to unhealthy food choices, particularly cheap processed foods. Poor nutrition associated with obesity may even be having an impact at the population level on overall health statistics. The Dutch—one of the tallest populations in the world—recently experienced a slight decline in average height at the population scale. One theory is that poorer nutrition associated with processed foods may be leading to shorter stature at the population scale (Pannett 2021).

perspective, deaths in this age group are especially significant because they remove active workers from the population, as well as individuals in their prime childbearing years. In economic and demographic analyses, the loss of a 20-year-old therefore has greater repercussions than the loss of an 80-year-old.

War and conflict have historically provided another significant cause of injury and death at the population scale (box 8.5). Today, an individual's risk of dying in

BOX 8.5 WARFARE AND POPULATION HEALTH

War can have devastating impacts on populations trapped by conflict. In Angola, for instance, a civil war from 1975 to 2002 resulted in 800,000 casualties and 4 million people displaced from their homes. Beyond those directly killed in the war, the use of land mines in Angola was particularly devastating, with civilians killed or maimed by the devices, even long after the war ended. Today, Angola is one of the most heavily mined countries in the world, with over 88,000 Angolans registered as living with disabilities from land mines and unexploded bombs in 2014 (MAG, n.d.). In addition to direct injuries, mined land and conflict zones contribute to food insecurity as land is rendered unavailable for food production.

It is not just direct injuries or deaths that are associated with wars, however. Until recently, deaths associated with disease during wartime frequently outnumbered deaths from injuries. During the Napoleonic Wars, eight times as many British soldiers may have died from disease as from battle wounds; in the American civil war, two-thirds of war-related deaths were probably caused by disease (Connolly and Heymann 2002). World War I also led to more deaths from illness than injury and subsequently triggered the global flu pandemic of 1918. A variety of factors associated with war explain this connection, including extensive population movements, poor sanitation, overcrowding and lack of shelter, diminished nutrition, and the collapse of public health systems. Population displacement, as refugees move to camps or temporary settlements, has been associated with a 60-fold increase in base mortality rates, with outbreaks of infectious diseases such as cholera in crowded camps leading to high mortality in already vulnerable populations (Connolly and Heymann 2002). Today, refugee populations often remain dispersed rather than crowded into camps, but disease can still be a significant problem. In Yemen, for instance, conflict since 2014 has led to the breakdown of public health and sanitation systems (including direct airstrikes on water facilities), resulting in the largest cholera epidemic in recent history, with outbreaks since 2016 leading to more than 1 million infections (Lyons 2017).

conflict is very low, although this could easily change if a major conflict were to break out. When wars do occur, the impacts at the population scale are profound as individuals in their prime working and childrearing years are disproportionately removed from the population. This distorts the population's structure, with widows, orphans, and the elderly suddenly having to take on additional responsibilities to make up for lost family members. The loss of a large number of men also has an impact on marriage and birth rates well into the future, with demographic "echoes" of lost cohorts showing up in future generations.

The Industrial Revolution, urbanization, and the epidemiologic transition

Though malnutrition and injuries have remained constant burdens on human populations for millennia, the rise of industrial lifestyles from the 1700s onwards brought a whole range of additional health challenges. In particular, as populations became more urbanized, rising population densities led to more rapid transmission of many infectious diseases. Airborne and waterborne diseases were able to spread especially easily in the cramped and unsanitary conditions that developed as cities grew, raising infant mortality rates. Over time, people responded by investing in innovations and technologies to alleviate these problems, most notably sanitation and other hygiene-related efforts. Combined with better nutrition and a plethora of health-related policy changes, these improvements began a slow shift toward improving health conditions for industrial populations.

The **epidemiologic transition** describes this gradual shift from deaths being caused largely by infectious diseases to predominantly non-infectious causes, as societies begin to consciously tackle the spread of infectious disease (figure 8.10). This transition began in the early 1800s in Europe and in the mid-20th century in the Global South when urban conditions were particularly dire. Over time, societies developed strategies that directly reduced the impact of infectious diseases, initially sanitation measures but also medical interventions such as sterilization, antibiotics, and vaccinations. Broader improvements to living and working conditions as well as better nutrition also played a significant role, leading to a general decrease in infectious disease. Today, infectious disease takes its greatest toll in the Global South, particularly in sub-Saharan Africa (figure 8.11).

As rates of infectious diseases decline, individuals live longer and are therefore more likely to develop non-infectious conditions, particularly heart disease and cancer. The broad term "non-infectious disease" encompasses a variety of health conditions that are not caused by an infectious agent, including malnutrition, heart disease, and diabetes. To some extent, the rise of many non-infectious diseases simply reflects the aging process, as wear and tear on the body allows **degenerative conditions** like heart disease to take hold. However, an additional explanation is the greater burden of **chronic** exposures to toxic substances that build up over a longer life span. Many substances are dangerous to people, including water contaminants such as lead and arsenic and substances in the air we breathe such as traffic fumes or wood smoke. In this respect, the worsening environmental conditions associated with the Industrial Revolution took a heavy toll, with poor air quality, chemical-tainted water, occupational exposures, and contaminants in food all damaging human bodies.

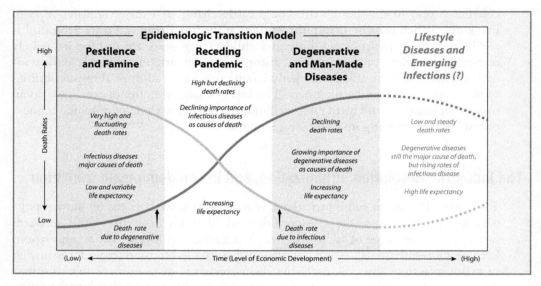

FIGURE 8.10 Epidemiologic transition model
The epidemiologic transition represents the shift from deaths occurring from largely infectious causes to mortality being primarily associated with non-infectious disease, as represented by the first three columns in the diagram. Many scholars now question whether we have reached the end of the epidemiologic transition and what might follow, with some suggestions that we may be entering a new era of infectious diseases as shown in column 4.
Source: Developed from Omran (1971).

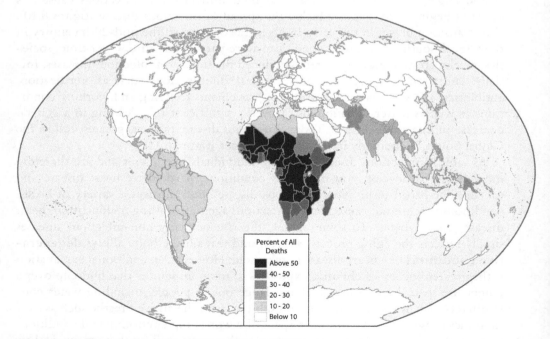

FIGURE 8.11 Proportion of deaths from infectious causes, 2019
Data source: WHO (2022).

Thankfully, public health measures have been put in place to protect us from many acute exposures that could be instantly life-threatening, such as industrial effluent in drinking water. Today, for people in affluent countries at least, what is more significant is a lifetime of low-level exposures to chemicals whose toxic impacts build up over time. Much of what **epidemiologists** do is assess the likelihood of harm from chronic low-level exposures to pollutants, using data sets from very large populations to identify even small increases in the risk of disease at the population level. Through epidemiological studies, we have discovered, for instance, that even low levels of mercury consumption can have neurological impacts on children and fetuses, leading to guidelines that encourage pregnant women to limit their consumption of species of fish that are often contaminated with mercury. Similarly, we know that lead (until recently found in white paint and vehicle fuel) contaminated the broader environment and was particularly damaging for small children, who could suffer neurological deficits after exposure. Recognition of this risk led to the banning of lead additives in many industries. Unfortunately, the sheer number of different chemicals in circulation today means that it is impossible to mitigate *all* possible exposures because many products are under-researched or hard to monitor or regulate. An oft-cited statistic suggests there may be as many as 800,000 chemicals of potential concern in widespread use, with the challenge of assessing the impacts of all of these chemicals, as well as potential combinations of them, near impossible. Furthermore, in low-income countries, where environmental legislation is often lax, many people remain exposed to high levels of *known* **carcinogens** and other pollutants. The campaign to remove lead from gasoline, for instance, took off in the West in the 1970s and most affluent countries had banned leaded fuel by the 1980s, but it was not until 2021 that Algeria stopped its use and the United Nations announced that lead had finally been removed from gasoline worldwide (BBC 2021). Today, the fact that it is often the poor and marginalized who bear the brunt of environmental exposures is well known and is explored in the field of **environmental justice**.

Even where pollutants have been brought under control, the toll of non-infectious disease remains, associated with other aspects of modern lifestyles. Indeed, a whole slew of non-infectious diseases are referred to as **lifestyle diseases** because their onset appears to be associated with aspects of modern Western lifestyles, particularly unhealthy diet, limited exercise, and stress. These diseases include heart disease, cancers, stroke, obesity, and diabetes. The Global North has now largely completed its transition from infectious to non-infectious causes of death. Many middle-income countries have also made huge progress in controlling infectious diseases and are now beginning to see rapidly rising rates of **chronic diseases** as people live long enough for these conditions to develop and the westernization of lifestyles generates conditions that favor these illnesses. Mexico now has obesity rates comparable to those in the Global North, for instance, and some countries in North Africa and the Pacific have even higher rates than most Western countries. With imported technologies and cultural preferences changing rapidly in response to images seen in the media, the epidemiologic transition is occurring much faster in many countries of the Global South than it did in the West, compressing it in ways that leave some communities at the beginning of the transition, whereas others are nearing the end. In such cases, we consider countries to be suffering from a **double burden of disease**—poorer communities continue to battle infectious diseases, even as rates of cancer and heart disease are rising in richer sub-populations. For the world's poorest countries, the epidemiologic transition is only just beginning, with large numbers of children still dying of infectious diseases, although rates of obesity, heart disease, and diabetes are starting to rise among an urban elite even in poor countries.

What might happen beyond the epidemiologic transition remains open to question. Many point to a recent resurgence in infectious disease since the late 20th century as evidence that we may be approaching a new era where infectious disease is once again becoming a major cause of death and disability. This has been associated with the rise of new epidemics of infectious diseases associated with ecological disruptions (e.g., COVID-19), rising rates of antibiotic resistance (e.g., drug-resistant strains of tuberculosis), and the more rapid spread of pathogens around our globalized world. We return to this topic in the next chapter.

In some populations, we have even seen unexpected declines in life expectancy that have complicated the supposedly unidirectional trajectory of the demographic transition model. In certain cases, this has been associated with increases in infectious disease, calling into question the epidemiologic transition as well. Some countries of sub-Saharan Africa lost 10 years of life expectancy during the HIV/ AIDS pandemic, for instance. Similarly, there was a 16 percent coronavirus-related increase in the death rate in the United States between 2019 and 2020 (CDC 2021). In other cases, social conditions have been held responsible—after the fall of the Soviet Union, for example, male mortality rose associated with high rates of alcohol use, smoking, and suicide. The idea of "deaths of despair" has become an increasing focus of health research, exploring how the breakdown of traditional social structures and rising economic inequalities are leaving some groups marginalized and at risk of destructive behaviors. Many of these newly emerging health crises have tight connections with how populations interact with their physical and social environments, and it is to this topic that we turn in the next chapter.

Conclusion

Global patterns of mortality and morbidity reveal stark inequalities in health at multiple scales, with more affluent individuals enjoying longer lives as well as greater periods in good health. This has important implications for individuals but also poses significant economic and social justice challenges at the population scale. Consideration of health indicators and patterns in health forces us to engage with the reality that the health of some populations is protected by global systems while others face occupational, environmental, and social circumstances that compromise health. A social justice perspective forces us to consider how best to tackle these persistent inequalities—we expand on this issue in the next chapter.

Discussion questions

1   Why is a country's infant mortality rate a better indicator of the overall "healthiness" of a society than its crude death rate? What factors are likely to influence the infant mortality rate vs the crude death rate?

2   What are some of the pros and cons of using numerical measures to summarize a community's health status? What other approaches could we use to complement the quantitative approach of health measures?

3   Why are maternal and peri-natal mortality critical social justice issues? How might a low-income country take an approach to addressing maternal and infant health that extends beyond access to healthcare?

4   Think of all the ways in which inequality can contribute to poor health that have been outlined in this chapter. What sorts of policy responses would you suggest to address these challenges?

Suggested readings

Aubrey, A. 2019. "Malnutrition Hits the Obese as Well as the Underfed." *NPR*, Goats and Soda, December 23, 2019. https://www.npr.org/sections/goatsandsoda/2019/12/23/785566796/malnutrition-hits-the-obese-as-well-as-the-underfed

Connolly, M., and D. Heymann. 2002. "Deadly Comrades: War and Infectious Diseases. *The Lancet* 360: s23–4. https://www.thelancet.com/pdfs/journals/lancet/PIIS0140-6736(02)11807-1.pdf

Gawande, A. 2021. "Costa Ricans Live Longer than We Do. What's the Secret?" *The New Yorker*, August 23, 2021. https://www.newyorker.com/magazine/2021/08/30/costa-ricans-live-longer-than-us-whats-the-secret

Ruggeri, A. 2018. "Do We Really Live Longer than Our Ancestors?" BBC Future, April 2, 2021. https://www.bbc.com/future/article/20181002-how-long-did-ancient-people-live-life-span-versus-longevity

Glossary

age-adjusted death rate: a death rate that has been standardized to reflect the age structure of a reference population

carcinogens: substances known to cause cancer

child mortality rate: deaths among children between birth and 5 years per 1,000 live births

chronic: persisting over a long period of time

chronic disease: a disease that persists for a long period of time; for example, heart disease, diabetes

congenital anomalies: structural or functional abnormalities that occur in babies before birth

crude death rate: number of deaths per 1,000 population; crude death rates control for population size but are not age-adjusted

degenerative conditions: illnesses that involve a progressive decline in bodily function

double burden of disease: a community suffering from both infectious and non-infectious diseases at the same time is said to be experiencing a "double burden of disease"

environmental justice: efforts to ensure that communities with less power do not have greater exposure to unhealthy environments

epidemiologic transition: the historic shift from primarily infectious to non-infectious causes of death

epidemiology: the study of the incidence and distribution of health concerns at the population scale; epidemiologists typically use large data sets and statistical techniques

food insecurity: unreliable access to food, leading to the possibility of malnutrition

infant mortality rate: deaths among infants under one year per 1,000 live births

life expectancy: the average number of years that an individual from a particular population is expected to live

lifestyle diseases: health conditions that are known to be related to lifestyle factors such as poor diet and lack of exercise

life table: statistical calculations of the likelihood of an individual of a particular age dying before their next birthday

malnutrition: poor nutrition—includes undernutrition, nutrient deficiencies, and obesity

maternal mortality rate: the number of women dying from complications of pregnancy or childbirth within 42 days of childbirth per 100,000 live births

morbidity: illness or injury

mortality: death

mortality rate: the number of people dying from a particular condition, or from all causes, relative to overall population size

neo-natal mortality rate: deaths in the first 28 days after birth per 1,000 live births

outlier: a data point that is significantly different from the others in the data set

patriarchy: social system dominated by men, marginalizing women

peri-natal mortality rate: the number of infant deaths between 28 weeks gestation and 7 days after birth per 1,000 live births

primary healthcare: healthcare focused around providing basic healthcare such as antibiotics and vaccinations in a community-based manner

social determinants of health: social, political, and economic conditions that influence people's health

undernutrition: absolute lack of food that can lead to stunting (low height-for-age) and wasting (low weight-for-height)

universal healthcare: government-funded healthcare available to all residents of a country

Works cited

AIHW [Australian Institute of Health and Welfare]. 2021. "Deaths in Australia." Australian Government, June 25, 2021. https://www.aihw.gov.au/reports/life-expectancy-death/deaths/contents/life-expectancy

Arias, E., and J. Xu. 2020. "United States Life Tables, 2018." *National Vital Statistics Reports* 69 (12). Hyattsville, MD: National Center for Health Statistics. https://www.cdc.gov/nchs/data/nvsr/nvsr69/nvsr69-12-508.pdf

Baker, R., and V. Gallicchio. 2020. "How United States Healthcare Can Learn from Costa Rica: A Literature Review." *Journal of Public Health and Epidemiology* 12 (2): 106–13. doi:10.5897/JPHE2020.1212

Bartlett, L., A. LeFevre, L. Zimmerman, S. Saeedzai, et al. 2017. "Progress and Inequities in Maternal Mortality in Afghanistan (RAMOS-II): A Retrospective Observational Study." *Lancet Global Health* 5: e545–55. doi:10.1016/S2214-109X(17)30139-0

BBC. 2021. "Highly Polluting Leaded Petrol Now Eradicated from the World, says UN." BBC, August 31, 2021. https://www.bbc.com/news/world-58388810

Caja Costarricense de Seguro Social. 2012. "CCSS Ofrece Cobertura con EBAIS al 94% de la Población." http://www.ccss.sa.cr/noticias/index/32-/379-ccss-ofrece-coberturacon-ebais-al-94-de-la-poblacion

CDC [Centers for Disease Control and Prevention]. 2020. "Infant Mortality Statistics from the 2018 Period Linked Birth/Infant Death Data Set." *National Vital Statistics Reports*. Table 2. https://www.cdc.gov/nchs/data/nvsr/nvsr69/NVSR-69-7-508.pdf

———. 2021. "Provisional Mortality Data—United States 2020." *Morbidity and Mortality Weekly Report*, April 9, 2021. https://www.cdc.gov/mmwr/volumes/70/wr/mm7014e1. htm#:~:text=In%202020%2C%20approximately%203%2C358%2C814%20deaths,828.7 %20deaths%20per%20100%2C000%20population

Connolly, M., and D. Heymann. 2002. "Deadly Comrades: War and Infectious Disease." *The Lancet*, 360: s23–4. https://www.thelancet.com/pdfs/journals/lancet/PIIS0140-6736(02)11807-1.pdf

FAO [Food and Agriculture Organization]. 2020. *The State of Food Security and Nutrition in the World*. Rome: FAO, IFAD, UNICEF, WFP, WHO. https://www.fao.org/documents/card/en/c/ca9692en

———. 2022. "Hunger and Food Insecurity." https://www.fao.org/hunger/en/

Gawande, A. 2021. "Costa Ricans Live Longer than We Do. What's the Secret?" *The New Yorker*, August 23, 2021. https://www.newyorker.com/magazine/2021/08/30/costa-ricans-live-longer-than-us-whats-the-secret

Geronimus, A. 2001. "Understanding and Eliminating Racial Inequalities in Women's Health in the United States: The Role of the Weathering Conceptual Framework." *Journal of the American Medical Women's Association* 56 (4): 133–6, 149–50.

Gladstone, R. 2021. "Famine Hits 350,000 in Ethiopia: Worst Hit Country in a Decade." *New York Times*, June 10, 2021. nytimes.com/2021/06/10/world/africa/ethiopia-famine-tigray.html

Hadi, A., M. Mujaddidi, T. Rahman, and J. Ahmed. 2007. "The Inaccessibility and Utilization of Antenatal Health-Care Services in Balkh Province of Afghanistan." *Asia-Pacific Population Journal* 22 (1): 29–42.

Höpflinger, F. 2012. *Bevölkerungssoziologie. Eine Einführung in demographische Prozesse und bevölkerungssoziologische Ansätze*. Weinheim and Basel: Beltz Juventa.

Jung, E., and H. Maroof. 2021. "Giving Birth under the Taliban." BBC, September 20, 2021. https://www.bbc.com/news/world-asia-58585323

Kent, M. 2010. "Young U.S. Adults Vulnerable to Injuries and Violence." PRB, July 19, 2010. https://www.prb.org/resources/young-u-s-adults-vulnerable-to-injuries-and-violence/

Lyons, K. 2017. "Yemen's Cholera Outbreak Now the Worst in History as Millionth Case Looms." *The Guardian*, October 12, 2017. https://www.theguardian.com/global-development/2017/oct/12/yemen-cholera-outbreak-worst-in-history-1-million-cases-by-end-of-year

MAG [Mines Advisory Group]. n.d. "Angola." Accessed March 23, 2022. https://www.maginternational.org/what-we-do/where-we-work/angola/

Marmot, M., C. Ryff, L. Bumpass, M. Shipley, and. N. Marks. 1997. "Social Inequalities in Health: Next Questions and Converging Evidence." *Social Science and Medicine* 44 (6): 901–10.

Marmot, M., G. Smith, S. Stansfeld, C. Patel, et al. 1991. "Health Inequalities among British Civil Servants: The Whitehall Study II." *Lancet* 337 (8754): 1387–93. doi:10.1016/0140-6736(91)93068-k

Mayer, S., and A. Sarin. 2005. "Some Mechanisms Linking Economic Inequality and Infant Mortality." *Social Science & Medicine* 60 (3): 439–55. doi:10.1016/j.socscimed.2004.06.005

Najafizada, S., I. Bourgeault, and R. Labonté. 2017. "Social Determinants of Maternal Health in Afghanistan: A Review." *Central Asian Journal of Global Health* 6 (1): 240. doi:10.5195/cajgh.2017.240

Omran, A. 1971. "The Epidemiologic Transition. A Theory of the Epidemiology of Population Change." *Milbank Memorial Fund Quarterly* 49: 509–38.

OurWorldinData. n.d. "GDP Per Capita, 2020." Accessed April 3, 2022. https://ourworldindata.org/grapher/gdp-per-capita-worldbank

Pannett, R. 2021. "The World's Tallest Populace Is Shrinking, and Scientists Want to Know Why." *The Washington Post*, September 19, 2021. https://www.washingtonpost.com/world/2021/09/19/netherlands-average-height-shrinking-tallest/

Pearce, J., and D. Dorling. 2009. "Tackling Global Health Inequalities: Closing the Health Gap in a Generation." *Environment and Planning A: Economy and Space* 41 (1): 1–6. doi:10.1068/a41319

Pesec, M., H. Ratcliffe, A. Karlage, L. Hirschhorn, et al. 2017. "Primary Health Care That Works: The Costa Rican Experience." *Health Affairs* 36 (3): 531–8. doi:10.1377/hlthaff.2016.1319

PRB [Population Reference Bureau]. 2020. "World Population Datasheet 2020." PRB, July 2020. https://www.prb.org/wp-content/uploads/2020/07/letter-booklet-2020-world-population.pdf

Rodríguez, A. 2005. *La Reforma de Salud en Costa Rica*. Santiago: CEPAL.

Roser, M., and H. Ritchie. 2019. "HIV/AIDS." OurWorldInData.org. https://ourworldindata.org/hiv-aids

Ruggeri, A. 2018. "Do We Really Live Longer than Our Ancestors?" BBC Future, October 2, 2018. https://www.bbc.com/future/article/20181002-how-long-did-ancient-people-live-life-span-versus-longevity

Sen, A. 1983. *Poverty and Famines: An Essay on Entitlement and Deprivation*. Oxford: Oxford University Press.

UNDESA [United Nations Department of Economic and Social Affairs]. 2019. "World Population Prospects 2019." https://population.un.org/wpp/

UNFPA [United Nations Population Fund]. 2019. "Trends in Maternal Mortality, 2000 to 2017." WHO, UNICEF, World Bank, and UNFPA, September 19, 2019. https://www.unfpa.org/featured-publication/trends-maternal-mortality-2000-2017

———. 2020. "Midwives on the Front Lines Working to Reverse Afghanistan's High Maternal Death Rate." UNFPA, October 21, 2020. https://www.unfpa.org/news/midwives-front-lines-working-reverse-afghanistans-high-maternal-death-rate#:~:text=Afghanistan%20has%20one%20of%20the,die%20per%20100%2C000%20live%20births

———. 2021. "As Women and Girls Bear the Brunt of the Crisis, UNFPA Urgently Seeks $29.2 Million to Save and Protect Lives in Afghanistan." UNFPA, September 13, 2021. https://www.unfpa.org/press/women-and-girls-bear-brunt-crisis-unfpa-urgently-seeks-292-million-save-and-protect-lives

UNICEF [United Nations Children's Fund]. 2019. "Maternal Mortality." UNICEF, September 2019. https://data.unicef.org/topic/maternal-health/maternal-mortality/

———. n.d. "Costa Rica Country Profile." Accessed September 4, 2021. https://data.unicef.org/country/cri/

Van Egmond, K., M. Bosmans, A. Naeem, P. Claeys, et al. 2004. "Reproductive Health in Afghanistan: Results of a Knowledge, Attitudes and Practices Survey among Afghan Women in Kabul." *Disaster* 28 (3): 269–82.

WHO [World Health Organization]. 2021. "Malnutrition." WHO, June 9, 2021. https://www.who.int/news-room/fact-sheets/detail/malnutrition

———. 2022. "Global Health Estimates 2019 Summary Tables." https://www.who.int/data/global-health-estimates

———. n.d. "Maternal Mortality in 2000–2017: Afghanistan." Accessed October 4, 2021. https://www.who.int/gho/maternal_health/countries/afg.pdf

Xu, J., S. Murphy, K. Kochanek, and E. Arias. 2021. "Deaths: Final Data for 2019." *National Vital Statistics Reports* 70 (8). CDC, July 26, 2021. https://www.cdc.gov/nchs/data/nvsr/nvsr70/nvsr70-08-508.pdf

Emerging trends in population health

After reading this chapter, a student should be able to:

1 discuss the dynamic nature of health patterns;

2 explain how environmental change influences patterns in infectious disease;

3 explain how changes to the social environment are affecting the incidence of chronic and degenerative diseases.

In the previous chapter, we focused on how **mortality** and **morbidity** show significant patterns at the population scale and the insights we can glean from global variations in health indicators. At the close of that chapter, we noted some potentially troubling trends on the horizon, related to the rapidly changing environmental and social contexts in which health is situated. Given our focus on sustainability and social justice, these topics require further discussion. There are numerous issues that we could raise in this respect, but here we focus on two key ideas. In part I, we focus on the natural environment, considering the ways in which disruptions to global ecological systems are influencing the range and incidence of infectious diseases. In part II, we turn to non-infectious diseases, focusing on how changing social environments are influencing disease patterns.

PART I: POPULATION HEALTH AND ENVIRONMENTAL CHANGE

The ecology of infectious disease

Infectious diseases are, by definition, caused by living organisms, including viruses, bacteria, fungi, and worms. These organisms have their own **ecology**, which ties them intimately to their environment via the need for food, shelter, and breeding opportunities. Whereas for some diseases it may only be the direct

DOI: 10.4324/9781003143253-9

environment of the body hosting them that is significant, for many others conditions in the broader environment are also critical. The value of geographic approaches, which focus on the importance of physical and social environments, is clear here.

For any infectious disease to circulate, the **pathogen** that causes it (e.g., the virus) must be transmitted effectively from one **host** to another. The new host may then become sick with the disease before potentially spreading it to others. Here we are most interested in diseases that affect human hosts, although all animals host their own unique range of pathogens. **Airborne diseases**, such as measles or tuberculosis, are passed directly from one host to another through the air, either in tiny aerosol particles that hang in the air to be inhaled by another host or via larger droplets that are exhaled and settle on surfaces, which can then get into another host when a person touches that surface and then their mouth, eyes, or nose. Since the coronavirus pandemic, we have all become experts on subtleties of transmission, with early advice suggesting that we should wipe down surfaces to avoid transmission soon replaced by evidence suggesting that COVID-19 is predominantly spread via aerosolized particles, refocusing our attention toward masks and adequate ventilation. Particularities of diseases such as this illustrate the huge importance of understanding routes of transmission to devising effective public health strategies. They also illustrate how subtle changes in the environment, or people's interactions within their environment, can greatly influence disease transmission. For airborne diseases, the likelihood of another person passing by and inhaling infectious particles or touching a contaminated surface while the pathogen is still viable must be high enough for the disease's transmission cycle to continue. Thus, many airborne diseases spread more effectively in crowded conditions.

In other cases, pathogens can persist in the environment (e.g., in water or soil) for longer periods of time before being picked up by another host. **Waterborne diseases** like cholera and giardia are caused by pathogens that can survive in water and are then ingested by a host via drinking water or water on uncooked foods such as salads. Waterborne pathogens may remain viable in water for long periods, but population density is still important because high-density human populations are more likely to contaminate local water supplies with fecal matter. Many waterborne diseases therefore show greater transmissibility in urban populations than in rural populations in conditions of poor hygiene. The dire disease circumstances of early industrial cities such as Victorian London, where perhaps one-third of children died before reaching adulthood, illustrate these circumstances well. It took many years for sanitation systems and environmental health acts to catch up with explosive population growth in Europe and make cities a healthy place to live. Today, we can see these same issues being replayed in the **megacities** of the Global South, where rapid urbanization has pushed large numbers of people into close quarters faster than sanitation infrastructure can keep up. This additional disease risk that comes with living in cities is sometimes referred to as the **urban penalty** (see also chapter 12).

A third group of diseases require another organism, called a **vector**, to transmit the pathogen from one host to another. **Vector-borne diseases** comprise about 17 percent of all infectious diseases and cause more than 700,000 deaths annually (WHO 2020). The best-known example is malaria, but many vector-borne diseases exist, most spread by mosquitos, flies, or ticks (table 9.1). People have developed a variety of behavioral adaptations to limit their exposure to vector-borne diseases,

TABLE 9.1 Significant vector-borne diseases

Disease	Pathogen; vector	Geographic range	Human toll (per year)
Malaria	*Plasmodium* parasite; *Anopheles* mosquitos	Widespread across the Tropics; highest disease burden in sub-Saharan Africa	Globally, over 200 million cases and 400,000 deaths (mostly in children)
Dengue fever	Virus; *Aedes* mosquitos	Widespread across the Tropics	96 million symptomatic cases; 40,000 deaths
Zika fever	Virus; *Aedes* mosquitos	Tropical Africa, Asia, South Pacific, and Americas	Variable: the 2015–16 epidemic led to at least a million cases; disease has since receded
Yellow fever	Virus; *Aedes* and *Haemogogus* mosquitos	Tropical and sub-tropical Africa and South America	A highly effective vaccine has greatly reduced the disease burden, which persists mostly in Africa
West Nile fever	Virus; *Aedes* mosquitos	Africa, North America, Europe, Middle East, and West Asia	Periodic outbreaks; only 20–30% of people infected are thought to have symptoms
Lyme disease	Virus; several tick species	Parts of Eurasia and North America	476,000 diagnosed cases in the United States alone; can be treated with antibiotics
Japanese encephalitis	Virus; mosquitos	Southeast and East Asia, Pacific Islands	68,000 clinical cases; no cure

Data sources: WHO (2017, 2019a, 2019c, 2020, 2022c) and CDC (2021).

including avoiding swampy areas where pests thrive, staying indoors at dawn and dusk to avoid mosquitos, or wearing long pants to avoid picking up ticks.

Vector-borne diseases are found where environmental conditions favor the vector, with the year-round warm temperatures of the tropics providing particularly amenable conditions for many vectors. The environmental conditions needed are often much more specific than just warm temperatures, however. The *Anopheles* mosquito that causes malaria, for instance, needs standing water in which to breed and is often associated with regions with ponds and drainage ditches. The human habitats created by irrigated agriculture are often ideal, sometimes even preferred over natural areas, contributing to malaria's persistent disease threat (Hinne et al. 2021). By contrast, the species of mosquito that is the main vector of dengue fever, *Aedes aegypti*, is a forest species that evolved to breed in small pockets of water trapped in leaves. Unfortunately, *Aedes aegypti* has adapted to use the small pools of water trapped in urban refuse as a substitute for these leaf pools, giving the mosquito an abundance of new breeding grounds. Today, *Aedes aegypti* infests many tropical cities and dengue fever is now a primarily urban disease.

Epidemics, pandemics, and environmental change

The intimate connections between vectors and specific environmental conditions mean that vector-borne diseases are particularly prone to fluctuations in incidence and range associated with landscape change. This can be used to actively control disease; for example, by draining swamps that host high densities of disease-bearing mosquitos. Today, in an era of rapid human-induced landscape change, many disease experts are concerned about the ways in which environmental change might inadvertently be *increasing* the range and incidence of vector-borne diseases, however. Given that changes to the landscape could theoretically improve or worsen conditions for the spread of a disease, it is reasonable to consider whether the impacts of multiple environmental changes might actually balance out, leaving conditions different but no worse. Unfortunately, the consensus among disease ecologists is that the net impact of current environmental disruptions will likely increase the spread of disease. Pathogens and vectors typically reproduce very rapidly, with short generation times, and are thus able to evolve to adapt to new conditions much faster than higher organisms like humans, whose slower breeding patterns are better adapted to constant conditions. This gives pathogens and vectors the upper hand in rapidly changing environments. Furthermore, our public health systems are carefully honed to current conditions—as conditions change, our health-related policies, behaviors, and infrastructure will need to be rapidly updated. As Walters (2004, 11–12) put it, "Over the past century or more, humans have so disrupted the global environment and its natural cycles that we risk evicting ourselves from our shelter of relative ecological stability. ... If the upsurge in new diseases is any indication, microscopic predators are taking full advantage of the instability."

In summary, infectious diseases have intimate connections with environmental conditions, as well as population density, distribution, and behaviors. In this context, disturbances such as urbanization and landscape conversion are recognized as key drivers of **epidemics** of infectious disease (Morse et al. 2012). A large literature has developed over the past 20 years detailing the ways in which infectious diseases have been re-emerging since the late-20th century, after huge success in controlling infectious diseases with basic interventions like sanitation, vaccines, and antibiotics in the early- to mid-20th century. Epidemics and **pandemics** have occurred periodically throughout human history, but many fear that the frequency of these events is increasing today as urbanization, globalization, and encroachment on wildlands increase our exposure to new pathogens and the speed at which they can be transported around the world.

Globalization is a key aspect of this link between human activities and disease, with increased mobility of people and goods creating the potential for the rapid movement of pathogens and vectors around the world. The introduction of a new disease to a previously unexposed population potentially leads to a high **case fatality rate** because the unexposed population has little immunity to the new disease. This is not a new process: global interconnectivity has been increasing steadily for hundreds of years, often with terrible epidemiologic consequences for newly exposed populations, as was forcefully illustrated when Native American and Pacific Islander communities were devastated by diseases brought from Europe in the colonial era (box 9.1). However, global interconnectivity has reached new heights in recent years, with the first quarter of the 21st century being marked by a string of epidemics, including SARS (sudden acute respiratory syndrome), several strains of influenza (flu), and Zika virus—plus, of course, COVID-19. Although only COVID-19 generated a global pandemic, many disease ecologists were concerned about the pandemic potential of several of these previous outbreaks.

Since the turn of the 21st century, the sheer magnitude of global interactions has reached new levels, with people and goods moving around the world at

BOX 9.1 THE "COLUMBIAN EXCHANGE"

The introduction of Old World diseases to the New World with colonists illustrates just how deadly a disease can be when a pathogen is brought to a population with little immunity. Prior to European contact, the Americas had been cut off from the disease pool of the rest of the world for thousands of years. As a result, New World populations had little immunity to European diseases at the time of first contact. Epidemics of smallpox, measles, tuberculosis, and other European diseases struck down Native American populations, leading to one of the most devastating depopulation events ever recorded (Mann 2006).

Although this is sometimes referenced as part of a broader "Columbian Exchange" of plants, animals, and pathogens between the continents, "exchange" is a misnomer with respect to disease, because few pathogens appear to have moved from the Americas to Europe. During the era of rapid colonization, Africa and Eurasia—with their large interconnected human populations, as well as numerous species of domesticated animals to seed new human diseases—had a large diversity of diseases established in human populations and, as a result, high levels of immunity to them. By contrast, far fewer pathogens were probably circulating among smaller Native American populations, who also had few species of domestic animals to provide a source of pathogens. As such, few diseases probably moved from the New World to the Old (Diamond 1997). The result was huge loss of life in Native American communities from European diseases but few diseases recorded to have moved in the opposite direction.

The precise impacts of this event in the Americas are still being investigated, with lingering disagreements over how many people lived in the Americas prior to contact, as well as what proportion may have died. Traditionally, the Americas were thought of as sparsely populated, but more recent research suggests that the pre-Columbian Americas may have been densely settled in places, with some estimates suggesting human populations of perhaps 50 million when Europeans arrived (Denevan 1992). As much as 90 percent of some Native American communities may have died from European-introduced diseases, with deaths numbering in the millions overall. Early European explorers sometimes recorded already depopulated landscapes when they first arrived in a new region, suggesting that waves of disease probably moved across the Americas ahead of the colonists (Mann 2006). Regardless of exact death rates, the implications for Native American communities were devastating, destroying whole communities and their cultures.

dizzying speed, carrying pathogens along with them. Additionally, we know that vectors can be transported in aircraft, leading to phenomena such as "airport malaria," where people living near airports have contracted malaria from accidentally imported mosquitos. Rapid transportation also increases the likelihood that localized epidemics will spread widely to become global pandemics—the speed at which COVID-19 spread across the world, despite travel bans and stay-at-home orders, reminded the global community of its vulnerability in this respect. Ebola virus disease provides another poignant example. During the late 20th century, Ebola periodically emerged in isolated outbreaks in remote parts of central Africa when the disease spilled over from bats into the human population. Though devastating for the local communities infected (Ebola has had a 50 to 90 percent case fatality rate in recorded outbreaks), the sheer remoteness of these communities shielded global populations for many years. This changed in 2014–16 when an outbreak in West Africa spread along local and then global trade routes, leading to over 28,000 cases and 11,000 deaths in West Africa (figure 9.1), as well as isolated infections in populations as far afield as Europe and North America (CDC 2019b).

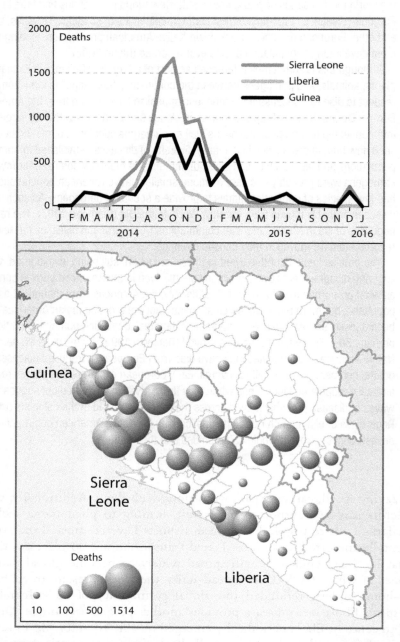

FIGURE 9.1 Trends in the number of Ebola deaths over time and by province in Sierra Leone, Guinea, and Liberia during the Ebola epidemic of 2014–16
Data source: AmeriGEOSS (n.d.).

Ebola also raises a second critical issue: our vulnerability to diseases from other animals. A variety of human activities are putting us in closer contact with animal pathogens, ranging from deforestation to the sale of wild animals for food to intensive agriculture. Although most animal pathogens are not adapted to infect humans, occasionally a pathogen is able to spread from an animal to a human in a so-called **spillover event**. Once the pathogen is in human bodies, it is then possible for it to evolve better mechanisms for spreading directly from human to human, presenting the potential for an epidemic. Despite the rarity of this sequence of events, the frequency of contact between people and animals means that this is probably the path by which most human infectious diseases have emerged over thousands of years. In this respect, the agricultural revolution almost certainly led to a significant increase in diseases circulating in the human population as people began living in close contact with livestock. Tuberculosis, measles, smallpox, and diphtheria may have emerged from cattle; influenza from pigs and ducks; the common cold from horses; and leprosy from water buffalo (Walters 2004; Lieberman 2013). This process continues to this day: around 60 percent of the infectious diseases that emerged between 1940 and 2004 likely came from animals (Jones et al. 2008). Infectious disease experts continue to worry about the potential for new human flu strains to emerge from exposure to pigs or poultry, as famously occurred in 1918 with the "Spanish flu" pandemic (box 9.2). Indeed, prior to the recent coronavirus pandemic, many disease experts believed that a strain of flu would cause the next major pandemic owing to the influenza virus's ubiquity and ability to mutate rapidly.

It is not just domestic animals that are cause for concern, however. The COVID-19 pandemic brought to the public's attention the potential dangers of exposure to pathogens of wild animals because it appears that the COVID-19 virus probably originated in bats. The notion of "One Health" emphasizes how human health is intricately tied to diseases circulating in other animals, as well as its dependence on broader environmental stability (Cunningham, Daszak, and Wood 2017). Encroachment on wildlands for agriculture, logging, and mining, as well as the bush meat trade and wild animal markets, provide ways in which people encounter pathogens from wild animals. Since the year 2000, an alarming number of epidemics have emerged in this way, with outbreaks of SARS, MERS (Middle East respiratory syndrome), Ebola, and COVID-19 all associated with pathogens crossing from wild animals to humans in spillover events. Additionally, several flu strains evolved from some combination of flu strains from wild and domestic birds as well as pigs (table 9.2). As noted above, diseases new to a human population often have a high case fatality rate because limited human exposure to the disease means little immunity to it—MERS (a disease associated with camels) kills 35 percent of people who contract it (WHO 2019b; figure 9.3); the H5N1 strain of avian flu killed about half of those infected between 2003 and 2015 (WHO 2015).

COVID-19 illustrates many of these patterns well; indeed, many infectious disease experts have been warning of the dangers of exactly this sort of disease outbreak since the early 2000s. The coronavirus that causes COVID-19 circulates naturally in bats and may have infected an intermediary species of animal before crossing over into the human population. The dense human populations of eastern China, many of which have close contact with wild animals, provide excellent conditions for wild viruses to cross into the human population. Wildlife farms, where wild animals are raised to be sold for human consumption, as well as markets that sell wild animals,

BOX 9.2 THE 1918 FLU PANDEMIC

Sometimes known as the "Spanish flu," the influenza pandemic that spread in the wake of World War I was the most devastating pandemic of recent history, ultimately infecting perhaps one-third of the global population and killing 50–100 million people (Barry 2017; CDC 2018). The virus responsible originated in birds before mutating into a strain that could affect humans. The geographic origin of the virus is still contested—with France, China, Vietnam, and the United States all suggested as possible sites of the disease's emergence— but the high mobility of soldiers at the time undoubtedly allowed the virus to spread rapidly, turning what might have been a localized outbreak into a pandemic (Barry 2017; figure 9.2).

From a population perspective, an intriguing aspect of this pandemic was the characteristics of the population affected. The challenging circumstances of war, including stress, injury, and forced movement of populations, left many with weakened immune systems and therefore more susceptible to disease. The unique age structure of those killed by the virus is perplexing, however. Typically, seasonal flu viruses tend to cause greatest mortality in elderly populations. In this case, however, the virus was notable for causing high mortality in otherwise healthy people, particularly those 20 to 40 years—a group not normally seriously affected by flu (CDC 2018). The exact reason for this finding is still debated. Perhaps the huge stress of the war put a particular toll on these middle age groups, who were most likely to be actively involved in the conflict and at greatest danger of infection. Researchers have hypothesized that further aspects might also be significant, however. Some older people may have had partial immunity to the pandemic flu strain owing to a previous

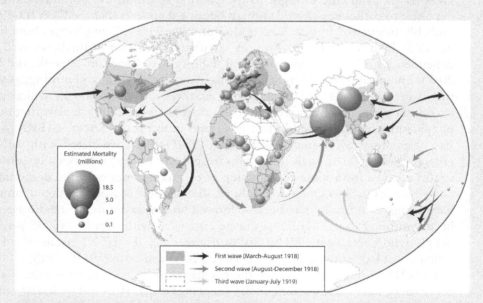

FIGURE 9.2 Spread of influenza during the 1918–19 flu pandemic
The map shows the spread of three waves of the pandemic as well as estimated deaths by country. Deaths are reported by 1918 national boundaries.
Data sources: Johnson and Mueller (2002) and Patterson and Pyle (1991); map drawn after a similar image at Africa Center for Strategic Studies (2020).

exposure to a similar strain of flu circulating earlier in their life, explaining their better than expected survival. Another hypothesis suggests that it may actually have been the very robustness of the immune response among the healthy 20-to-40-year-old cohort that was the problem, via a damaging over-reaction to the virus by the body's own immune system.

have been highlighted as posing a particular risk for introducing viruses into the human population (WHO 2021b). No one knows when this particular coronavirus evolved the ability to spread directly from human to human, but it is likely that direct human–to–human transmission was occurring for several months (if not years) prior to the outbreak at the Wuhan seafood market in December 2019 that marked the beginning of global recognition of the disease (Woodward 2021). Although concerns have been raised that the virus that causes COVID-19 could have escaped from a lab, the idea that it emerged naturally in the way described above is perfectly plausible and acts as an important cautionary tale.

Another environmental factor raising considerable concern is the potential for climate change to alter the incidence and distribution of infectious diseases. This could occur by changing key factors for disease transmission such as the length of season

TABLE 9.2 Infectious disease events that emerged from animals during the 21st century

Year	Disease	Transmission route	Geographic scope	Cases/deaths
2002–04	SARS (severe acute respiratory syndrome)	Bats to intermediate animal host (civet?) to humans; then human-to-human transmission	Emerged in China; most cases in China and Hong Kong; some exported cases	Approx. 8,000 cases and 800 deaths
2003 onwards	Avian flu (H5N1)	Wild waterfowl to domestic poultry; bird-to-human transmission	Emerged in Southeast Asia, most cases in Southeast Asia but other isolated outbreaks; e.g., Egypt	Approx. 800 cases and 400 deaths (2003–15)
2009–10	Swine flu (H1N1)	Pigs to humans; then widespread human-to-human transmission	Probably emerged in Mexico; worldwide spread	Estimated 0.7–1.4 billion cases; 150,000–575,000 deaths
2012 onwards	MERS (Middle East respiratory syndrome)	Camels to humans (may have originated in bats); occasional human-to-human transmission	Emerged in Middle East; most cases in Middle East with isolated exported cases	Approx. 2,500 cases; 870 deaths (2012–20)
2013 onwards	Avian flu (H7N9)	Poultry to humans; no human-to-human transmission to date	China	Approx. 1,500 cases; 600 deaths (2013–18)
2014–16	Ebola	Bats to humans; widespread human-to-human transmission	West Africa, isolated exported cases	Approx. 38,600 cases; 11,000 deaths
2019 onwards	COVID-19	Bats to intermediate animal host to humans?; widespread human-to-human transmission	Emerged in China; worldwide spread	As of April 2022, over 500 million cases and over 6 million deaths, and counting ...

Data sources: Johns Hopkins Coronavirus Resource Center (2022), WHO (2015, 2021a), CDC (2017b, 2019a, 2019b), and Roos (2011).

FIGURE 9.3 Camel market, United Arab Emirates
Dromedary camels act as a reservoir of the MERS (Middle East respiratory syndrome) virus. Although the exact role of camels in transmission of the disease is not well understood, close contact between camels and people at markets such as this may have introduced the disease to the human population. Most cases of MERS transmission to date have occurred in healthcare facilities through direct human-to-human contact, however (WHO 2019b).

over which vectors can survive or the speed at which pathogens can breed as temperatures rise or the amount of standing water in an environment as precipitation patterns change. The distribution and abundance of vectors may be particularly profoundly affected, given their very specific environmental requirements. In this respect, there is considerable concern that warming temperatures may increase the latitudinal and altitudinal range of many disease vectors, and changes to precipitation patterns could lead to better breeding conditions for vectors like mosquitos. The poleward spread of mosquitos, for instance, may introduce vector-borne diseases such as West Nile virus, dengue fever, and Chikungunya fever to Northern Europe (Fischer et al. 2009). Changing weather and vegetation patterns will, of course, make conditions *less* favorable for the transmission of a disease in some cases. Once again, however, there is reason to believe that changes will be negative overall with respect to disease. Warmer conditions are likely to lead to more places where vectors and pathogens can thrive, as well as increase the speed of their breeding cycles. Meanwhile, more extreme weather conditions such as stronger storms have been linked to the potential for more disease outbreaks associated with standing water, as well as disruptions to infrastructure such as piped water. We expand on these ideas in chapter 12.

Though the idea of environmental disruptions increasing rates of infectious disease may seem bleak, a better understanding of the pathways by which this occurs increases our chances of finding solutions. The recent COVID-19 crisis has illustrated how quickly people can make behavioral changes given appropriate information (e.g., wearing masks, avoiding crowded settings). Beyond this, we have started to put in place some of the structural changes that could better protect us in the future (e.g., improved ventilation, better-funded public health surveillance systems). We increasingly have the scientific know-how to make great strides in other arenas too, such as strategies to mitigate climate change. At this point, it is not so much lack of understanding as lack of political will and economic barriers that prevent us from investing in mitigation approaches. It is aspects such as these that form the social environment—the topic to which we now turn.

PART II: NON-INFECTIOUS DISEASES AND THE SOCIAL ENVIRONMENT

It is not just the physical environment that is influencing recent shifts in population health. Economic and political shifts, as well as broader cultural changes, have also triggered some concerning declines in health status. These changes suggest that we should not assume that we are on a one-way path toward better health. In this section, consistent with our social justice focus, we are interested in aspects of the social environment that influence health, focusing particularly on rising inequalities and the destabilization of structures of society. We turn our attention to non-infectious diseases, many of which have close ties to the social environment.

Chronic, degenerative, and lifestyle diseases

Over time, more and more lifestyle factors have been associated with negative health consequences, from more sedentary activity patterns, to processed foods, to long workdays. Although the influence of each individual factor may be small, the combined effect of multiple factors can be large. For instance, stress associated with a challenging job may be manageable, but if it is combined with poor diet and using cigarettes to deal with stress, significant health impacts may be incurred. The impact of Western lifestyles has received particular attention in terms of this sort of combination of multiple stressors. The heart disease, cancers, and strokes that cause most deaths in affluent countries are all believed to have numerous lifestyle triggers, many of which build up over many years, with diet, exercise, and stress often identified as key underlying issues. Many of these factors also show complex interactions. Obesity is such a significant risk factor for the development of type II diabetes, for example, that the term "diabesity" has been coined (Kalra 2013). This combination of factors makes it challenging to tease apart the relative impact of different contributory factors, but overall the influence of the social environment on our health seems to be significant.

The particular importance of increasingly sedentary lifestyle and rich diet to our health has led to the term **diseases of affluence** being applied to many of the **chronic diseases** that typify old age in affluent contexts. Today, chronic diseases (ailments that develop and persist over a long time such as cancer and heart disease) cause more than 80 percent of deaths in most rich countries (figure 9.4). As affluence increases globally, the incidence of these diseases is rising in poorer populations, too—more people now suffer from diabetes in the low-income world than in the Global North, for instance, although rates remain lower (WHO 2022a). In the

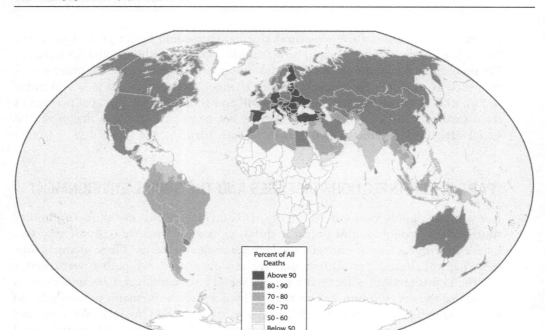

FIGURE 9.4 Proportion of deaths due to chronic disease by country, 2019
Data source: WHO (2022b).

affluent world, we are now seeing a shift, however, as the most educated populations are increasingly taking protective actions against diseases of affluence (going to the gym, reducing red meat consumption, giving up smoking, etc.). As a result, in rich countries, it is now often the poor who suffer from lifestyle diseases at the highest rates, presenting an intriguing paradox. How did "diseases of affluence" shift to become diseases of poverty in the Global North?

Obesity provides a particularly instructive case in this regard. Obesity is a leading contributor to many causes of death, including the two biggest killers among chronic diseases: cancer and heart disease. Obesity has traditionally been considered a disease of affluence because its onset is closely linked to the rich diet and sedentary occupations of the wealthy. Today, however, obesity is a global epidemic, with half of all people overweight or obese in the hardest-hit countries (figure 9.5). The distribution of obesity is highly uneven, however. In the world's poorest communities—where people still undertake heavy agricultural labor and have restricted diets—obesity remains uncommon. However, with growing affluence in the Global South, an emerging obesity crisis is now hitting many middle-income countries. A shift in diet, with Western-style processed foods often associated with modernization, has seen a huge rise in the consumption of fast foods and snack foods in many middle- and even low-income countries. In communities where increasing affluence may still not equate to large amounts of disposable income, processed snacks are some of the easiest small luxuries for people to afford, being both cheap and widely available (figure 9.6). At the same time, populations are becoming increasingly sedentary as traditional jobs in agriculture are replaced by less active urban employment.

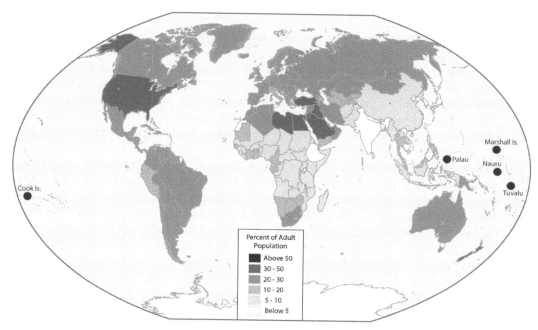

FIGURE 9.5 Obesity rates by country, 2016
Data source: OurWorldinData (2016).

FIGURE 9.6 Snack vendor, Cusco, Peru
Unhealthy foods such as soft drinks and processed snacks provide a cheap treat for populations of many parts of the world, including in the Global South.

Meanwhile, in the Global North, where health movements have taken off among the affluent, obesity and diabetes are increasingly concentrated in poorer populations. In such cases, "wellness culture" can be seen as a luxury for those with the time and disposable income to be able to afford more expensive wholefood diets and active leisure activities. For the poor, by contrast, less healthy choices often retain considerable appeal. Just as in the Global South, they offer an affordable treat for populations who may feel that other luxuries are unaffordable. Processed foods and fast foods are also relatively cheap and are easier to prepare than many whole food alternatives—of great significance to the working poor, who may have little energy to prepare meals from scratch after long workdays. Lack of time and energy may also explain why low-income communities are less likely to exercise during leisure time. Ultimately, the connections between obesity and affluence are extremely complicated and seem to have deep-rooted psychological connections, leaving us, to date, lacking effective interventions. Indeed, some commentators have deemed the connections between obesity and poverty to be so complicated that it might be easier to tackle obesity indirectly by improving people's socio-economic status via education rather than trying to directly address obesity itself (Lee 2013).

Examples such as these illustrate how psycho-social aspects of health are increasingly being recognized as highly significant, making the social environment a key influence on population health. Whereas disease was once seen as something that struck down the unlucky, today our scientific world increasingly depicts disease as an avoidable event that can be circumvented if we make the right healthy choices. Much preventative medicine is based on trying to improve people's health via campaigns that nudge people toward healthier lifestyles, including anti-smoking strategies, vaccination programs, safe sun campaigns, or healthy eating promotions. Although personal choices undoubtedly influence health outcomes, it is important to acknowledge that our broader social environment constrains or enables particular choices for individuals. In this context, social scientists distinguish between **structural constraints** (aspects that constrain an individual's choices—often political, economic, or cultural factors) and the **agency** of individuals (the power a person has to make their own decisions). We know, for instance, that far more people from economically disadvantaged backgrounds smoke in the Global North, even though the health problems of smoking have been widely publicized to all socio-economic groups. Using a structural framework, we could consider what aspects of the broader social environment might make it more likely that a poorer individual takes up smoking than a richer person and/or less likely to give up once they have started. This might include stressful conditions associated with uncertain employment, living in communities where it is more likely that other people around you smoke, or a marginalized economic situation that offers little money to pursue other pleasurable activities.

Studies of the political and economic environments in which we live suggest that the social environments of many disempowered communities—whether marginalized by socio-economic status, race, or other factors—compromise health in numerous ways. In the United States, for example, where health inequalities are particularly pronounced, poorer communities tend to suffer from higher rates of heart disease and cancers, earlier onset of many degenerative conditions, higher rates of accidental death, and higher rates of suicide. Similar patterns appear if we compare African American health outcomes with those of non-Hispanic whites (box 9.3). Unfortunately, these health disparities seem to have been widening in

BOX 9.3 HEALTH INEQUALITIES IN THE UNITED STATES

Though the United States has some of the best healthcare in the world, the United States is also notable for having poorer health indicators than would be expected for its high income. The United States spends more per capita on healthcare than other comparable countries (approximately 17 percent of GDP or nearly twice as much as most other countries of the Global North), and yet the United States has a lower life expectancy, higher infant mortality rate, higher chronic disease burden, higher obesity rates, higher suicide rate, higher number of hospitalizations from preventable causes, and larger numbers of avoidable deaths than most countries at a similar level of economic development (Tikkanen and Abrams 2020). Young Americans, in particular, experience significantly worse health than their peers in other high-income countries, with Americans aged 15 to 24 more than twice as likely to die as their peers in other high-income countries such as Germany, Japan, and France. Meanwhile the United States' infant mortality rate is about three times that of other affluent nations (Rogers et al. 2022). A variety of factors contribute to this situation.

The United States is unique among high-income countries in not having a publicly funded, universal system of healthcare—less than 1 percent of the population lacks health insurance in most European countries; the rate is about ten times this in the United States (Hummer and Hamilton 2019). This leaves many Americans paying out of pocket for their healthcare, reducing their use of preventative care and leaving many conditions untreated until they trigger a crisis situation. Although the Affordable Care Act of 2010 ("Obamacare") showed some success in expanding the reach of publicly funded healthcare, the number of people lacking health insurance remains significant.

High rates of income inequality more broadly also contribute to poorer health, given abundant evidence that poverty takes a significant toll on health. Almost 15 percent of American children were growing up in poverty in 2019, giving the United States one of the highest rates of child poverty in the high-income world (Kearney 2021). Poverty has been associated with poorer diet, higher rates of risk-taking behaviors, as well as higher stress and less power in social situations—all of which are associated with chronic disease.

Health inequalities extend beyond socio-economic status to other marginalized populations, too. If we look at data by race, broadly speaking, whites and Asians have the best health outcomes in the United States, whereas African Americans and Native Americans have the worst, with Latino populations presenting a mixed picture (low infant mortality rates and high life expectancy but high rates of obesity, diabetes, and disability). Recent immigrants tend to have better health outcomes than the US-born population, perhaps helping to explain why Latino populations have better-than-expected health given their minority status. African American populations, by contrast, experience higher rates of infant mortality, premature birth, obesity, and diabetes than US averages, as well as the onset of heart disease, diabetes, and stroke at significantly younger ages than white Americans (CDC 2017a; figure 9.7).

One factor explaining health differences by race and ethnicity is differences in socio-economic status among ethnic groups, with poverty responsible for a proportion of the poorer outcomes in populations of color. The confounding effect of lower average socio-economic status does not tell the whole story, however. Even affluent African Americans experience worse health than their white counterparts, suggesting that there is also something specific to the African American experience that generates poor health statistics.

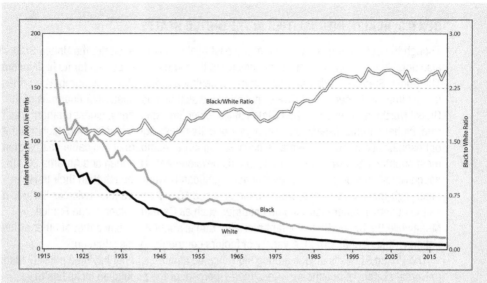

FIGURE 9.7 Black/white racial disparities in infant mortality rates in the United States
Infant mortality rates have declined over the past 100 years for both black and white populations. However, the ratio between these two rates has increased over the same time period, indicating that infant mortality rates in black populations have worsened relative to those for whites.
Data source: CDC (2022).

Geronimus (2001, 133) argued that for African Americans in the United States, the stress associated with "repeated experience with social, economic, or political exclusion" can lead to something akin to accelerated aging via a phenomenon known as "weathering." As a result, policies aimed at addressing health inequalities will not erase inequalities while racism persists (Hummer and Hamilton 2019).

many countries in recent years, leading health scholars to explore which aspects of modern society might be compromising health. One factor that is receiving significant attention is increasing inequality associated with changing political and economic structures.

Declining health and the social environment

Historical examples illustrate how political and economic upheavals can have profound influences on population health. One of the most famous cases was the declining health associated with the breakup of the Soviet Union in the early 1990s. This decline was particularly significant for young and middle-aged men, who saw a decrease in life expectancy at this time (figure 9.8). One estimate suggests 2.5 to 3 million excess deaths may have occurred among young and middle-aged adults between 1992 and 2001 (Men et al. 2003). Although scholars continue to debate exactly what triggered this decline in health status, the transition from a carefully

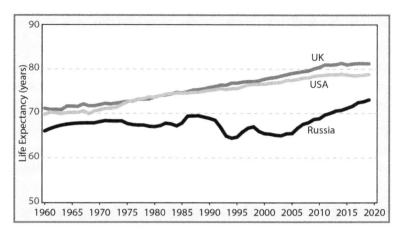

FIGURE 9.8 Life expectancy in Russia, compared with the UK and United States, 1960–2020
Notice the decline in life expectancy experienced in Russia in the early 1990s, coinciding with the breakup of the Soviet Union. Since the early 2000s, life expectancy has begun to increase again but is still not comparable to that in countries such as the UK and United States, which saw relatively steady increases in life expectancy throughout this period.
Data source: World Bank (n.d.).

controlled communist economy to a capitalist one undoubtedly generated economic instability, which is frequently associated with health declines. At the same time, the state reeled back many state-run programs, generating considerable uncertainty over the future for a population accustomed to significant government intervention in many aspects of life. Problems such as the collapse of the healthcare system, poor diet, high HIV rates, and high rates of alcohol and tobacco use undoubtedly explain the decline in life expectancies, but psychosocial stress experienced during and after the breakdown of the Soviet Union may be the most critical underlying factor (Heleniak 2010). These broader uncertainties and fears for the future also contributed to the significant decline in fertility rates recorded at the time, with fertility dropping from 2.1 babies per woman in 1988 to 1.2 in 1996 (Barkalov 1999).

More recently, deteriorating health has been recorded in the United States in low-income white populations and similarly attributed to economic instability. The year 2015 marked the first decline in life expectancy in more than 20 years in the United States, followed by further declines in 2017 and 2018 (Acciai and Firebaugh 2017). White, non-Hispanic, middle-aged Americans without a college degree experienced some of the greatest declines in life expectancy, even as older and more educated groups continued to see life expectancy improvements, suggesting that disadvantages associated with low socio-economic status were pivotal. The exact cause of these deteriorating health outcomes is still a matter of debate, but one suggestion is that diminishing job opportunities for those with limited education in an increasingly competitive globalized economy may be partly responsible (Case and Deaton 2015). By this argument, globalization of the economy is leading to growing divides between "winners" and "losers," as affluent professional classes in the Global North reap the rewards of globalization while many blue-collar workers have seen their jobs move overseas. This has led not only to unemployment and economic hardship for blue-collar workers but also a loss of status, which seems to be affecting white men in particular, given their traditionally high status in American society. For many white men of limited education, the well-paid, benefited jobs in manufacturing that their fathers held are simply no longer available, and the jobs now on offer involve either the need for significant retraining or a loss of status.

The resulting health crisis has been noteworthy, with the term **deaths of despair** coined to recognize increasing mortality from a group of causes linked to psychological distress. This includes accidental poisoning (particularly drug overdose), suicide, and alcohol-related liver disease (Case and Deaton 2015). In the United States, opioid overdoses present a significant aspect of this, with prescriptions for opioid painkillers reaching crisis proportions in recent years. Although the opioid crisis cuts across American society, those of lower socio-economic status are particularly likely to be affected. Many poorer communities have traditions of manual labor, making them especially likely to have received a prescription for an opioid painkiller because of an occupational injury; furthermore, many low-income Americans are underinsured, meaning that prescriptions for painkillers may be offered to them rather than more expensive treatments such as physiotherapy. This exposure to opioids, combined with stressful and unstable economic conditions, has contributed to high rates of opioid misuse in many low-income communities. Some scholars have gone so far as to argue that the opioid crisis is the primary factor explaining rising death rates in low-income white populations in the United States.

It is not just low-income Americans who are experiencing deteriorating health, however. Common with many other parts of the Western world, broader changes to American society are also being held responsible for health declines across society. Social isolation has received particular attention in recent years. Increasingly fluid family structures and aging societies have both been associated with larger numbers of people living alone. In the United States, one-third of people over 65 live alone and 40 percent of Americans over 45 say that they are lonely (AARP 2010). Loneliness has been associated with a variety of negative health outcomes, including heart disease, stroke, arthritis, and type II diabetes. Mental health problems are also associated with loneliness, including higher rates of depression, anxiety, dementia, and suicide. Certain sub-populations are particularly at risk of social isolation related to longstanding social marginalization, including immigrants, LGBTQ+ populations, ethnic minorities, and older adults (NASEM 2020). Isolation is not just a problem in the Global North. In China, for instance, the suicide rate among those over 70 is more than four times that of the general Chinese population (and twice the global average for the elderly). Notably, suicide decreases during Chinese New Year when families reunite, suggesting that social isolation is a key contributory factor (Fang et al. 2021). Some countries have experimented with trying to improve social connections among community members—for example, through social clubs—as an indirect way to improve community health. For instance, Kelly et al. (2021) documented the positive impacts of "men's sheds" on the health of older men—a group that is particularly likely to experience social isolation on retiring from the workforce. Originating in Australia in the 1990s but now also used in other Western countries, men's sheds provide a workshop environment for men to engage in meaningful practical activities with the goal of also fostering positive social interactions and have been found to lower rates of depression and increase confidence among participants (Kelly et al. 2021).

Health and wellbeing

Though the social environment can trigger negative health outcomes, it is equally possible to enhance people's lives with improvements to the social environment, as the example of men's sheds illustrates. This includes efforts as diverse as improving access to healthcare via policy changes, reducing social inequalities via economic restructuring, or fostering a positive cultural environment via education. These approaches view health as influenced by broader structures of society and use a **holistic** framework to think beyond clinical solutions to health problems.

This holistic framework suggests that a wide variety of factors need to be considered to estimate the relative "healthiness" of different communities. The Bloomberg Global Health Index, for instance, provides a healthiness ranking based on measures as diverse as access to healthcare, quality of sanitation, relative incidence of risk-taking behaviors, pollution levels, and rates of chronic conditions such as obesity. In 2020, the Bloomberg Global Health Index concluded that Spain had the healthiest population in the world, owing to factors such as a Mediterranean diet rich in good fats and low in processed foods, high-quality universal healthcare, and a high proportion of people walking when possible (WorldHealth.Net 2020). According to these sorts of criteria, the healthiest countries are all affluent and mostly in Europe and Asia (table 9.3). The United States falls behind many other Western countries at position 35, primarily due to high rates of obesity (WorldHealth.Net 2020).

Beyond the obvious impact of affluence, these rankings also reinforce the importance of equitable societies, with political systems focused on providing high levels of social care generating better population health indicators than those of more individualistic societies. A variety of other cultural factors, particularly related to diet and family structure, are also shown to be significant, with aspects as diverse as a tradition of eating family meals to how likely kids are to walk to school contributing to societal health.

Although we have so far focused on physical measures of health such as obesity rates, population health goals are increasingly focused around less tangible measures

TABLE 9.3 The world's "healthiest" countries, 2020

Ranking	Country
1	Spain
2	Italy
3	Iceland
4	Japan
5	Switzerland
6	Sweden
7	Australia
8	Singapore
9	Norway
10	Israel

Source: WorldHealth.Net (2020).

as well. The idea of **wellbeing** is now commonly used to move us beyond simple physical health toward the idea that we also need to strive toward broader life satisfaction. Wellbeing focuses attention on facets such as sound mental health, strong inter-personal connections, and even ideas like hope for the future. Though the value of wellbeing as a goal for a population is probably one we can all agree upon, it is a challenging concept to define and often hard to address with policy measures. Nonetheless, there are efforts to try to measure these sorts of aspects. The World Happiness Report, for instance, combines countries' economic standing (measured by GDP) with information on social support, healthy life expectancy, the freedom to make your own life choices, perceptions of corruption, and measures of generosity to produce "happiness" rankings. European countries tend to score highest, with Finland topping the rankings in 2020, owing to its high GDP and strong social safety net, as well as a strong sense of community combined with personal freedoms. Afghanistan, with extreme political instability and significant poverty, scored lowest (World Population Review 2021).

Conclusion

Although we have made huge strides toward improving population health over the last 100 years, a variety of issues urge us to consider whether health can also worsen over time. From a sustainability perspective, we are forced to engage with the uncomfortable reality that many aspects of modern lifestyles are not sustainable from an environmental or health perspective. Rising rates of infectious disease associated with landscape change are perhaps the clearest illustration of how health, environment, and economic development are intricately interconnected, as well as one of the strongest arguments for promoting sustainability more broadly. A "One Health" approach asks us to acknowledge that epidemics of infectious disease might be an inevitable consequence of unsustainable forms of economic growth and modernization that have sacrificed broader ecological sustainability and the health of human communities for short-term profit.

At the same time, our social environment is being called into question as the decline of many community structures coincides with increased globalization of the economy and rising inequality. In some contexts, this is leading to a breakdown of traditional health-promoting structures such as strong community ties and government safety nets. The result is a rise in negative health outcomes, particularly for those at the bottom of the socio-economic hierarchy, as well as those otherwise marginalized by society. It is important to remember, however, that social structures can also support health and that rebuilding opportunities for positive interactions within communities may be as important for promoting health as traditional health policies like vaccination or anti-smoking campaigns.

Discussion questions

1   In what ways is human health intimately tied to the natural environment? How might we use a "One Health" approach to rethink how we address infectious disease?

2   Reflect on some of the ways in which globalization can influence the spread of infectious disease. In what ways might our current globalized economy be particularly vulnerable to the development of a pandemic? How do we see these vulnerabilities illustrated in the case of the COVID-19 pandemic?

3   "Deaths of despair" and the obesity crisis both reflect situations of worsening health for those of lower socio-economic status in affluent countries, even as richer members of society are living longer and healthier lives. Might health policy be better served by policies that address economic inequality and education, rather than focusing narrowly on health promotion? Why or why not?

4   Think about your own experiences with health and healthcare. What *structural constraints* have you experienced in trying to lead a healthy life? In what circumstances did your own personal *agency* allow you to overcome these barriers? How might people living in other circumstances be more or less likely to have experienced these barriers?

Suggested readings

The Commonwealth Fund. 2020. "U.S. Health Care from a Global Perspective, 2019: Higher Spending, Worse Outcomes?" The Commonwealth Fund, January 30, 2020. https://www.commonwealthfund.org/publications/issue-briefs/2020/jan/us-health-care-global-perspective-2019

Cunningham, A., P. Daszak, and J. Wood. 2017. "One Health, Emerging Infectious Diseases and Wildlife: Two Decades of Progress?" *Philosophical Transactions of the Royal Society B* 372: 20160167. http://dx.doi.org/10.1098/rstb.2016.0167

Lee, H. 2013. "The Making of the Obesity Epidemic." *The Conversation*, March 20, 2013. https://thebreakthrough.org/journal/issue-3/the-making-of-the-obesity-epidemic

Morse, S., J. Mazet, M. Woolhouse, C. Parrish, D. Carroll, et al. 2012. "Prediction and Prevention of the Next Pandemic Zoonosis." *Lancet* 380 (9857): 1956–65. https://doi.org/10.1016/S0140-6736(12)61684-5

Glossary

agency: the ability of an individual to make their own choices; used in contrast to structural constraints

airborne disease: an infectious disease that is spread through droplets in the air or exhaled and picked up from surfaces

case fatality rate: the number of deaths from a particular cause of death as a proportion of all cases of that disease

chronic disease: a disease that develops over and persists for a long period of time; for example, cancer, obesity

deaths of despair: term coined to reflect deaths from causes such as alcoholism, drug abuse, and suicide that are potentially connected to rising inequality and declining economic fortunes for poorer communities in affluent societies

diseases of affluence: diseases traditionally associated with aspects of affluent Western lifestyle, such as rich diet and sedentary lifestyle; for example, heart disease

ecology: the study of the interactions of communities of organisms with their environments

epidemic: an outbreak of infectious disease; now sometimes applied also to non-infectious diseases; for example, obesity epidemic

holistic: taking a broad approach that emphasizes complex interconnections; with respect to health, this implies considering more than physiological aspects to health, particularly the influence of the social environment on health, including mental health

host: in the context of disease, an organism that is providing a place for another organism to live

infectious disease: disease caused by a living agent of disease such as a virus or bacteria

megacity: a very large city, currently usually defined as having at least 10 million inhabitants

morbidity: illness or injury

mortality: death

pandemic: an infectious disease outbreak of global scope

pathogen: a disease-causing organism (e.g., virus)

spillover event: the spread of a disease from an animal population into the human population

structural constraints: structures of society, often economic and political, that limit individuals' choices

urban penalty: compromised health related to living in an urban area associated with factors such as overcrowding and contaminated water

vector: an organism that transmits disease from one host to another; for example, a mosquito or tick

vector-borne disease: a disease that is spread from host to host by a vector

waterborne disease: a disease that is spread in contaminated water

wellbeing: a state of health that considers holistic ideas of fulfillment, contentment, and psychological soundness, in addition to physical health

Works cited

AARP [American Association of Retired Persons]. 2010. "Loneliness among Older Adults: A National Survey of Adults 45+." AARP, September 2010. https://assets.aarp.org/rgcenter/general/loneliness_2010.pdf

Acciai, F., and G. Firebaugh. 2017. "Why Did Life Expectancy Decline in the United States in 2015? A Gender-Specific Analysis." *Social Science & Medicine* 190: 174–80. doi:10.1016/j.socscimed.2017.08.004

Africa Center for Strategic Studies. 2020. "Lessons from the 1918–1919 Spanish Flu Pandemic in Africa." ACSS, May 13, 2020. https://africacenter.org/spotlight/lessons-1918-1919-spanish-flu-africa

AmeriGEOSS. n.d. "Sub-national Time Series Data on Ebola Cases and Deaths in Guinea, Liberia, Sierra Leone, Nigeria, Senegal and Mali since March 2014." Accessed March 9, 2022. https://data.amerigeoss.org/dataset/rowca-ebola-cases/resource/9b68ab69-e0b3-4ff5-b2d8-90bf446acd52

Barkalov, N. 1999. "The Fertility Decline in Russia, 1989–1996." *Genus* 55 (3/4): 11–60. https://www.jstor.org/stable/29788609

Barry, J. 2017. "How the Horrific 1918 Flu Spread across America." *Smithsonian Magazine*, November 2017. https://www.smithsonianmag.com/history/journal-plague-year-180965222/

Case, A., and A. Deaton. 2015. "Rising Morbidity and Mortality in Midlife among White Non-Hispanic Americans in the 21st Century." *Proceedings of the National Academy of Sciences* 112 (49): 15078–83. doi:10.1073/pnas.1518393112

CDC [Centers for Disease Control and Prevention]. 2017a. "African American Health." https://www.cdc.gov/vitalsigns/aahealth/index.html

———. 2017b. "SARS Basics Fact Sheet." CDC, December 6, 2017. https://www.cdc.gov/sars/about/fs-sars.html

———. 2018. "History of 1918 Flu Pandemic." https://www.cdc.gov/flu/pandemic-resources/1918-commemoration/1918-pandemic-history.htm

———. 2019a. "2009 H1N1 Pandemic." CDC, June 11, 2019. https://www.cdc.gov/flu/pandemic-resources/2009-h1n1-pandemic.html

———. 2019b. "2014-2016 Ebola Outbreak in West Africa." CDC, March 8, 2019. https://www.cdc.gov/vhf/ebola/history/2014-2016-outbreak/index.html

———. 2021. "Why Is CDC Concerned about Lyme Disease?" https://www.cdc.gov/lyme/why-is-cdc-concerned-about-lyme-disease.html

———. 2022. "Linked Birth and Infant Death Data." National Vital Statistics System, January 4, 2022. https://www.cdc.gov/nchs/nvss/linked-birth.htm

Cunningham, A., P. Daszak, and J. Wood. 2017. "One Health, Emerging Infectious Diseases and Wildlife: Two Decades of Progress?" *Philosophical Transactions of the Royal Society B* 372: 20160167. http://dx.doi.org/10.1098/rstb.2016.0167

Denevan, W. 1992. "The Pristine Myth: The Landscape of the Americas in 1492." *Annals of the Association of American Geographers* 82 (3): 369–85. https://doi.org/10.1111/j.1467-8306.1992.tb01965.x

Diamond, J. 1997. *Guns, Germs, and Steel: The Fates of Human Societies.* New York: W.W. Norton.

Fang, H., Z. Lei, L. Lin, and P. Zhang. 2021. "Family Companionship and Elderly Suicide: Evidence from the Chinese Lunar New Year." March 2021, NBER Working Paper Series. https://www.nber.org/system/files/working_papers/w28566/w28566.pdf

Fischer, D., R. Stahlmann, S. Thomas, and C. Beierkuhnlein. 2009. "Global Warming and Exotic Insect-Borne Diseases in Germany." *Geographische Rundschau International Edition* 5 (2): 32–8.

Geronimus, A. 2001. "Understanding and Eliminating Racial Inequalities in Women's Health in the United States: The Role of the Weathering Conceptual Framework." *Journal of the American Medical Women's Association* 56 (4): 133–36, 149–50.

Heleniak, T. 2010. "Mortality Trends in the Former USSR." *Geographische Rundschau International Edition* 6 (3): 56–62.

Hinne, I., S. Attah, B. Mensah, A. Forson, and Y. Afrane. 2021. "Larval Habitat Diversity and *Anopheles* Mosquito Species Distribution in Different Ecological Zones in Ghana." *Parasites and Vectors* 14, 193. https://doi.org/10.1186/s13071-021-04701-w

Hummer, R., and E. Hamilton. 2019. *Population Health in America.* Oakland: University of California Press.

Johns Hopkins Coronavirus Resource Center. 2022. "COVID-19 Dashboard." Center for Systems Science and Engineering, April 17, 2021. https://coronavirus.jhu.edu/map.html

Johnson, N., and J. Mueller. 2002. "Updating the Accounts: Global Mortality of the 1918–1920 'Spanish' Influenza Pandemic." *Bulletin of the History of Medicine* 76 (1): 105–15.

Jones, K., N. Patel, M. Levy, A. Storeygard, et al. 2008. "Global Trends in Emerging Infectious Diseases." *Nature* 451 (Feb): 990–4. https://www.nature.com/articles/nature06536

Kalra, S. 2013. "Diabesity." *Journal of Pakistan Medical Association* 63 (4): 532–4. https://pubmed.ncbi.nlm.nih.gov/23905459/

Kearney, M. 2021. "Child Poverty in the U.S." Econofact, February 5, 2021. https://econofact.org/child-poverty-in-the-u-s

Kelly, D., A. Steiner, H. Mason, and S. Teasdale. 2021. "Men's Sheds as an Alternative Healthcare Route? A Qualitative Study of the Impact of Men's Sheds on User's Health Improvement Behaviours." *BMC Public Health* 21: 553. https://doi.org/10.1186/s12889-021-10585-3

Lee, H. 2013. "The Making of the Obesity Epidemic." The Conversation, March 20, 2013. https://thebreakthrough.org/journal/issue-3/the-making-of-the-obesity-epidemic

Lieberman, D. 2013. *The Story of the Human Body: Evolution, Health, and Disease*. New York: Knopf Doubleday Publishing Group.

Mann, C. 2006. *1491: New Revelations of the Americas before Columbus*. New York: Random House.

Men, T., P. Brennan, P. Boffetta, and D. Zaridze. 2003. "Russian Mortality Trends for 1991–2001: Analysis by Cause and Region." *British Medical Journal* 327 (7421): 961. doi:10.1136/bmj.327.7421.964

Morse, S., J. Mazet, M. Woolhouse, C. Parrish, et al. 2012. "Prediction and Prevention of the Next Pandemic Zoonosis." *Lancet* 380 (9857): 1956–65. https://doi.org/10.1016/S0140-6736(12)61684-5

NASEM [National Academies of Sciences, Engineering, and Medicine]. 2020. *Social Isolation and Loneliness in Older Adults: Opportunities for the Health Care System*. Washington, DC: The National Academies Press.

OurWorldinData. 2016. "Share of Adults That Are Obese, 2016." https://ourworldindata.org/obesity

Patterson, K., and G. Pyle. 1991. "The Geography and Mortality of the 1918 Influenza Pandemic." *Bulletin of the History of Medicine* 65 (1): 4–21. https://www.jstor.org/stable/44447656

Rogers, R., R. Hummer, E. Lawrence, T. Davidson, and S. Fishman. 2022. "Dying Young in the United States." *Population Bulletin* 76 (2). https://www.prb.org/resources/dying-young-in-the-united-states/

Roos, R. 2011. "Study Puts Global 2009 H1N1 Infection Rate at 11% to 21%." CIDRAP, August 8, 2011. https://www.cidrap.umn.edu/news-perspective/2011/08/study-puts-global-2009-h1n1-infection-rate-11-21

Tikkanen, R., and M. Abrams. 2020. "U.S. Health Care from a Global Perspective, 2019: Higher Spending, Worse Outcomes?" The Commonwealth Fund, Issue Briefs, January 30, 2020. https://www.commonwealthfund.org/publications/issue-briefs/2020/jan/us-health-care-global-perspective-2019

Walters, J. 2004. *Six Modern Plagues and How We Are Causing Them*. Washington, DC: Island Press.

WHO [World Health Organization]. 2015. "Cumulative Number of Confirmed Human Cases for Avian Influenza A (H5N1) Reported to WHO, 2003–2015." WHO, March 31, 2015. https://www.who.int/influenza/human_animal_interface/EN_GIP_201503031cumulativeNumberH5N1cases.pdf

———. 2017. "West Nile Virus." WHO, October 3, 2017. https://www.who.int/news-room/fact-sheets/detail/west-nile-virus

———. 2019a. "Japanese Encephalitis." WHO, May 9, 2019. https://www.who.int/news-room/fact-sheets/detail/japanese-encephalitis

———. 2019b. "Middle East Respiratory Syndrome Coronavirus (MERS-CoV)." WHO, March 11, 2019. https://www.who.int/news-room/fact-sheets/detail/middle-east-respiratory-syndrome-coronavirus-(mers-cov)

———. 2019c. "Yellow Fever." WHO, March 7, 2019. https://www.who.int/en/news-room/fact-sheets/detail/yellow-fever

———. 2020. "Vector-borne Diseases." WHO, March 2, 2020. https://www.who.int/news-room/fact-sheets/detail/vector-borne-diseases

———. 2021a. "MERS Situation Update, October 2021." WHO, October 2021. http://www.emro.who.int/health-topics/mers-cov/mers-outbreaks.html

———. 2021b. "Reducing Public Health Risks Associated with the Sale of Live Wild Animals of Mammalian Species in Traditional Food Markets." World Organisation for Animal Health, WHO, and UNEP, April 12, 2021. WHO/2019-nCoV/Wet_Markets/2021.1

———. 2022a. "Diabetes." https://www.who.int/health-topics/diabetes#tab=tab_1

————. 2022b. "Global Health Estimates 2019 Summary Tables." https://www.who.int/data/global-health-estimates

————. 2022c. "Yellow Fever." https://www.who.int/health-topics/yellow-fever#tab=tab_1

Woodward, A. 2021. "Suspicions Mount that the Coronavirus Was Spreading in China and Europe as Early as October, Following a WHO Investigation." *Business Insider,* February 10, 2021. https://www.businessinsider.com/coronavirus-circulated-europe-china-before-wuhan-outbreak-2020-12

World Bank. n.d. "Life Expectancy at Birth, Total (Years)—Russian Federation." Accessed March 1, 2022. https://data.worldbank.org/

WorldHealth.Net. 2020. "Bloomberg's Global Health Index for 2020." Bloomberg, June 18, 2020. https://worldhealth.net/news/bloombergs-global-health-index-2020/

World Population Review. 2021. "Happiest Countries in the World 2021." https://worldpopulationreview.com/country-rankings/happiest-countries-in-the-world

10

Migration patterns and theories

After reading this chapter, a student should be able to:

1 describe major historic and current migration flows;

2 discuss the major types of migration and their motivations;

3 recognize major migration theories.

Migration is one of the most contentious population issues of our times because of the multiple effects on the countries of origin and destination, as well as on the migrants themselves. For migrants, a migration is often a mixture of hope and hardship: hope for a better life elsewhere but also hardships associated with leaving behind friends, family, and country. Hope can become reality when a migrant finds a job abroad that allows them to support their family or achieve their career goals. Unfortunately, all too often, achievements are tempered by negative migration experiences, including exploitation, racism or discrimination, and even deportation. In the worst-case scenario, migration can be deadly, especially for people who try to reach other countries by boat, as stowaways, or over dangerous land crossings.

The effects of migration are equally diverse for the hosting communities. Economically, migrants can provide a significant boost to a destination country by filling labor shortages, transferring knowledge, or creating jobs through entrepreneurship. However, migrants may also be blamed for undercutting wages or taking jobs from native workers, as well as posing a financial burden on the destination country if they need support, although many scholars argue that immigrants contribute more economically to host countries than they take out. For the country of origin, outmigration may help relieve unemployment or population pressure, as well as provide an economic boost through **remittances** (money migrants send back to their families). On the other hand, it is often well-qualified workers who are most able to migrate elsewhere, which can result in a **brain drain** of some of the most dynamic and accomplished members of society.

From a demographic perspective, hosting countries tend to benefit from gaining young, economically active people because most migration streams are dominated

DOI: 10.4324/9781003143253-10

by young adults, whereas countries of origin often suffer from problems associated with large dependent populations if children and the elderly are left behind. Immigrants change the demographic profile of a country not just by swelling the numbers through their own immigration but also because immigrants tend to arrive in their childbearing years and may bring patterns of higher fertility from their country of origin, thereby raising fertility rates.

Migration can also substantially shape the ethnic and racial composition of a country. Cultural diversity is often viewed as a positive impact of migration, but the extent to which diversity is valued varies widely. For countries with a traditionally unified ethnic identity, such as many European countries, migration may be seen as diluting national identities. In Germany, for example, the Pegida movement (Patriotic Europeans against the Islamicisation of the Occident) is based on the premise that immigrants are threatening native culture (Murray 2017). This movement and other expressions of **Islamophobia** are now evident in a variety of European countries (box 10.1). On the other hand, many European citizens keenly support the value of cultural diversity, particularly after the racist atrocities of World War II, and strongly condemn racial discrimination (figure 10.1). At another extreme, countries that have been shaped by large-scale immigration may see diversity as an explicit part of their national identity. Canada, for instance, defines itself as multicultural. **Multiculturalism** is based on the idea that immigrants retain many of the cultural characteristics of their home country and that the broader community benefits from a diversity of cultural practices. Another approach is the idea of countries being a cultural **melting pot**, where people of different cultures "melt" into the culture of the host country. The United States officially endorses the melting pot ideal, although in reality it is more multicultural. Despite these diversity ideals, cultural, racial, and ethnic tensions continue to simmer even in countries with long histories of immigration.

Politically, migration is extremely controversial as well. Migration policies are a compromise between competing interests, with governments wanting to appear to be protecting their own citizens by carefully moderating immigration and cracking down on illegal migrations, while at the same time allowing the entry of certain migrants, particularly those who are highly skilled or who are willing to take jobs that the native population is unwilling to do. Increasingly, some economists are also urging governments to see migration as a potential solution to aging populations and rising **dependency ratios**. However, there has been a recent backlash against immigrants in many countries, associated with what many perceive as cultural threats, as well as immigrants being blamed for job losses (that are arguably a result of broader processes of global economic restructuring rather than immigration). Governments are left treading a fine line between these diverging perspectives on immigration. The United States provides a particularly strong example of this in recent years. President Trump fueled negative opinions and even fear of immigration, imposed a ban on migrants from several Muslim countries, and vowed to seal the Mexican border by extending the border wall. As soon as he took office, President Biden reversed many of the migration policies implemented during the Trump administration, showing how politically divided the United States is on migration matters. Over the last few decades, Europe has also seen the rise of anti-immigrant sentiment, with nationalist parties such as the National Rally

BOX 10.1 BURQA BANS AND ISLAMOPHOBIA IN EUROPE

Tensions over cultural differences arise frequently between migrants and the societies to which they migrate. Conservative migrant groups may oppose the more liberal ways of the countries they move to, or destination societies may object to values and traditions that immigrants bring with them. In Europe, some of the strongest tensions emerge between native populations with Christian traditions and recent Islamic immigrants.

Though anti-Islamic sentiment sometimes leads to acts of outright racism, including violent attacks, in other cases the controversies are subtler. One well-known case is *l'affaire du foulard* (the headscarf affair) that began in France in the 1980s. Three female Muslim students were suspended from their French school because they refused to remove their headscarves and the wearing of headscarves was seen as incompatible with the French value of *laïcité* (secularism). Ultimately, France passed a law forbidding the wearing of "ostentatious religious symbols" in 2004. The law prohibits the wearing of not only Islamic veils but also Jewish kippas and large Christian crosses (discreet crosses and Jewish Stars of David are allowed) but is widely perceived as targeting Muslim traditions (Auslander 2000).

The wearing of veils (hijabs) and other headcoverings by Muslims is viewed differently by different people. Some feminists and human rights advocates interpret the hijab as a sign of female oppression and argue that women wear it because of social pressure. They point out that though Islam requires women to dress modestly, it does not mandate that a woman cover her face, suggesting that veils are more of a cultural preference than a religious requirement (Ali-Karamali 2008). Others argue that wearing a headcovering or veil is a woman's choice and should be protected as a religious practice. In this view, forbidding women to veil not only targets Muslims unfairly but is also a violation of human rights that protect religious practices.

Since 2011, France has prohibited full-face veils in public places such as public transportation, parks, and even streets, and in 2021 an even stricter ban was proposed (for example, forbidding girls under 18 to wear a headscarf in public), triggering a fierce debate between people who see this as reinforcing France's commitment to secularism and those who interpret this as Islamophobia (Lang 2021). "Burqa bans" (a burqa is a flowing dress with a mesh over the eyes—one of a variety of Islamic head coverings) have now spread to many other European countries, despite the very small number of women wearing burqas in Europe (about 300 each in Belgium and the Netherlands, for instance). Belgium passed a burqa ban in 2011, the Netherlands in 2016, Austria in 2017, Denmark in 2018, and Switzerland in 2021, and Germany and Norway are considering following suit (Müller 2019). Some non-European countries have also banned burqas and some other forms of veils (such as Sri Lanka) or ban them in certain settings such as schools (even in some Muslim countries). One of the reasons cited for these bans is public safety, because a person cannot be recognized under a full-face veil.

(National Front) in France, Alternative für Deutschland (Alternative for Germany), and Freiheitliche Partei Österreichs (Austrian Freedom Party), even as other parties try to maintain a public face that is welcoming to immigrants.

Numerous other examples could be listed, but even this brief overview clearly illustrates that the impact of migrations on both origin and destination countries is complex. It also illustrates how migration can be perceived as both a solution to

FIGURE 10.1 "Many Walls Have to Be Torn Down"
This mural by Ines Bayer and Raik Hönemann on Berlin's East Side Gallery—an open-air gallery covering more than a kilometer of the remaining stretch of the Berlin Wall—references not only the Berlin Wall but also the metaphorical walls between people of different origin. It is a monument to freedom and diversity.

problems (such as labor shortages or aging societies), as well as a problem (such as creating ethnic tensions or taking jobs from native workers), depending on the lens through which people view migration. In the next few pages we provide an overview of contemporary migrations, different types of migration, and migration theories. In the next chapter, we develop the idea of voluntary versus forced migrations to focus on human rights issues arising in the context of migration.

A global overview of current migration flows

We live in an "age of migration" (de Haas, Castles, and Miller 2020). In 2020, there were over 280 million international migrants in the world (UN 2021), representing a significant increase since the mid-20th century (figure 10.2). Globally, population growth and migration have generally kept in step, with the percentage of international migrants fluctuating around 3 percent of global population (de Haas, Castles, and Miller 2020). It is important to remember, however, that these numbers only refer to *international* migrants, not *internal* migrants—people who move within their own countries. It is estimated that the number of internal migrants far exceeds that

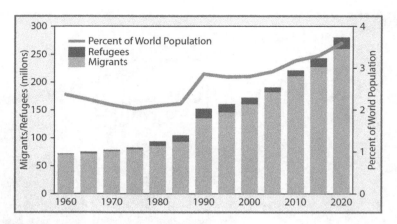

FIGURE 10.2 World migrant and refugee population, 1960–2020
Data source: UNHCR (2022) and UN (2021).

of international migrants, especially in the most populous countries of the world such as India, China, Indonesia, and Nigeria.

In regards to international migration, current flows are dominated by movements toward the affluent regions of North America, Europe, and the Gulf States, with significant regional flows also across the Russian border and in South Asia. As such, a relatively small number of countries host a large share of international migrants—the United States, Russia, Saudi Arabia, and Germany currently top the list with over 10 million immigrants each, followed by the UK, United Arab Emirates, France, Canada, and Australia (Pison 2019). If we consider countries of origin, the dense populations of Europe and Asia are the source of some of the largest migrant flows (figure 10.3).

To contemplate the impact of immigrants, however, it is often more meaningful to consider the percentage of the total population that are immigrants, rather than absolute numbers (figure 10.4). Here the United Arab Emirates (UAE), with immigrants comprising 87.3 percent of its population, stands out, followed by the other Gulf countries. Broadly speaking, countries with high proportions of immigrants can be divided into five groups: (1) sparsely populated countries with large oil reserves such as the UAE, Kuwait, and Qatar; (2) microstates or territories with special tax rules such as Macao, Monaco, and Singapore; (3) classic immigration countries (those that have actively attracted immigrants for a long time and continue to encourage at least some forms of immigration) with large territories and low population densities such as Australia and Canada; (4) rich western democracies such as Western European countries; and (5) countries of first asylum such as Lebanon and Chad, which host large numbers of refugees fleeing from a neighboring country (Pison 2019).

Several countries that host large numbers of migrants, such as the UK, Germany, and Russia, have relatively large numbers of their own citizens living in other countries, illustrating that countries can simultaneously be countries of origin and destination. Many immigrants to these countries come from less well-off countries (so-called vertical migration), but the citizens leaving these countries migrate primarily to other rich countries (horizontal migration), such as Germans going to the UK or Brits going to Australia.

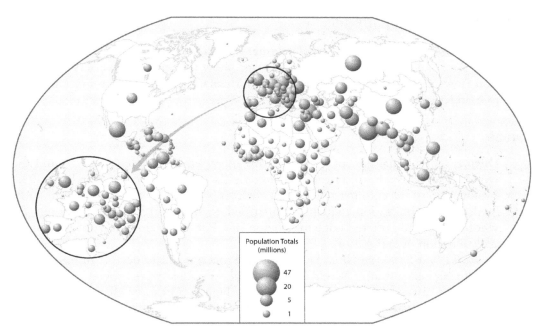

FIGURE 10.3 Distribution of migrants by place of origin, 2020
Data source: UN (2021).

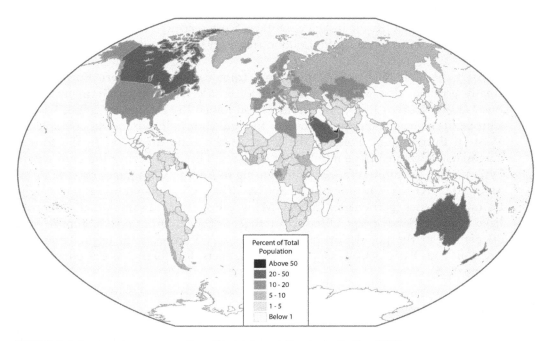

FIGURE 10.4 Immigrants, as a proportion of the total population, by destination, 2020
Data source: UN (2021).

Regional migration patterns

We now provide a short historical summary of migration flows by world region to help place the cases discussed in context and show the enormous variety of migration flows and patterns. We then turn to a thematic approach to migration in the remainder of the chapter.

Europe

During the colonial era, many Europeans migrated to their colonies around the world to serve as administrators, missionaries, soldiers, sailors, or merchants. The exploitation of the colonies not only made Europe rich but was also a driving force behind industrialization, which, in turn, led to impoverished wage laborers for whom emigration offered hope for a better life. For a long time, Europe was essentially a region of emigration.

World War I and the Great Depression led to a significant decline in migrations, but during and just after World War II countless people were moved against their will, as we discuss in detail in chapter 11. The biggest shift came after World War II, when Europe was transformed from a region of emigration to a major destination for migrants. In the immediate post-war years, migration was driven by decolonization, which led to the return of citizens from former colonies, as well as new immigrants from these colonial ties. This included people from the Caribbean and the Indian sub-continent moving to the UK, North Africans to France, and Indonesians to the Netherlands. Rapid economic growth also resulted in the active recruitment of thousands of migrant workers to Western European countries up to the 1970s. In the UK, for instance, workers were recruited in large numbers from former colonies to work in the London transport system and the National Health Service.

After the Cold War ended, migrations from Eastern to Western Europe were finally possible, first with the migration of ethnic Germans to Germany. Later, once several Eastern European countries had joined the European Union, Eastern Europe became an important source of foreign labor for a number of countries in Western Europe (e.g., Poles to the UK). At the same time, war in the former Yugoslavia generated large numbers of refugees. The Arab Spring and civil war in Syria created yet more refugees, contributing to Europe's "migrant crisis" of 2015. Since 2022, Russia's invasion of Ukraine has driven Ukrainian refugees into neighboring countries, particularly Poland. Many European (and some other) countries have been quick to accept Ukrainian refugees. Today, immigrants comprise over 30 percent of the population in some regions of Europe, with major cities like London and Paris acting as particular magnets for foreign-born populations (figure 10.5).

North and South America

The New World has been profoundly shaped by migration. During the colonial era, an estimated 12 million Africans were forcibly brought to the New World as slaves, resulting in sizeable communities of African origin in Brazil, the Caribbean, and the southeastern United States. After slavery was abolished, slaves were

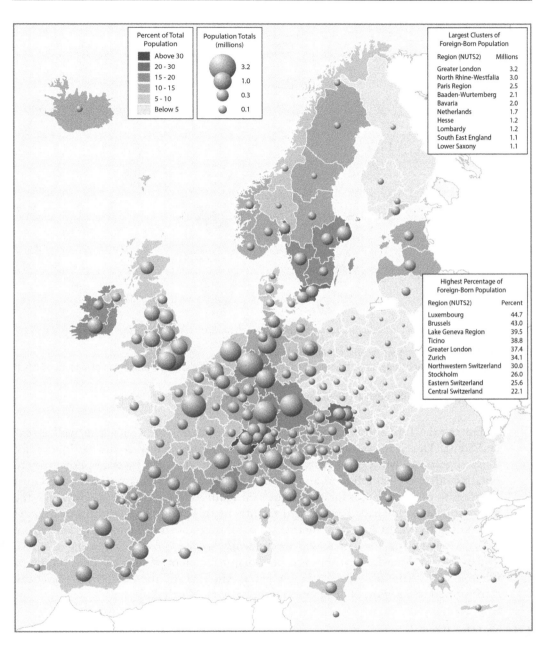

FIGURE 10.5 Foreign-born population in Europe, 2015
Data source: UN (2021); OECDstat (n.d.).

increasingly replaced by **indentured laborers** who were brought to the New World by particular employers, to whom they were tied for a fixed period of time to pay for their passage (figure 10.6). Many indentured laborers came from Asia, particularly to parts of the Caribbean and South America, adding to the ethnic mixture of the Americas.

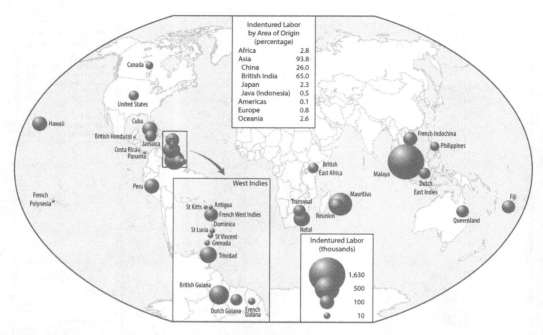

FIGURE 10.6 Indentured laborers by area of origin and destination
Symbols show the number of indentured laborers by destination. The table shows the proportion of indentured laborers by area of origin.
Data sources: Northrup (1995), Sandhu (1969), Roberts and Byrne (1966), and Unger (1944).

During the 19th and 20th centuries, the United States and Canada, as well as several South American countries such as Brazil and Argentina, were the destination for mass migrations. Most migrants were from Europe, attracted by the opportunity to obtain land, as well as religious and political freedom. After a hiatus during the Great Depression of the 1930s, migration picked up again after World War II. However, the migration landscape had changed significantly by this point. Though Europeans kept on coming to North America, their share of all migrants declined as migration from Latin America and Asia increased dramatically to the United States and Canada. Today, North America continues to be a major recipient of international migrants, whereas Latin America—particularly Central America and the Caribbean—has been transformed to a region of emigration, especially to North America.

Asia and the Pacific

Fueled by economic development and social change, Asia has become a major player in migration, both as a source region and for interregional migrations. Asian countries still vary widely in terms of economic development and therefore also migration movements, making it very difficult to draw out overarching trends. Since World War II, the two main migration flows have been post-colonial migrations to Europe, facilitated by preferential access for people from ex-colonies (e.g., Indians, Bangladeshis, and Pakistanis to the UK), as well as refugee flows during the Cold War (e.g., from Korea and Vietnam). Since the 1970s, labor migrations have gained in importance, including migration streams to North America and

Australia after exclusionary policies against Asians ended in those countries. In addition to affluent Western countries, the Gulf states have become major destinations for Asian labor migrants because oil wealth has generated more jobs than could be filled by native populations. Recently, wars and persecution have created renewed refugee flows in Asia, including conflicts in Afghanistan and Iraq and the persecution of the Rohingya who have been fleeing Myanmar.

The absolute number of Asian migrants is high due to the large population of Asia, but emigrants account for only a relatively small percentage of the population in most Asian countries. One exception is the Philippines, which is the largest exporter of labor in the region, having pursued labor export as a deliberate strategy (box 10.2). The Pacific Islands have also been characterized by high rates of outmigration, owing to limited economic opportunities in many of these tiny countries. Though Australia and New Zealand were once part of the European migration system, receiving large

BOX 10.2 THE PHILIPPINES—A LABOR EXPORTER

Very few countries in the world have as many citizens living abroad as the Philippines—over 10 million, or about 10 percent of the population. These migrants make up a sizeable proportion of global migrants in a number of industries, particularly nursing, domestic workers, and seafaring—a quarter of the world's mariners are Filipino, for example (Almendral 2018). The importance of migration for the Philippines cannot be overstated: the remittances that migrants send home account for about 5 percent of the country's GNP and a fifth of its export earnings, as well as being a lifeline for many families. Additionally, emigration helps relieve domestic unemployment. Migration is highly valued and celebrated in the Philippines—migrant workers are labeled as *bagong bayani* (heroes) and awards are given to the best migrant workers.

The situation is the outcome of a deliberate government policy. In the 1970s, the government began to promote labor migration to Gulf countries when rising oil prices increased demand for migrant labor. Both private agencies and a government agency, the Philippines Overseas Employment Administration (POEA), recruited labor for foreign employers and shipping companies. In the 1990s, the execution of a Filipina worker in Singapore extended the role of the POEA beyond the recruitment of workers to also promoting workers' wellbeing and rights (O'Neil 2004).

Originally, many Filipinos worked in construction and as mariners, but the migrations have diversified over time. Today, Filipinos and Filipinas also work in the health sector and other professional jobs, as cleaners or nannies, and in factories. Perhaps most famously, the Philippines is a major source of nursing staff, with almost 20,000 nurses leaving the Philippines to work abroad every year (Almendral 2018). In the UK, almost 20,000 Filipinos work for the National Health Service, equivalent to 10 percent of all Filipinos in Britain (Quinn 2020). Though male migrants originally outnumbered females, today women account for roughly half of all emigrants from the Philippines. Destinations have also diversified—many emigrants still work in Gulf countries, but many are now also employed in richer Asian countries, Europe, and North America (figure 10.7). Though many remain on temporary contracts, others have become permanent immigrants, especially in Europe and North America (O'Neil 2004).

Most migrants from the Philippines migrate as part of official programs, so they have some protection in case of mistreatment (for example, the Filipino government stepped in

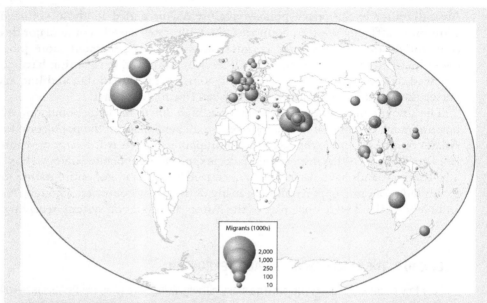

FIGURE 10.7 Destinations of Filipino migrants
Data source: UN (2021).

after several cases of abuse in Hong Kong), as well as access to insurance and pension plans. These programs also provide training specific to the destination country (for example, nannies going to the United States may be taught how to bake brownies!). This extensive preparation for their jobs abroad makes Filipino workers very attractive to employers.

Overall, the Philippines has succeeded in managing orderly migration flows to the economic benefit of its domestic population, while offering some protection to its citizens as workers. Many see it as an exemplary model to be emulated by other countries (Mendoza 2015). However, critics argue that the protection of workers is insufficient and that the government largely ignores the plight of undocumented migrants and victims of trafficking. Other concerns include the impact of migrations on families (especially children, who are often raised by grandparents while their parents work overseas), whether the export of laborers can sustain the economy in the long term, and the reliance on migration as the main form of social advancement in the Philippines.

numbers of settlers from Europe (particularly the UK), they are now increasingly integrated in the Asia–Pacific migration system as destination countries.

Africa

Africa has experienced interregional migration for a long time due to the large number of African countries, the artificial borders imposed during colonialism, and the large number of porous borders, especially in the Saharan region. In recent times, numerous conflicts have also led to large numbers of refugees within the continent. Though public discourse is often dominated by images of North and West Africans fleeing to Europe, the vast majority of African migrants continue to move *within* Africa (*The Economist* 2021), with South Africa as the main destination country and other migration streams flowing mostly from the interior of the continent towards the coasts.

Of all major world regions, Africa has the lowest levels of intercontinental migration because many countries remain relatively poor and isolated. Nonetheless, some migration flows have developed from Africa to other world regions. North African countries have been important senders of labor migrants to Europe since the 1950s. In addition, the migration of highly skilled Africans to Europe and North America has resulted in a significant brain drain from Africa, which is losing many of its doctors, engineers, and other skilled workers. In response to widely publicized examples of illegal boat migrations across the Mediterranean, Europe is increasingly cracking down on migrations from Africa.

The Middle East

Migration rates from the Middle East are higher than those from Africa. Some of the world's largest refugee groups come from this region, including over 5 million Palestinians who fled from Israel in various waves since the 1940s and over 6 million Syrians who escaped the turmoil of the Arab Spring and the civil war in Syria since 2011. Many of these refugees stay in the region, making it also one of the world's major refugee destinations. The oil-rich countries surrounding the Persian Gulf have also become major destinations for labor migrants from South and Southeast Asia since the 1970s (figure 10.8). The Middle East is now the third most important destination for migrant laborers after North America and Europe (Cummings and Christmann 2020).

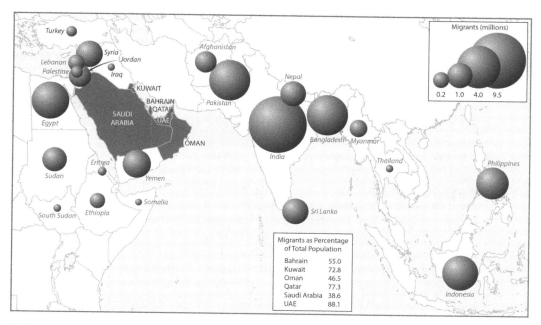

FIGURE 10.8 Major sources of labor migrants to the Gulf countries
The bubbles show the number of emigrants leaving each country to migrate to the Gulf States.
Data source: UN (2021).

Israel constitutes a special case in this region. As the Jewish homeland, Israel was designed to be a country of immigration. Israel has always been very open to the immigration of Jews, which it sees as returning co-ethnics, but fairly restrictive toward non-Jews (Raijman 2020). Until just after World War II, immigrants were mostly from Europe, many fleeing anti-Jewish policies promoted by the Nazis, but migrants now increasingly arrive from other world regions. Among the major immigration destination countries in the world, Israel is by far the smallest country.

Types of migration

The experiences of migrants, as well as the impact on countries of origin and destination, vary substantially by the type of migration. As we consider different types of migration, it is important to acknowledge that categories and labels can be highly contentious. To cite just a few examples: some forms of migration defy easy categorization if different factors are at work simultaneously; academics, policymakers, and the general public may use terms and categories in different ways; and certain categories or labels can have important, sometimes derogatory, connotations. For instance, the term "guestworker" clearly suggests that migrants are temporary *guests* and not future immigrants, which has been used to justify excluding them from full participation in destination societies. Similarly, the term "illegal alien," used in the United States to refer to migrants who are in the United States illegally, is often considered more hurtful than the term "undocumented migrant." Nonetheless, defining and categorizing migrants can be useful in gaining a more nuanced understanding of migration.

One of the first distinctions to underscore is the difference between mobility and migration. The term **mobility** refers to any human movement—regardless of distance covered, the duration of the movement, or whether the movement crossed an administrative border. This includes tourism, business travel, commuting, evacuating a city because of an approaching hurricane, and pilgrimages, for instance. The term migration is used when a person changes residence across an administrative border for an extended period of time, usually a year or more. **International migration** refers to people moving from one country to another; **internal migration** is moving within a country, such as moving to a different state or province. Though migrations across international borders often have more significant impacts (such as a different language, culture, and way of life, as well as bureaucratic implications), in some cases even a move across town can radically change a person's life—for example, when an African American moves from a predominantly African American neighborhood to a predominantly white one or a Catholic in Northern Ireland relocates to a predominantly Protestant neighborhood. Furthermore, some international migrations do not radically change circumstances for a migrant (de Haas, Castles, and Miller 2020). For example, a Dutch person moving to the Flemish-speaking part of Belgium, an Austrian moving to Germany, or a Moroccan moving to Algeria can continue to speak their language, and cultural differences may be relatively small. The administrative burden of moving also varies significantly. Moving from one country to another within the European Union may not require more paperwork or permissions than moving from Minnesota to Colorado in the United States, for instance. That being said, in most cases the distinction between international and internal migration is meaningful, because countries are still the primary entities that control who can enter their national territory and under what conditions.

A second major distinction is the duration of the migration, with the two most important categories being **permanent** and **temporary**. Some migrants relocate with the intention to stay in another country permanently. Others plan to stay abroad for a limited amount of time and then return home, including international students, expatriate workers, or labor migrants. However, it is quite common for migrations that were intended as temporary to ultimately become permanent. For example, a refugee may find that conditions in their country of origin do not improve sufficiently to allow a return, or an international student may obtain employment in the country where they studied or a partner who encourages them to settle permanently. Governments often stress that certain visas are only temporary to allay fears around immigration within the native population, while being cognizant of the fact that many temporary migrants will ultimately attempt to adjust their status to permanent residency. In the United States, for instance, international students must declare that they do not intend to immigrate permanently to obtain a student visa and yet it is clear that the United States benefits enormously from the influx of skills and expertise that come with international students, providing little incentive for the United States to force students to leave when their period of study is complete. Indeed, there are a number of ways in which international students can extend their stay or try to stay permanently.

Some migrations can best be described as periodic, circular, seasonal, or recurrent. For example, a person serving in the military who is deployed multiple times but periodically returns home can be said to engage in periodic migration. A student moving from their home to campus for the semester and back home during vacations engages in a circular migration. Another example of circular migrations can be found in the extractive industries, particularly mining and oil. Much of this work takes place in remote and often dangerous locations, so the workforce is rotated in and out on a regular basis.

Migrations can also be classified according to their primary motivation. **Labor migrations** are the most common migration and are often subdivided into highly skilled migrations (e.g., of doctors, researchers, or engineers) and low-skilled migrations (e.g., of manual laborers or farmworkers). However, this distinction is problematic as well because at least some forms of manual labor require significant skills, and some highly qualified migrants cannot find jobs in their destination countries that match their skills or have difficulties getting qualifications transferred. For example, a doctor may find that their qualifications are not recognized in their new host country and turn to working as a nurse. **Family migrations** occur to reunite families—for example, when a family follows a migrant who had originally arrived as a labor migrant—and **love** or marriage **migrations** bring together partners. **Lifestyle migrations** occur when people move because of lifestyle choices that encourage them to seek out a particular type of community (e.g., a gay community, senior community, or golfing community; box 10.3). Some people might even make finer distinctions among types of migrations, but it is important to remember that these categories are all based on the *primary* motivation, when in reality most migrations are driven by multiple motivations.

BOX 10.3 LIFESTYLE MIGRATIONS

Lifestyle migrations (also called amenity migrations) are voluntary migrations where people move in search of a particularly appealing community, environment, or experience. Some people move to participate in activities that are associated with specific places. For example, young people are often attracted to the "bright lights" of a large city. Members of the LGBTQ+ community may want to move to a large city because their lifestyle is more accepted there than in rural areas or small towns or because they want to live in a community of people who are similar to them. Within cities, people may move to particular neighborhoods for lifestyle choices such as students moving into student neighborhoods characterized by affordable housing and entertainment such as bars or young professionals moving into gentrifying inner-city neighborhoods with good access to amenities and short commutes.

Whereas some lifestyle migrations are directed toward cities, others go in the opposite direction. Since the 1970s, **counterurbanization** trends have developed in many affluent countries as people tired of city life make a conscious choice to live in rural areas to enjoy a slower-paced lifestyle and pleasant environment. Often the destinations chosen are on the outskirts of big cities so that the services and benefits of the city can still be accessed. Improvements in telecommunications have increasingly enabled some working-age adults to move out of the city to smaller communities that offer the benefits of rural life while technology allows them to stay connected to city jobs, giving rise to so-called Zoom towns. Some people also move to rural areas to engage in specific activities such as skiing or surfing. As with many other migration streams, counterurbanization is highly selective, with middle-aged and older people of higher socio-economic status predominating. In some cases, these rural migrations have revitalized areas that had been suffering population losses, such as mountain villages in the European Alps (Steinicke and Löffler 2019). On the other hand, amenity migrations become a significant problem if affluent newcomers price local house buyers out of the market. In the desirable ski towns of the Rocky Mountains, for instance, many long-term residents can no longer afford housing in resorts and workers employed in the ski industry often have to commute long distances.

Retirement migrations are another important form of lifestyle migration, which can be permanent (e.g., retirees moving from Northern to Southern Europe in search of peaceful village life) or temporary (e.g., "snowbirds" in the United States who leave the cold north for the southern states every winter). In Europe, retirement migrations often cross international borders to reach warmer climes and coastal locations. We even see migration channels, such as Brits preferring the Costa del Sol, the French Catalonia and Valencia, and Germans the Balearic and Canary Islands in Spain. These places, in turn, often cater to these nationals—the island of Mallorca is jokingly called the 17th federal state of Germany because 60,000 Germans spend their winters there (and many more visit as tourists throughout the year) and many services are available in German. Though providing special services to migrants is, of course, appreciated by these groups, in some cases it may lead to resentment if locals feel that their communities are being taken over by newcomers.

FIGURE 10.9 Refugees then and now
Sources: Hoke (2018) and Anonymous (1914).

Migrants are also frequently categorized according to their legal status, resulting in additional categories such as refugees and asylum seekers and illegal, undocumented, or irregular migrants. **Refugees** have long been part of the migration landscape (figure 10.9). According to the 1951 United Nations Refugee Convention (also called the Geneva Convention), a refugee is a person "who is unable or unwilling to return to their country of origin owing to a well-founded fear of being persecuted for reasons of race, religion, nationality, membership of a particular social group, or political opinion" (UNHCR, n.d.). It is important to note that people who flee persecution must cross an international boundary to qualify as refugees; those who flee within a country are labelled **internally displaced people** (box 10.4). The two groups nonetheless experience many of the same challenges and hardships. The term refugee is also often confused with **asylum seeker**—a person who has applied for refugee status but has not yet been granted it. Once a person has been granted refugee status, they are entitled to the basic civil rights that all other legal immigrants of a country have, including access to services such as healthcare and education. However, many asylum seekers' applications are unsuccessful and they are never recognized as refugees.

With respect to refugees, we can distinguish between acute flows and anticipatory flows. For example, when Rwandans fled from Rwanda during the genocide in the early 1990s, the violence had already begun, so this was an acute flow. In other cases, people fear that conditions may worsen in the future and flee while they are still able to leave in anticipatory flows. Some Ukrainians began moving toward Ukraine's western borders in anticipation of a Russian invasion in early 2022, for instance, once it became clear that Russia was building up forces at Ukraine's eastern border. The crisis subsequently became acute when Russia invaded Ukraine, leading to millions more leaving the country.

Persecution based on race, ethnicity, religion, and nationality are probably the most common reasons to flee, but people may also be persecuted because of other characteristics—in some countries, homosexuality is still illegal, and in others gender-based violence is common, for example. Some policymakers and scholars argue that the definition of a refugee should be expanded to account for this greater diversity of potential persecutions. Others argue against lengthening the list and note that most persecuted groups could already receive refugee status through being persecuted as members "of a particular social group," which is already referenced in the current definition (Copeland 2003).

Apart from in the case of family reunification and marriage, migrants who cannot claim refugee status are generally assumed by authorities to be **economic migrants**—those seeking a better life from their migration. Today, economic migrants are typically only welcomed if they bring skills to contribute to the local economy. A variety of visas are available to individuals arriving to fill specific jobs or able to contribute particular skills or even to take up training opportunities in the destination country. For those who do not fit these criteria for legal migration, another option may be to attempt an **illegal migration**. Whether these individuals should be labeled **illegal migrants** is controversial. Some people argue that the term "illegal" should never be applied to a person, because it is degrading as well as inaccurate because it is the acts that people commit, not the people themselves, that are illegal. Others believe that it is precisely their legal status (or lack thereof) that determines people's experiences and that it is therefore appropriate to use the term illegal. Alternative terms such as "irregular," "unauthorized," or "undocumented" have been suggested and may seem more value-free at first sight but also carry implicit biases (de Haas, Castles, and Miller 2000). In this book we avoid the term illegal migrants but use the terms illegal migration and **undocumented**

BOX 10.4 INTERNALLY DISPLACED PEOPLE

Internally displaced people (IDPs) are migrants who have to flee their homes because of conflict, human rights violations, or disasters but do not leave their country. Because IDPs have not crossed an international border, they are not protected under international law. IDPs far outnumber asylum seekers and refugees but receive much less attention. In addition to being less visible to outsiders, providing international aid to IDPs can be tricky because it can be seen as violating a country's sovereignty.

There were about 46 million IDPs in the world in 2020 (figure 10.10), the vast majority concentrated in just ten countries. Syria (6.5 million) and the Democratic Republic of the Congo (5.5 million) have by far the highest numbers, but IDPs are also numerous in Ethiopia, Burkina Faso, Afghanistan, El Salvador, Yemen, Mali, South Sudan, and Nigeria (IDMC 2021a). Generally speaking, about 80 percent of those who are forced to move are women and children.

In Syria, protests against President Bashar al-Assad since the beginning of the Arab Spring in 2011 quickly became an armed conflict that resulted in about 5.5 million refugees and an even larger number of internally displaced people (IDMC 2021b). More recently, in 2021, the number of IDPs is soaring in Afghanistan after the takeover by the Taliban. At the end of 2020, about 2.9 million Afghans were already internally displaced, with another 400,000 or so added as the Taliban once again swept through the country.

Recently, natural disasters have also caused many displacements. In 2019, the top 20 disasters that displaced peopled were all weather related such as a monsoon that led to severe flooding in India; several cyclones in Bangladesh, India, and China; and wildfires in the United States.

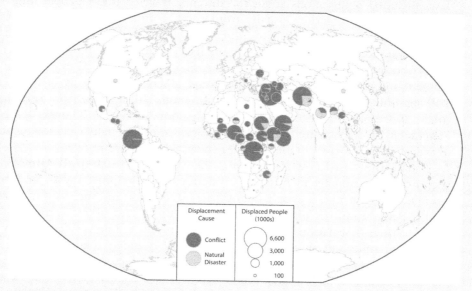

FIGURE 10.10 Internally displaced people, 2020
Data source: IDMC (2022).

migration interchangeably, while acknowledging that these terms are not problem free either.

In the context of illegal migrations, it is important to distinguish between illegal entry and illegal stay. In the media we hear frequently of illegal entry, such as people moving covertly across a land border or arriving clandestinely by boat, but in fact it is more common for people to enter a country legally—for example, on a tourist or student visa—and then remain beyond their entitled stay: **visa overstayers**. Asylum seekers may also find themselves ending up in a country illegally if their application for refugee status is rejected (box 10.5). The experiences of these groups are quite different from those of people entering illegally because their trip is usually not dangerous and by the time they lose their legal status they already have some experience with the host country. Nevertheless, lacking proper documentation, they are highly vulnerable to exploitation and may be forced to work under difficult conditions, as well as struggle with access to services such as health and education.

Migration theories

As has become clear in the previous section, there are many different types of migrations, influenced by a huge number of contextual factors. Additionally, many migrations are motivated by more than one factor, and one type of migration can turn into or trigger another type of migration. It is therefore extremely difficult, maybe impossible, to develop a single coherent theory of migration (McLeman 2014). As a result, scholars employ a number of different migration theories. Here we present some of the most frequently discussed theories. The first well-known attempt to develop a theory of migration was Ravenstein's 1889 "Laws of Migration." His laws included observations about the distance traveled (most people migrate only short distances), direction (usually from rural to urban areas), migrant selectivity (urban dwellers are more likely to migrate than rural people, women are more likely to migrate short distances, and men more likely to migrate internationally), motivations (mostly economic), and factors increasing migration (level of economic development and improvements in transportation). Remarkably, many of these laws are still applicable today. For example, most migrants still migrate from rural to urban areas, and most migrations cover relatively short distances. Though Ravenstein's statement that women are less likely to migrate internationally is no longer accurate, Ravenstein was correct in identifying gender as an important factor.

Today, the best-known migration theory is the **push–pull model**, which explains migration by identifying economic, demographic, social, and environmental factors that push people out of their places of origin (such as war, unemployment, persecution, or environmental degradation) and other factors that attract them to their destination (such as job opportunities, religious freedom, availability of land, or a pleasant environment). This model is widely used because it is simple and intuitive and can incorporate a wide range of factors. However, this theory does not shed light on *how* these factors work or how they interact with one another. It therefore cannot explain why some people who experience a certain set of factors migrate while others in the same situation do not or why some places experience in- and outmigration at the same time. Despite its obvious shortcomings, the basic ideas behind the push–pull model continue to permeate thinking about migration.

BOX 10.5 FROM ASYLUM SEEKER TO UNDOCUMENTED MIGRANT

Migrants perform large amounts of agricultural work in Europe and North America, both legally and illegally. In many cases migrants cross borders precisely to engage in seasonal agricultural labor, but in some cases asylum seekers who failed to receive refugee status end up working illegally as migrant laborers. As Reckinger (2018) described in his book *Bittere Orangen* (bitter oranges), many of the agricultural workers in southern Italy originally entered Italy as asylum seekers. Many of these mostly African men had previously worked as guest workers in countries such as Syria and Libya and lived in simple but decent conditions but fled when war broke out. With few options, they traveled to Italy hoping for a better life. These once legal migrant workers first became asylum seekers but ended up in the country illegally after being denied refugee status.

To understand how this happened we have to understand the legal situation in Italy. Italy follows the Geneva Convention in regards to refugees: a recognized refugee is granted a 5-year renewable residence permit, which allows them to work legally and gives them access to social and health services. While their cases are working their way through the legal system, asylum seekers are housed in camps. Decisions are meant to be reached within a month, but in practice they often take much longer. In southern Italy, the system works particularly poorly, and asylum seekers often have to leave the camps and fend for themselves while they await a decision on their refugee status. Those who are denied receive a *foglia di via*, a deportation notice, but are not actually deported. Without papers or money to move elsewhere, for many the only option is to stay in the country illegally. Even those who are granted refugee status may find themselves in a precarious situation because little support is available for refugees. So, though their legal status varies, both groups often work as day laborers on farms for lack of alternatives.

The challenging circumstances of ex-asylum seekers are made even clearer when compared with the Eastern European seasonal laborers who often work alongside them. Although pay is comparably low across both groups, the Eastern Europeans, who are not only there legally but are also white, tend to enjoy higher status and better access to housing. Many of the African refugees, by contrast, survive in tent cities. This hierarchy is also visible in the work, with Eastern Europeans usually the supervisors or shift leaders and Africans the day laborers. Typically, agricultural labor moves seasonally (in southern Italy the cycle often starts with the olive harvest on Sicily, followed by the orange harvest around Rosarno, and finally the tomato harvest around Foggia), which means that the men move from one squatter settlement to another, further reducing their chances of obtaining decent housing or access to medical services.

Most of the people stuck in this extremely precarious situation are men. In Italy there is great resistance to refugee camps, and economically more powerful northern Italy grudgingly accepts camps for female refugees but refuses to house male refugees; camps for men are therefore in the south and often in peripheral locations. This gender split also mirrors local needs for laborers. In northern Italy there is a high demand for caregivers and sex workers, work typically performed by women, whereas in southern Italy there is a need for cheap agricultural labor. The intersection of locally different labor markets and uneven application of asylum procedures thus produces a differentiated geography of precariousness. More broadly, this case study illustrates many factors that determine the experience of a migrant worker: legal status, problematic and unequally applied policies and procedures, lack of adequate government support, as well as how nationality, race, and gender intersect to make some people more vulnerable than others.

Another well-known theory applied to migration is **neoclassical economic theory**. It interprets migration as the outcome of geographical differences in the supply and demand for labor and resulting wage differences. According to this theory, people migrate from low-wage areas with a surplus of laborers to higher-wage areas in need of labor. At the micro or personal level, this theory sees migrants as rational actors who carefully weigh the economic benefits of migration. In a similar vein, **human capital theory** interprets migration as being driven by people's quest to maximize their human capital (knowledge and skills). This theory explains migrant selectivity; for example, that young and more skilled people migrate at higher rates because they have the most to gain from migration. Both of these theories focus on economic factors (broadly conceived) and ignore other motivations or constraints (e.g., whether people are allowed to enter another country or whether their credentials transfer). They also assume that potential migrants are well informed about their destination countries and take rational decisions as individuals.

In contrast to these theories based on choice and rational decisions (or migrants' **agency**), **structuralist theories** see migrants' opportunities as being constrained by wider structures of society. For example, **world systems theory** interprets migration as shaped by macro-level structures such as economics and politics. Similarly, **globalization theory** views migration as an essential response to the global division of labor. Migrants are afforded limited agency in these theories and are basically considered to be large moveable labor resources. Migration in this framework may actually increase existing inequalities (rather than diminish them over time as neoclassical theories assume); for example, by draining some countries of their skilled workers and strengthening the pool of talent in countries that are already advantaged.

A final set of theories looks at migrants as embedded in family or other social networks. The **new economics of labor theory** sees migrations not as the actions of individuals but as a household strategy. This strategy might involve the desire to increase household income through remittances (money sent home by migrants) or be motivated by the wish to reduce risk by diversifying sources of income for the household. **Network theory** shows how the ties between migrants and their contacts in their home country can stimulate further migration by reducing the costs and risks of moving. Already settled migrants help new migrants by organizing travel, helping with finding housing and work, assisting with language and paperwork, and easing the transition to the new culture. The term **chain migration** is often used in this context, because the migration of one person can result in the migration of a chain of others. Notably, this term has come to take on politically negative connotations in some countries, implying that one migrant may lead to an influx of many more. Networks not only encourage and facilitate migration but also play a role in the development of **transnationalism**, where migrants remain actively involved in their home countries and participate in both societies.

Conclusion

Migration has occurred throughout human history, but since the end of World War II several new trends have emerged that have reshaped migrations patterns. First, migration has become a global phenomenon. Though the overall share of migrants has remained remarkably constant at around 3 percent over several decades, more countries are now part of the global migration system. Countries of origin are mostly those at lower levels of economic development, whereas destination countries are typically affluent countries (except with respect to refugees) and have become very ethnically diverse as a result; many middle-income countries now both send and receive migrants. Connected to this point, in some cases migration flows have changed direction and emigration countries have become immigration countries. This is most clearly the case for European countries, which sent millions of migrants abroad until about the 1960s but have now become major destinations for migrants. Furthermore, new places have emerged as destinations. Most notable, the Gulf region has become a hub for migrant workers owing to oil wealth, massively changing its population composition, with foreign-born populations outnumbering natives in several Gulf countries.

Second, the composition of migrants has changed. Though women have always migrated as part of family units, changes in the nature of labor migrations mean that more and more women migrate on their own, often to work as caretakers and in other service occupations. The composition of migrants has also been reshaped in terms of the skill levels of the migrants, as advanced economies participate in the global race for talent but also still need workers in low-wage occupations, resulting in distinct migration streams.

Third, migration has become a highly politicized topic and is increasingly interpreted not only as a population and social issue but also as a security issue, leading to the fortification of borders, restrictive policies against certain immigrant groups, and contentious debates about admission and integration policies. We explore these topics further in the next chapter as we consider the social justice implications of migration.

Discussion questions

1   In your home community or another place you are familiar with, is immigration seen as enriching the community or as a problem? Why? How could tensions be addressed?

2   Should more countries pursue a labor export policy like the Philippines? What are the advantages and disadvantages of deliberately sending people abroad for work?

3   Should the definition of a refugee be expanded to explicitly include specific groups that face persecution such as transgender individuals or girls threatened by female genital mutilation? Why or why not?

4   Should the international community be allowed (or even required) to intervene when countries cannot adequately take care of internally displaced persons?

Suggested readings

Asis, M. 2017. "The Philippines: Beyond Labor Migration, Toward Development and (Possibly) Return." Migration Policy Institute, July 12, 2017. https://www.migrationpolicy.org/article/philippines-beyond-labor-migration-toward-development-and-possibly-return

Bolter, J. 2019. "Explainer: Who Is an Immigrant?" Migration Policy Institute, February 2019. https://www.migrationpolicy.org/content/explainer-who-immigrant

Gill, P. 2019. "In Nepal, Out-migration Is Helping Fuel a Forest Resurgence." https://e360.yale.edu/features/in-nepal-out-migration-is-helping-fuel-a-forest-resurgence

Piela, A. 2022. "Burka Enforcement and Burka Bans: Where Extremist Policies Meet." *Religion and Politics*, May 24, 2022. https://religionandpolitics.org/2022/05/24/burka-enforcement-and-burka-bans-where-extremist-policies-meet/

Glossary

agency: the ability of an individual to make their own choices; used in contrast to structural constraints

asylum seeker: a person requesting the protection of a country; when an asylum application is successful, the person is granted official refugee status

brain drain: loss of people with high educational levels or skills through emigration

chain migration: the migration of one person resulting in the migration of others

counterurbanization: the movement of people from urban to rural areas, often in pursuit of more pleasant surroundings

dependency ratio: measure of the number of dependents under age 15 and over 65, compared with the population aged 15 to 64

economic migration: migration primarily motivated by economic factors such as higher wages or better jobs

family migration: migration occurring primarily to reunite families

globalization theory: theory that sees migration as part of the global division of labor where countries specialize in certain types of economic activities

human capital theory: theory that sees migration as motivated by a person's desire to maximize their skills

illegal migrant: contested term for a person who has crossed into another country without legal authorization; *see also* **undocumented migrant**

illegal migration: migration that occurred without legal authorization

indentured labor: a form of contract labor where a person has to work for a specified amount of time (usually multiple years) to work off the costs of travel or money advanced

internally displaced person (IDP): a person forced to relocate within their own country, usually as a result of conflict or environmental disaster; often IDPs are unable to cross international borders to reach safety

internal migration: migration within the borders of a country

international migration: migration across international borders

Islamophobia: fear, hatred, or prejudice against Islam or Muslims

labor migration: migration occurring primarily to work

lifestyle migration (amenity migration): migration motivated by wanting to live a particular lifestyle or enjoy particular amenities

love migration: migration motivated primarily by the desire to be with a partner

melting pot: the idea of people of different origins assimilating into one united culture over time; most commonly applied to the United States

mobility: any form of human movement, regardless of duration

multiculturalism: a policy that encourages people to retain their original culture when joining a new society

neoclassical economic theory: migration theory that sees migration as primarily motivated by wage differentials or the availability of jobs

network theory: migration theory that emphasizes the importance of connections between people in shaping migrations

new economics of labor theory: migration theory that emphasizes decision making at the household or family scale in migration

permanent migration: migration involving a long-term change of residence

push–pull model: model that sees migration as the outcome of some factors pushing people out of their region or country of origin and other factors pulling them to another location

refugee: a person who has been forced to leave their country to escape war, persecution, or natural disaster

remittances: money sent by migrants to family members in their country of origin

structuralist theories: theories that emphasize the importance of broader structures of society such as the economy as influences

temporary migration: migration where there is an intention to return to the home country

transnationalism: in the context of migration, active in more than one country

undocumented migrant: a person who has crossed into another country without legal authorization, *see also* **illegal migrant**

visa overstayer: a person who entered a country legally on a visa but fails to leave the country before the expiration of this visa

world systems theory: theory that sees processes such as migration as determined by global economic and political forces

Works cited

Ali-Karamali, S. 2008. *The Muslim Next Door: The Qur'an, the Media, and That Veil Thing*. Ashland, OR: White Cloud Press.

Almendral, A. 2018. "Why 10 Million Filipinos Endure Hardship Abroad as Overseas Workers." *National Geographic* 12. https://www.nationalgeographic.com/magazine/article/filipino-workers-return-from-overseas-philippines-celebrates

Anonymous. 1914. "Belgian Refugees in 1914." https://commons.wikimedia.org/wiki/File: Belgian_refugees_1914.jpg

Auslander, L. 2000. "Bavarian Crucifixes and French Headscarves." *Cultural Dynamics* 12 (3): 283–309.

Copeland, E. 2003. "A Rare Opening in the Wall. The Growing Recognition of Gender-Based Persecution." In *Problems of Protection. The UNHCR, Refugees, and Human Rights*, edited by N. Steiner, M. Gibney, and G. Loescher, 101–15. New York: Routledge.

Cummings, V., and L. Christmann. 2020. "Internationale Arbeitsmigration in die arabischen Golfstaaten." *Geographische Rundschau* 4: 28–32.

de Haas, H., S. Castles, and M. Miller. 2020. *The Age of Migration. International Population Movements in the Modern World*. London: The Guilford Press.

The Economist. 2021. "Many More Africans Are Migrating within Africa than to Europe." October 30, 2021. https://www.economist.com/briefing/2021/10/30/many-more-africans-are-migrating-within-africa-than-to-europe

Hoke, Z. 2018. "Rohingya Refugees Entering Bangladesh after Being Driven Out of Myanmar, 2017." Screenshot. Voice of America, February 14, 2018. https://commons.wikimedia.org/wiki/File:Rohingya_refugees_entering_Bangladesh_after_being_driven_out_of_Myanmar,_2017.JPG

IDMC [Internal Displacement Monitoring Centre]. 2021a. "Global International Displacement Database." https://www.internal-displacement.org/database/displacement-data

———. 2021b. "Syria." https://www.internal-displacement.org/countries/syria

———. 2022. "Global Internal Displacement Database." https://www.internal-displacement.org/database/displacement-data

Lang, C. 2021. "Who Gets to Wear a Headscarf? The Complicated History Behind France's Latest Hijab Controversy." *Time*, May 19, 2021. https://time.com/6049226/france-hijab-ban/

McLeman, R. 2014. *Climate and Human Migration. Past Experiences, Future Challenges.* New York: Cambridge University Press.

Mendoza, D. 2015. "Human Capital: The Philippines' Labor Export Model." *World Politics Review.* https://www.worldpoliticsreview.com/insights/15998/human-capital-the-philippines-labor-export-model

Müller, M. 2019. "Where Are 'Burqa Bans' in Europe?" *Deutsche Welle.* https://www.dw.com/en/where-are-burqa-bans-in-europe/a-49843292

Murray, D. 2017. *The Strange Death of Europe: Immigration, Identity, Islam.* London: Bloomsbury Continuum.

Northrup, D. 1995. *Indentured Labor in the Age of Imperialism, 1834–1922.* Cambridge University Press

OECDStat. n.d. "Database on Migrants in OECD Regions." Accessed January 30, 2022. https://stats.oecd.org

O'Neil, K. 2004. "Labor Export as Government Policy: The Case of the Philippines." Migration Policy Institute, January 1, 2004. https://www.migrationpolicy.org/article/labor-export-government-policy-case-philippines/

Pison, G. 2019. "Which Countries Have the Most Immigrants?" World Economic Forum, March 13, 2019. https://www.weforum.org/agenda/2019/03/which-countries-have-the-most-immigrants-51048ff1f9/

Quinn, B. 2020. "Coronavirus Exerts Heavy Toll on Filipino Community in UK." *The Guardian*, April 17, 2020. https://www.theguardian.com/world/2020/apr/17/coronavirus-exerts-heavy-toll-on-filipino-community-in-uk

Raijman, R. 2020. "A Warm Welcome for Some: Israel Embraces Immigration of Jewish Diaspora, Sharply Restricts Labor Migrants and Asylum Seekers." Migration Policy Institute, June 5, 2020. https://www.migrationpolicy.org/article/israel-law-of-return-asylum-labor-migration

Reckinger, G. 2018. *Bittere Orangen: Ein neues Gesicht der Sklaverei in Europa.* Bonn: Bundeszentrale für politische Bildung.

Roberts, G., and J. Byrne. 1966. "Summary Statistics on Indenture and Associated Migration affecting the West Indies, 1834–1918." *Population Studies* 20 (1): 125–45.

Sandhu, K. 1969. *Indians in Malaya: Some Aspects of Their Immigration and Settlement.* Cambridge, UK: Cambridge University Press.

Steinicke, E., and R. Löffler. 2019. "'New Highlanders' in den Alpen." *Geographische Rundschau* 3: 32–7.

UN [United Nations]. 2021. "International Migrant Stock." UN Population Division. https://www.un.org/development/desa/pd/content/international-migrant-stock

Unger, L. 1944. "The Chinese in Southeast Asia." *Geographical Review* 34 (2): 196–217.

UNHCR [United Nations High Commissioner for Refugees]. 2022. "Refugee Data Finder." https://www.unhcr.org/refugee-statistics/download/?url=wP9Ny8

———. n.d. *The 1951 Refugee Convention.* Accessed July 21, 2021. https://www.unhcr.org/1951-refugee-convention.html

Social justice issues in migration

After reading this chapter, a student should be able to:

1 consider the spectrum of migration types from forced to voluntary;

2 apply the concepts of structure and agency to migration;

3 discuss how social equity issues are intertwined with migration.

In the previous chapter, we discussed a variety of different types and theories of migration. Though there are numerous topics that we could have expanded on in this chapter, we have chosen to explore the particularly important distinction between **forced** and **voluntary migration**, consistent with our focus on social justice. At one extreme are those migrants making lifestyle choices such as moving to be with a partner or choosing to live in a more pleasant climate, which can be considered to be genuinely voluntary. By contrast, many migrations occur when people's choices are limited by challenging political, social, economic, or environmental situations, potentially leaving little choice but to move. It is here that migration raises a host of social justice issues. The argument we make in this chapter is that forced and voluntary migrations form a spectrum, rather than discrete categories, with many migrations that are commonly assumed to be voluntary actually influenced by significant constraints. In this respect, we use a **structuralist perspective** to emphasize the social justice inequities that characterize global migration flows.

Structure and agency in migration

The notion of **agency** in the social sciences refers to individuals' ability to make their own life decisions. This is often set in contrast to **structures** of society such as economic circumstances or political frameworks that constrain this agency. Affluent communities and those with privileged status in society typically have the greatest agency to make their own migration choices, with many borders more porous to elites than to the poor. The United States has a special visa category for

DOI: 10.4324/9781003143253-11

"individuals with extraordinary ability or achievement," for instance, providing a route of entry to the United States for groups such as athletes, models, and actors. Similarly, many countries have special visas that allow the very rich to enter the country with few criteria beyond their wealth, in the hope that they will use their affluence to generate jobs.

It is not just elites who receive preferential treatment in migration flows, however. For those in the middle classes who have had the privilege of education or training, there may also be the option to move overseas to study or get a work visa based on skills in areas such as technology or healthcare. Some classic immigration countries also offer opportunities for tradespeople to migrate—for instance, Australia and New Zealand have prioritized immigrants from a variety of trades, including hairdressers and chefs in recent years. It is understandable that countries would prioritize migrants with skills that are in demand but also undeniable that this makes borders less porous to those who have not had the opportunity to develop these skills. For those with limited education or training, borders are now generally closed. This is in stark contrast to the large-scale migrations of the 19th and early 20th centuries when countries like the United States and Canada offered land to those prepared to farm it.

At the other end of the spectrum are situations where people are forced to migrate against their will. At the greatest extreme, millions of Africans were brought to the New World literally in chains during the era of slavery. Today, few people are physically forced to move, but migrations can still be forced in other ways; for instance, when people are driven from their homes by conflict or persecution because of their race, religion, or other social characteristics. In the middle of these two extremes is a large group classified as **economic migrants**. Subcategories within this group fall at opposite ends of the forced–voluntary continuum. Economic migrants may undertake voluntary migrations if they move to take up a better paid or more desirable job. By contrast, poor peasant farmers may be considered forced migrants if they move to flee circumstances in which they can no longer make a living—for instance, where climate change has reduced the productivity of their land or their property has been subdivided to the point where it can no longer support the household.

With these diverse examples in mind, some researchers classify migrants according to the degree of agency they have. In this framework, a lifestyle migrant has the greatest agency; economic migrants fall somewhere in the middle of the spectrum; people displaced by environmental disasters, war, or persecution have limited agency; and enslaved people have the least agency of all (figure 11.1). We use this as an organizing framework for the examples we provide in this chapter, focusing particularly on those migrations that include some degree of force or lack of agency. We begin with migrations that are clearly forced before delving into the gray areas in the middle of this spectrum.

Slavery

One of the most widely known examples of forced migration and human rights abuses is the transatlantic slave trade. The conquest of the Americas after 1492 originally resulted in the movement of only a small number of Europeans. The Spanish

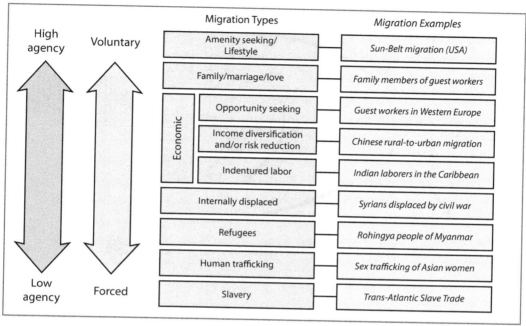

FIGURE 11.1 Degree of agency in migration process
Source: Drawn after a similar image in McLeman (2014).

and Portuguese colonizers saw the newly discovered territories primarily as places to exploit, and the annual flotilla brought back mineral resources, above all silver from Zacatecas in Mexico and Potosí in Bolivia. To exploit these natural resources as well as agricultural products, a large workforce was needed. Originally indigenous people were forced to provide labor, but their numbers were soon diminished by disease. As a result, European colonizers turned to African slaves, who had immunity not only against the diseases the Europeans brought with them to the Americas but also against many tropical diseases (Ottmer 2016).

From the 16th to the 19th century, about 12.5 million Africans were forcibly taken to the New World, with an estimated 12 percent dying en route due to the horrendous conditions (Gates 2014). Once they reached the Americas, the enslaved were sold to plantation owners, merchants, and other slave traders at auctions. The conditions varied somewhat throughout the New World. In British colonies such as the US South, enslaved people were bought and sold and treated as property. They had few rights and were usually enslaved for life. It is often argued that slavery in British colonies was harsher and crueler than in Spanish colonies, where those enslaved were at least viewed as people and had the right to complain about conditions, could marry without their master's consent, and could purchase their freedom, although few ever managed this.

In terms of sheer numbers, the transatlantic slave trade generated one of the largest forced population movements ever and the most significant forced movement between continents. The largest numbers of enslaved people were brought to Brazil and the Caribbean to work on plantations, with only a small fraction ending up in the British colonies of North America. The slave trade was part of a broader

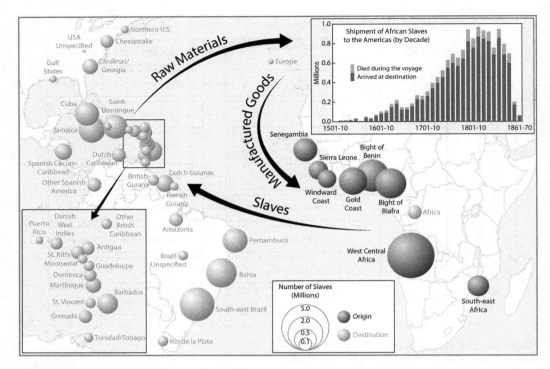

FIGURE 11.2 The transatlantic slave trade
Data source: SlaveVoyages (n.d.).

trade system, with manufactured goods being transported from Europe to Africa, enslaved people from Africa to the Americas, and resources such as sugar and cotton from the Americas to Europe (figure 11.2). Europeans tapped into pre-existing slave trades that had persisted in Africa for centuries—for example, Muslim Arabs and Berbers from North Africa had a well-established trade route transporting enslaved West Africans across the Sahara—but the scale of the transatlantic trade was unprecedented.

Today, slavery has officially been abolished around the world, but many people argue that new forms of slavery have emerged. Some estimates suggest that as many as 40 million people may be trapped in some form of slavery today, more than at any other time in human history in terms of absolute numbers (Hodal 2019). As opposed to traditional forms of state-sponsored slavery, there is no longer legal ownership of people. Instead, people are enslaved in the sense that another human being has control over them and may resort to violence to keep them under control. Though in some cases ethnic or racial discrimination still contributes to enslavement, more often people are enslaved simply because they are poor and lack alternatives, especially in countries experiencing rapid population growth, where "an immense underclass of systematically impoverished and vulnerable people" exists (Kara 2012, 210). As such, most enslaved people today are found in poor countries, particularly in Asia and Africa (figure 11.3). Ironically, because people are not literally purchased any more, the cost of obtaining slave labor is now lower than in the past, so people are more likely to be disposed of when they are no longer considered profitable (Bales 1999; box 11.1).

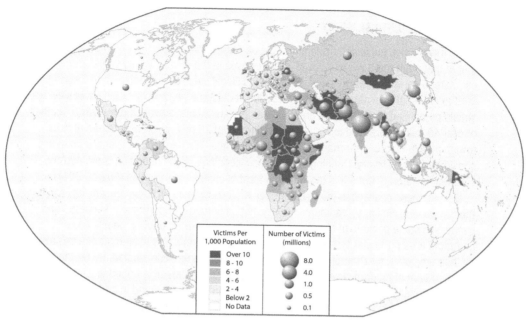

FIGURE 11.3 Victims of modern slavery, 2018
Data source: Walk Free (2018b).

Forced migration and ethnic cleansing

Another historic form of forced migration involved the removal of native peoples from traditional lands as colonial settlement expanded. Although this occurred in numerous countries, perhaps the best-known effort began in the United States in 1830 when the US Congress passed the Indian Removal Act. This act required several Native American tribes (the Choctaw, Cherokee, Creek, Seminole, and Chickasaw) to give up their lands in the southeastern United States to white settlers for cotton production and to relocate to so-called Indian Territory west of the Mississippi (in today's Oklahoma), which was not part of the United States then. Despite huge resistance, most groups were eventually coerced into moving. Some Cherokee refused to relocate and were ultimately forcibly removed from their homelands by the US army. According to estimates, a quarter of the Cherokee died during this enforced trip west by foot, horse, wagon, or steamboat, earning the march the name "Trail of Tears" (NPS 2020; figure 11.4). Today we would label this forced removal of around 100,000 Native Americans **ethnic cleansing** (Rückert 2020).

More examples of historic forced migrations could be given, but in many ways the 20th century is "the century of forced migrations" and deserves special attention. At this time, strong nationalism and racism propelled many countries to seek an ethnically homogeneous population, with forced migrations frequently seen as a way to achieve this. During peace time it would have been difficult to implement these ethnic cleansings, but the World Wars and their aftermath provided the opportunity to forcefully relocate numerous groups. Forced migrations not only

BOX 11.1 MODERN SLAVERY

When Kevin Bales published his landmark book *Disposable People* in 1999, he estimated that at that point there were about 27 million slaves in the world. Twenty years later, estimates vary widely, partly because slavery is illegal everywhere in the world (making it hard to obtain official data) and partly because of disagreements over exactly how slavery should be defined in the modern era. For example, the Global Slavery Index gives a number of over 40 million enslaved in 2016, comprising about 25 million forced laborers and 15 million people in forced marriages (Walk Free 2018a)—a category that is only sometimes considered to be slavery. Defining slavery is challenging—if the definition is too narrow, suffering is ignored, but if it is too wide the term becomes meaningless (Androff 2011).

Despite these challenges, three categories are widely recognized as present-day slavery. **Chattel slavery**, similar to traditional slavery, puts a person into permanent servitude and survives in some pockets of Mauritania and Sudan. In **contract slavery**, people have contracts for a particular type of work such as fishing, farming, or mining but are enslaved by their employer; for example, by having their identity papers confiscated. By far the most common type of modern slavery is **debt bondage**, where a person is bound to a slaveholder by debt. Globally, the largest number of bonded laborers can be found in agriculture, but they are also common in brickmaking, mining, jewelry, textiles and carpetmaking, shrimp and fish harvesting and processing, commercial sex, and domestic service. About 80 percent of bonded laborers live in South Asia, accounting for about 1 percent of the South Asian population. The highest numbers of bonded laborers are employed in agriculture, but the proportions are much higher in other industries, such as brickmaking and shrimp harvesting, where 50 to 60 percent of workers may be bonded laborers (Kara 2012). Though the vast majority of bonded laborers are enslaved in their own country, there are examples of bonded labor tied to migration and **human trafficking**, as we discuss later in this chapter. For example, many Cambodian and Burmese men work in terrible conditions on Thai shrimp boats and may not be allowed to leave the boats for months at a time (Murphy 2018).

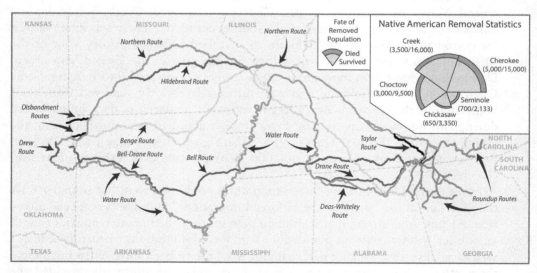

FIGURE 11.4 Trail of Tears
Date sources: NPS (2022), History in Charts (2020), and Pauls (n.d.).

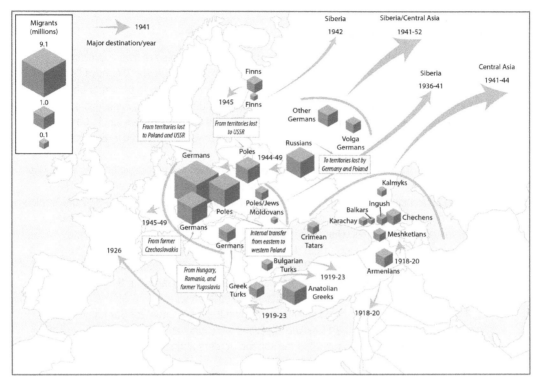

FIGURE 11.5 Major European forced population movements in the pre- and post-World War period
Data source: Kaya (2002).

became common but were widely accepted as a way to reshape the population composition of European countries (Schwartz 2021). To cite just a few examples: during World War I the Jewish population and some other groups in Czarist Russia were seen as internal enemies who collaborated with the Germans and Austro-Hungarians and were deported to the east. After the end of World War I, Germany, Austria, and Hungary received about 2 million people who were expelled from the territories they had lost in the war. In southeastern Europe, Greeks had to leave Turkish territory, and Turks had to leave Greece. Figure 11.5 illustrates the numerous forced migrations occurring at this time.

The sheer volume of forced migrations during and after World War II surpassed previous waves, with estimates of the number who had to flee or were expelled in Europe alone as high as 60 million people, or about 10 percent of Europe's population (Ottmer 2016). Nazi Germany recruited about 8 million forced laborers (both civilians and prisoners of war) from over two dozen different countries. The Nazis also forcibly resettled about 9 million people as Germans from outside German territories were brought back into the country, while non–Germans were expelled. Between them, Hitler's and Stalin's policies resulted in the forced relocation of about 30 million people between 1939 and 1944, and another 31 million people became victims of forced migrations between 1944 and 1948 after the defeat of Nazi Germany. At the end of the war, the advance of the Soviet army forced 14 million Germans out of their original areas of settlement (box 11.2). Poles lost

BOX 11.2 GERMANY'S EXPERIENCES WITH REFUGEES

Germany is a country shaped by refugees more than most people are aware. Millions of Germans were expelled from former German territories after World War II or have family members who were, so German history is much more interwoven with the experience of refugees than many other countries' histories (Kossert 2021). This may explain in part why the German Basic Law (constitution) has guaranteed the right to seek **asylum** in Germany since 1949 and why Germany welcomed refugees more warmly in the 2015 European "migrant crisis" than many other countries.

About half of the 25 million people who had to leave their homes during or after World War II were Germans—the largest forced population movement in Europe (figure 11.6). These mass expulsions were the direct result of the brutal war and crimes committed by the Nazis. At the end of the war, Germany lost about a quarter of its national territory and Germans were forced to leave areas where they had lived for hundreds of years in eastern and southeastern Europe (Kossert 2008). Many of the German expellees from places like Pomerania, Bohemia, Masuria, and Silesia were not welcomed in what remained of Germany. They were German citizens and spoke the same language, but they were considered foreigners in the West owing to cultural differences.

The question of whether the expulsion was legitimate punishment for Germany's actions during the war or whether the 14 million expellees were innocent victims is controversial. Historically, the Allied forces tolerated the shifting borders and associated ethnic cleansing

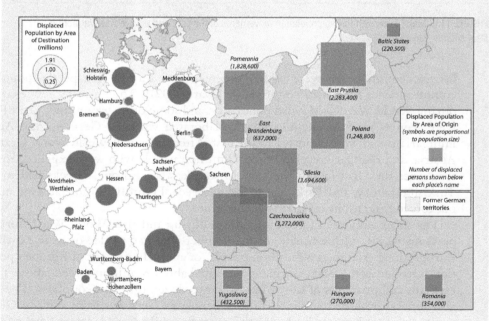

FIGURE 11.6 Expulsion of Germans from parts of Eastern Europe after World War II
As the boxes on the map show, the vast majority of Germans were expelled from Czechoslovakia and the formerly German territories of Silesia, Pomerania, and East Prussia. The size of the circles indicates the numbers of people who migrated to each destination state.
Data sources: Berliner Zeitung (2014) and Bundesminister für Vertriebene (1953).

of Germans or even saw them as a way to permanently solve ethnic tensions, given the cruelties Germany had inflicted during the war. Churchill, for example, saw the mass expulsions as an ugly but inevitable part of the reorganization of Europe (Kossert 2008). Even today the topic is often ignored, and expellees report feeling that they have not been recognized as victims. Only with German reunification and ethnic cleansings in the Balkans in the 1990s did the topic receive some attention. In 2021 a center finally opened in Berlin to document the suffering of the *Vertriebene* (expellees).

their homes in eastern Poland, after it became Soviet territory; many thousands of Ukrainians and Belarussians were forcibly resettled from Poland to the Soviet Union; Finns had to flee when the Soviet Union annexed parts of Karelia. These examples are just a short list of the forced migrations during this time period.

Although we have focused on Europe so far, forced migrations have taken place in other parts of the world as well. Perhaps most notable, in 1947, British India became independent and split into India and Pakistan, which then consisted of West Pakistan (modern-day Pakistan) and East Pakistan (now Bangladesh). The split resulted in a wave of migrations of Muslims from India into West and East Pakistan and Hindus from West and East Pakistan to India. Over 300,000 Muslims left the city of Delhi, India, alone and 600,000 Hindu refugees came to the city, completely reshaping the population composition. Similarly, every second inhabitant of Karachi, Pakistan, was a refugee. About 3 million people throughout the region lived in hastily built camps, and it is estimated that about 1 million people died. Never before was there such a large wave of refugees and expulsions over such a short time span (Ottmer 2016). This and other cases made the 1940s the darkest period of forced migrations.

Unfortunately, forced migrations continue. To list just a few examples, in the 1990s Christian Armenia occupied the Armenian enclave of Nagorno-Karabagh in Muslim Azerbaijan and forced about a million Muslims out of the occupied territories; at the same time, half a million Armenians had to leave Azerbaijan. Similarly, thousands of Georgians were forcibly removed from Abkhazia, an autonomous region in Georgia (Kossert 2021). The list of forced migrations included in this chapter seems overwhelming, but it is important to note that we have included only a small selection of cases. It also crucial to remember that there is a lot of overlap between forced migrants and refugees, the migrant group we turn to next.

Refugees

As noted in the previous chapter, the term **refugee** is applied to individuals who have crossed an international border owing to well-founded fear of persecution. According to the United Nations, refugees should be given special consideration, and many countries have formal immigration processes that define refugees as distinct from other migration streams. Obtaining refugee status has become more challenging in recent years, however. Many countries demand significant evidence of persecution from **asylum seekers** before giving them refugee status, even where this may not be available, arguably denying many asylum seekers the refugee status

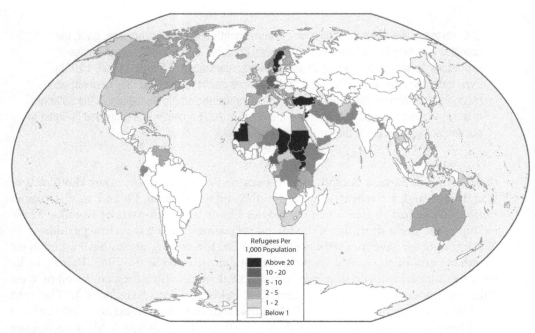

FIGURE 11.7 Distribution of refugees, 2019
Data source: UNHCR (2022).

that they need. For those whose asylum applications are denied, deportation may be the ultimate penalty.

The vast majority of refugees today live in low-income countries, with countries like Pakistan, Lebanon, Jordan, and Turkey hosting the largest numbers (figure 11.7). In some countries, refugees account for a significant percentage of the population—for example, Syrian refugees comprise over 20 percent of the population of Lebanon (de Haas, Castles, and Miller 2020). Jordan also has a long history of providing refuge to people, from the Circassians and Chechens from the Caucasus who fled Czarist Russia in the 1860s, to the Palestinians who left newly founded Israel in 1948 and during the Six Days' War, and from people fleeing Kuwait when Iraqi troops invaded, to Iraqis fleeing the 2003 American invasion of Iraq. Today roughly half of the population of Jordan is Palestinian. Jordan also currently houses 1.4 million Syrian refugees. This means that one in five of Jordan's inhabitants is Syrian, comparable to all Belgians suddenly moving to the UK. More than a quarter of Jordan's budget goes toward supporting refugees, given limited financial contributions from the international community (Saleh 2016).

Although most refugees live in the Global South, it is in the Global North that many of the high-profile debates over refugees occur. From World War II until the 1970s, richer countries were generally willing to accept refugees and provide them with assistance, and the United Nations High Commissioner for Refugees (UNHCR) was established in 1950 to protect refugees. Beginning in the 1970s, a surge in refugee numbers began to generate tensions, however. In the 1970s and '80s, people fled the US-supported anti-communist military regimes in Latin America and war in Vietnam, Cambodia, and Laos. In the 1990s the collapse of the Soviet Union resulted in large numbers of people seeking asylum in Europe and

North America. To stem the flow, rich countries implemented more restrictive immigration policies for asylum seekers as well as other types of migrants.

Refugees have become a particular political flashpoint in Europe in recent years, despite Europe absorbing just 1 percent of global refugees (Meier-Braun 2015). Until the 1990s, asylum policies in Europe had been generous, but the growing number of people requesting asylum after the end of the Cold War strained these liberal approaches. Furthermore, the creation of the European Union (EU) meant that borders between EU countries were becoming largely meaningless, focusing greater attention on the EU's outer borders (Gottschlich and am Orde 2011). In response, a number of changes for asylum seekers were implemented by the EU at the time (de Haas, Castles, and Miller 2020). For example, to stem the flow of migrants to the rich nations of the West, a number of Eastern European countries were declared to be "safe countries"; asylum seekers were required to apply for asylum in the first safe country they reached rather than in the country in which they would prefer to live. Additionally, efforts were made to prevent people from coming to Europe in the first place; for example, by requiring visas for entry and imposing carrier sanctions that punish airlines for transporting passengers without valid paperwork. Additionally, more of the burden of proof shifted to the migrants themselves. All of these changes were thoroughly tested in 2015, during Europe's "migrant crisis" (figure 11.8). European governments had to react quickly to the challenge of huge numbers of refugees arriving from the countries affected by the democratization movements that swept the Middle East known as the "Arab Spring," as well as a surge in migrants from other poor countries. These refugees arrived rapidly, precluding the development of carefully crafted migration policy.

Changes in refugee policy over time have also been dramatic in the United States. Until 1980, the United States admitted people fleeing from communist countries, while often denying refugee status to people who fled violence and persecution in non-communist dictatorships. In 1980 the United States officially adopted the Geneva Convention definition of refugees, which considers people to be refugees if they have a well-founded fear of persecution. However, when faced with simultaneous waves of Cuban and Haitian migrants arriving by boat in 1980, the United States persisted with its preference for defining those fleeing communist regimes as refugees by admitting Cubans but denying entrance to Haitians. Haitians were defined as economic migrants, even though many faced persecution from the Haitian government. Cubans have enjoyed special status in the United States since the 1960s, when the most generous refugee program ever was created for them, and the Cuban Adjustment Act allowed Cuban immigrants to obtain US citizenship faster than any other nationality (Masud-Piloto 1996). More recently, a "wet foot, dry foot" policy was in place for Cubans from the 1990s until 2017. If the US Coast Guard caught Cubans trying to reach the United States in the waters between Cuba and the United States, they were returned to Cuba. However, if they managed to reach dry land they were allowed to stay in the United States. This policy was extraordinary because it applied to people from only one country and because a couple of inches of water decided whether a Cuban could stay in the United States or not.

Beyond policies for admitting refugees, further controversy persists over how best to accommodate refugees once they have arrived. Refugee camps feature in the popular image of refugees, but camps are considered only a short-term solution

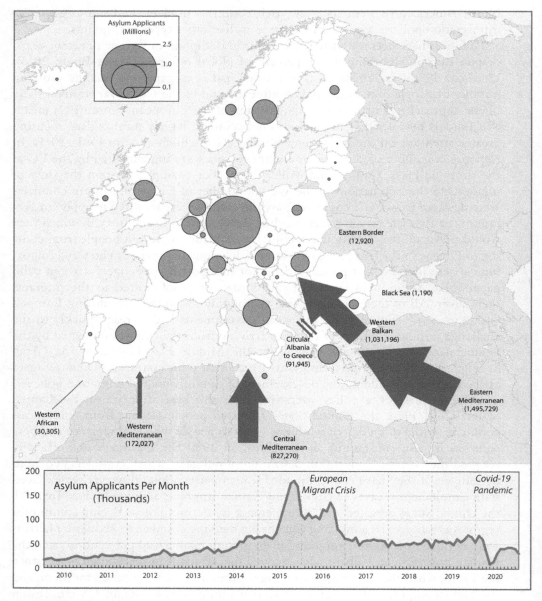

FIGURE 11.8 Europe's migrant crisis
Data sources: EUROSTAT (2022) and FRONTEX (2021).

(although many refugees have to endure them for extended periods). When refugees are concentrated in specific places, it is comparatively easy for aid organizations to provide for them from a logistical point of view, but living conditions are often poor and the cramped conditions of camps can create problems with disease and social tension. When refugees are integrated into a city it may be easier for them to find employment and housing, but it can be harder to provide services for them and integration may heighten conflicts with the local population if there is competition for scarce resources. In Jordan, for instance, about 20 percent of refugees live in camps while the rest live spread throughout the country, including 800,000 in the city of Amman and 160,000 in Mafraq. These large numbers put enormous pressure on local infrastructure in terms of providing energy, clean water, healthcare, and schools, generating significant social tensions (Saleh 2016).

Eventually "durable solutions"—those that allow refugees to return to a normal life—have to be found. The main durable solutions are (1) **repatriation** to the country of origin, (2) local settlement, and (3) **resettlement** in a third country. Many people consider voluntary repatriation to the country of origin to be the best solution because it allows people to remain within their culture and help rebuild their country. However, for this to be viable, the problems that led to the migration have to be resolved. If people fled from genocide, as occurred in Rwanda and Bosnia in the 1990s, a return may mean living in the same neighborhood as the people who perpetrated the genocide, for instance, and so ways to resolve tensions must be found. The country must also have recovered sufficiently to offer economic opportunities to live a decent life. Depending on how long people were outside their country of origin, they may also face reverse culture shock on their return, and children born in the host country may feel little connection to their parents' country of origin. When migrants do not repatriate voluntarily, some countries have resorted to forced repatriation, which raises human rights issues. For example, in the early 2000s, Congolese refugees were forcibly repatriated from Rwanda, and more recently Kenya was accused of sending Somali refugees back to Somalia against their will.

If people cannot or do not want to return, local settlement may be a viable option. Local settlement in the host county is often challenging as well, though, especially when host countries are struggling to provide infrastructure and social services to their host population. Local settlement is nonetheless often preferable to resettlement in a third country if refugees can stay in a culturally similar environment. For example, it is arguably easier to integrate Arabic-speaking Muslim migrants in an Arabic-speaking Muslim host country than in a far-away European country with a very different culture and language. On the other hand, cultural differences may create tensions even for populations crossing into a neighboring country. The Lebanese government, for instance, has been apprehensive about accepting Muslim immigrants owing to concerns about maintaining a balance between Lebanon's own Muslim and Christian populations. Resettlement is typically the most difficult and costly solution, and only a small percentage of all refugees are resettled—less than 1 percent in most years (UNHCR 2021). The UNHCR defines resettlement as the selection and movement of refugees from the country to which they fled to a third country that is willing to extend permanent resident status to them. This means that resettled refugees have most of the rights enjoyed by citizens of the host country, as well as a path to citizenship. In 2018, 27 countries around the world accepted resettled refugees, with the United States taking by far the largest number (box 11.3), followed by Canada and several western European countries. Some countries have resettlement programs that admit a certain number of refugees every year, whereas others respond to particular crises.

Refugees are frequently seen as a burden on the host society. We often hear of "floods of refugees," suggesting that they are some sort of disaster that threatens the destination society. Integrating refugees into the local community can be difficult for both sides, especially when cultures and worldviews are different. Likewise,

BOX 11.3 REFUGEE RESETTLEMENT TO THE UNITED STATES

In the United States, the President in consultation with Congress determines the number of refugees it is willing to accept. The United States has a long history of taking in resettled refugees and has admitted more resettled refugees than any other country, especially in the 1990s when over 100,000 refugees arrived annually. However, numbers have since dropped, from 85,000 in 2016 (under the Obama administration) to 15,000 in 2021 (under the Trump Administration). The refugee system is highly politicized in the United States, with the government deciding which groups of people are of special humanitarian concern, typically reflecting US policy preferences.

The places of origin have fluctuated over time as conflicts have shifted, but we can speak of three broad phases in US refugee resettlement: a Cold War phase (until 1991) dominated by refugees from the Soviet Union as well as other communist countries such as Vietnam, Cambodia, Laos, and Cuba; a Balkans phase (1990s), in which many refugees from the war and genocide in the former Yugoslavia were admitted; and a civil war phase (since 2001), with refugees coming from a variety of African and Asian countries (Singer and Wilson 2011). Over the last 2 decades, the three most important countries of origin have been Myanmar, Iraq, and Somalia, but each individual year can look very different. In 2019, for instance, most refugees came from the DR Congo, Myanmar, Ukraine, and Bhutan.

The majority of refugees are resettled in large cities, with the states of California, Texas, New York, and Florida taking in the largest number of refugees over the past 20 years, resulting in a highly uneven geography of refugee resettlement (Krogstad 2019). During the Cold War and Balkans periods, New York and Los Angeles absorbed most of the refugees, but since then the locations have diversified. In recent years, refugees have also been resettled to smaller places, where they have a potentially more dramatic influence on the local economy and population composition. For example, 12 percent of the population of Wausau, Wisconsin—a city of less than 40,000—is Hmong. Roughly a quarter of the students in Wausau are Hmong, presenting significant challenges for schools dealing with so many non-native English speakers, and initially healthcare providers struggled with cultural differences in approaches to healthcare. The adjustment period was difficult for both the Hmong and the recipient society, but the Hmong in Wausau are now seen as a success story (Singer and Wilson 2011).

formulating and implementing fair and effective refugee policies is challenging and, when policies fail, refugees may find themselves in an even worse position than before. Particularly challenging is what to do when asylum seekers are denied refugee status, which can leave them in a particularly precarious situation. At this point, refugees become economic migrants, who typically receive less sympathy on the global stage.

Economic migrants

Whereas refugees migrate owing to persecution, economic migrants are defined by economic factors triggering their move. Numerous rural–urban migrants head toward the "bright lights" of the city, such as the flow of migrants from rural western China to the factories of the East. Others attempt international crossings in search of job opportunities in more affluent countries—in southern Africa, migrants seek economic opportunity in the mines and cities of South Africa; in central America, migrants flow northwards, often first to Mexico before attempting to reach the United States. These flows are fueled by global economic inequalities that mean there are huge differences in living standards across certain borders. For many living in poverty in poorer countries that have had to cope with histories of colonial exploitation as well as current unfair terms of trade, migration may be one of only a limited number of opportunities for advancement. For others, the decision to move may be pressured by factors such as debt, land degradation, political instability, and climate change—yet again blurring the distinction between forced and voluntary migration. In reality, therefore, the distinction between economic migrants and refugees can be fuzzy, with most migrations caused by multiple factors. In this section, we consider some of the historic and contemporary migrations that are typically considered to be economically motivated, while emphasizing how many of these migrations may be less voluntary than people believe owing to significant constraints acting on the migrants.

Settler migrations

The settler migrations of the 19th and 20th centuries represent a huge movement of people with economic motives. People left crowded European countries seeking opportunities for new land or employment in the New World in particular but also in Australia, South Africa, and New Zealand. Historically, the classic immigration countries like the United States and Canada defined themselves as nations of immigrants and took pride in providing economic opportunities as well as religious and political freedom, ultimately offering citizenship to newcomers. Whereas Spanish and Portuguese colonizers primarily sought to exploit the natural resources of their colonies, British colonies were interpreted as settler colonies from the beginning (Ottmer 2016). The number of European migrants to the New World was originally small—before 1820, only about 5 million people of European origin lived in North America, and far fewer in Australia, New Zealand, and Brazil—but these pioneers blazed the trail for the mass migrations that began around 1820 (box 11.4).

BOX 11.4 ECONOMIC OPPORTUNITY IN THE "CLASSIC IMMIGRATION COUNTRIES"

In the United States, the first mass migration wave, called the Old Migration, was dominated by northwestern Europeans (especially British, Irish, and Germans), who settled primarily as farmers in rural areas. Immigration controls were minimal as the US government used migrants to settle the frontier. Around 1880, northwestern Europeans were joined by southern and eastern Europeans (the New Migration). These migrants were more likely to come from urban areas and congregated in the large cities on the East Coast. They were also more likely to be Catholic, Orthodox Christian, or Jewish and to settle in visible ethnic neighborhoods. As a result, they often encountered prejudice from people who had arrived earlier because they were perceived as different and unwilling to assimilate. The earlier groups of migrants identified as White Anglo-Saxon Protestants (WASPs), with the exception of the predominantly Catholic Irish, and often discriminated against people who were perceived as not falling into these categories (including the Irish).

So many Europeans decided to try their luck in the New World at this time—about 12 million immigrants were processed at perhaps the most famous immigration center on Ellis Island in New York between 1892 and 1924—that immigration controls were eventually instituted. As the US frontier began to close, migrants increasingly sought out other destination countries. Whereas the United States took in a wide variety of European migrants during the Old and New Migrations, Latin American countries primarily attracted migrants from southern Europe. The United States was by far the most important destination country in terms of numbers, but both Argentina and Canada took in more immigrants in relation to their population size at this time.

The beginning of the era of immigration control in the 20th century is notable not only in terms of controlling the size of flows but also because many countries attempted to control *who* immigrated. In Australia, the highly discriminatory "White Australia Policy" prevented the immigration of people of non-European origin from 1901 until the 1970s. To similar effect, the United States imposed a quota law in the 1920s, which permitted 2 percent of those already in the country of a given nationality (based on 1890 census data) to immigrate every year. This law was discriminatory because it privileged those from the European nations who had initially settled the United States, namely, the British, Irish, and Germans (Orchowski 2015). Asians were explicitly excluded from the United States through special laws such as the Chinese Exclusion Act, which remained in effect from 1882 until the 1940s. These discriminatory policies were finally terminated by the 1965 Immigration and Nationality Act, which established family reunification as the guiding principle of US immigration and allowed the United States to become the multiethnic country that it is today (figure 11.9).

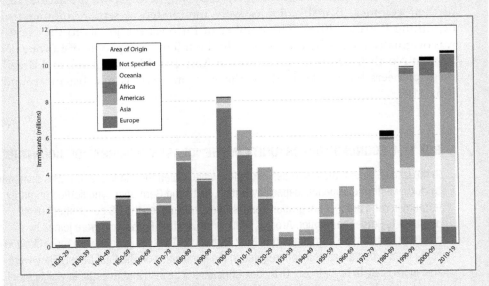

FIGURE 11.9 Immigration to the United States by area of origin since 1820
Data source: US Department of Homeland Security (2020).

The large-scale overseas migrations of the 19th and early 20th centuries are closely connected to the demographic transition in Europe. During this period, Europe developed from an agrarian society into an industrial one, and falling mortality rates combined with persistently high fertility led to population pressures that were relieved through overseas migrations (Bähr 2010). Though some migrants sought religious or political freedom, the majority were motivated by economic opportunity. Although these migrants came voluntarily, many of them faced considerable constraints in their home countries as rapid population growth in Europe had forced many people off the land, leading to rising unemployment in the rapidly expanding cities. Ireland provides a particularly interesting case in this respect because emigration was supported by the state as a way to relieve the pressures of the Irish potato famine of the 1840s, which cost about 800,000 lives. In the course of only 5 years, about a fifth of Ireland's population emigrated (Bähr 2010). Though the Irish who came to the New World are usually considered economic migrants, the presence of famine in Ireland blurs the boundary between forced and voluntary migration in this case.

Temporary labor migration programs

Subsequent to settler migrations, another major form of economic migration developed: temporary labor migration programs (TLMPs) or **guestworker** programs. Here workers are invited to work temporarily in a specific industry but without the expectation of permanent settlement or eventual citizenship in the host country. These programs became common in the mid- to late 20th century, particularly during the rebuilding of the post–World War II period. People were needed as workers but not settlers, which had important implications for how the migrants were viewed and treated (Bähr 2010).

TLMPs can be subdivided into two major types. The first category consists of temporary workers filling seasonal jobs for which there is an insufficient supply of domestic workers. The vast majority of these jobs are in agriculture, but they can also include people working in other seasonal jobs such as in the tourist industry. One example is Canada's Seasonal Agricultural Labor Program, which has been operating since 1966. This program allows people from participating countries (Mexico and several Caribbean countries) to work in Canada for 9 months in a limited number of agricultural activities (Government of Canada 2022). In the United States, the Bracero Program brought agricultural labor from Mexico from 1942 until 1964. Today the United States offers a special visa (H-2A) to temporary agricultural workers, most of whom still come from Mexico, although citizens of over 80 countries are eligible to apply.

The second category is temporary workers in permanent jobs in industry, agriculture, or other sectors. In this case, jobs are year-round and long term, but different workers rotate through the jobs to avoid the migrant workers turning into immigrants. Several European countries, including the UK, Belgium, the Netherlands, Germany, and France, put guestworker programs of this type in place after World War II, often using colonial ties to recruit labor. For example, Pakistanis and West Indians came to the UK, Moroccans and Algerians to France, and the Surinamese to the Netherlands. The importance of these programs cannot be overstated because they led to one of the largest economically motivated population movements in modern history, comparable in size only to the transatlantic migrations in the 19th and early 20th centuries.

Though these programs were often highly successful from an economic point of view, they have been criticized for using migrants' labor without guaranteeing formal immigration to the country or allowing workers to bring their families with them. The term "guestworker" has become controversial in this context because it emphasizes how migrants are guests in the host country with no permanent right of abode, even though they are contributing significantly to the host country's economy. Even carefully managed programs can result in human rights concerns because migrant workers are vulnerable to exploitation due to their precarious status as temporary workers (Costa and Martin 2018). Additionally, employers face the challenge of constantly retraining workers and are precluded from holding on to valuable employees.

The German experience illustrates many of these challenges. Without significant colonial ties, the German guestworker (*Gastarbeiter*) program (1955 to 1973) was unusual in relying on bilateral treaties. West Germany signed the first treaty with Italy in 1955, which became the model for subsequent treaties with Spain, Greece, Turkey, Portugal, and Yugoslavia. Most of the guestworkers were employed as skilled or semi-skilled labor in construction, mining, and metal industries. Though the treaties gave them job security for the duration of their contract, migrants frequently had to endure poor working conditions, shift work, and ill treatment from supervisors, as well as segregated, often dormitory-style, company housing. To limit the stay of each guestworker and discourage immigration, a rotation principle was implemented that moved different workers through each particular job.

By the 1960s, it had become clear that the rotation principle presented challenges for employers, so the German government increasingly granted extensions to work permits and did not enforce the requirement that guestworkers leave Germany at the end of their contracts. Criticism of the approach also developed in broader society—as Swiss playwright Max Frisch famously said of guestworkers in 1965: "We asked for workers, but we got people." By 1972, guestworkers comprised 12 percent of the German labor force, but in 1973 the oil crisis brought an abrupt end to the guestworker program. Guestworkers were encouraged to return home, but many instead decided to stay in Germany and apply for visas for their families to join them (Chin 2007). Ironically, therefore, the guestworker program ended up leading to significant immigration to Germany, with over 5 million immigrants living in West Germany by 1990, giving the country the largest foreign-born population in Europe (Chin 2007). This led to two inter-related changes: the composition of the foreign population changed as more women and children arrived, and the percentage of immigrants employed declined as a higher percentage of the new foreign population were dependents (Bähr 2010).

The situation became even more contentious in Germany after the end of the Cold War when 4.5 million ethnic Germans migrated to Germany from Eastern Europe in the 1990s. These groups were considered Germans in the communist countries in which they had found themselves in the post-war period and self-identified as German and took the opportunity to move to Germany when restrictions on movement lessened after the fall of the Iron Curtain. Because they were legally still considered German, they were given immediate German citizenship and offered financial support and language courses on arrival in Germany. As a result, a glaring contradiction arose: people born in Eastern Europe but considered ethnically German became German citizens immediately, even though many spoke little German and had never lived in Germany, whereas many children of guest-workers who had lived in Germany their whole lives and spoke fluent German were denied German citizenship. This discrepancy ultimately led to a reform of German citizenship law, allowing citizenship by place of birth to the children of foreigners, provided that one parent has resided in Germany for at least 8 years.

Though the German guestworker program ended half a century ago, guest-worker-style programs have not. Many of the same problems are now arising in contemporary programs in the Persian Gulf region, which has been a major desti-nation for labor migrants since the 1973 oil crisis. At this time, oil producers began to artificially raise oil prices as a bloc, creating a crisis for the rest of the world but a windfall for the oil producers. This sudden increase in oil wealth led many Gulf countries to embark on large infrastructure projects, often using foreign labor owing to their small domestic workforces. Initially most of the workers came from Arab countries such as Syria, Jordan, and Yemen. Over time, the Arab migrant workers protested against their exploitation by fellow Arabs, leading to a shift toward non-Arab workers from South and Southeast Asia. At the same time, the migrant labor force become more female, as Asian women were recruited as domes-tic servants and for nursing and other caretaking jobs (de Haas, Castles, and Miller 2020), leading to dramatic changes in the population composition of these countries.

In Jordan, Lebanon, and many Gulf countries, the **kafala system** supplies cheap labor to rapidly developing economies and regulates the relationship between migrant workers and their employers. Migrants' visas are linked to their employer under the kafala system, reducing their ability to demand better conditions. Arguably, the system is even worse for the laborers involved than traditional European guestworker programs, because the immigrants are typically not pro-tected by labor laws in the countries where they work. As such, the system is very one-sided—it helps economic development but has been widely criticized for exploiting workers, subjecting them to abuse, and being rife with racial and gender-based discrimination. If a worker leaves employment without permission, for example, the worker loses their legal status and may face imprisonment or deporta-tion (Mahdavi 2011). Because workers are threatened with imprisonment if they quit their jobs, even if they are escaping abuse, some scholars see this as a form of **modern slavery**. Additionally, employers may confiscate workers' passports/visas, or migrants may be forced to pay a "recruitment fee" that is so high that the worker can never pay it off (a form of "debt bondage"). Other human rights abuses associ-ated with the kafala system include confining workers to their dormitories and using deceptive or falsified contracts where workers have to do work that their ini-tial contract did not specify. In addition, many African and South Asian workers face racial discrimination (for example, in the form of lower pay or having to do the worst jobs) and many women have reported sexual violence (Robinson 2021). The kafala system once again blurs the distinction between forced and voluntary migra-tion because workers who undertook a migration voluntarily can find themselves locked into situations that are hard to leave.

Illegal migrations

Given that opportunities for legal migration are strictly controlled by immigration restrictions in most countries, many governments are now confronted with illegal migrations, and huge efforts are being made to prevent them in many countries. Many of the individuals involved are assumed to be economic migrants and yet may be experiencing political persecution or other experiences worthy of refugee status, yet again blurring the distinction between forced and voluntary migrations. On one hand, we must acknowledge that an illicit border crossing is an illegal act that sidesteps the legal migration channels that many other migrants are laboriously working their way through. On the other hand, given the dangerous lengths to which some migrants go to cross borders illicitly, we can also perhaps consider that the pressures to migrate must be genuinely high for a migrant to be willing to take such risks.

A further social justice issue is the human rights violations that have been recorded associated with some countries' responses to undocumented migrants. For example, Australia has been widely criticized for interning migrants arriving illegally for long periods without trial. During the Trump administration, the United States was reported to be separating children from their parents after illegal border crossings, arguably as an effort to deter further migrants from attempting to cross. Even if migrants do succeed in entering a country illegally, lack of proper documentation leaves them living clandestinely, potentially for many years. This may leave migrants without access to medical care or other services, as well as open to exploitation owing to the constant threat of being reported to authorities. The children of undocumented immigrants are also in a precarious situation and one not of their making. If parents are deported, the only way for children to stay with their parents may be to "return" to a country they may never have even visited (Bhabha 2014).

Illegal crossings themselves are also often extremely dangerous, especially if they occur by sea or through remote locations like deserts. It is very difficult to track migrant deaths, but some experts estimate that there may be two deaths for each body recovered (Brian and Laczko 2014). Sea voyages are especially dangerous: the chance of boats capsizing is high and other dangers such as hypothermia, dehydration, infections, and even shark attacks cost many lives. For example, in the 1970s, thousands fled Vietnam by boat trying to reach Southeast Asian countries like Indonesia, Malaysia, and Singapore. Many of the boats were pushed back out to sea when they came close to the coasts of other countries, forcing people to stay at sea for weeks before they could land. Similarly, in the 1980s, both Cubans and Haitians made the treacherous journey to the United States by boat, and an additional wave of Cuban refugees tried to reach the United States in the 1990s. Many of the rafts used for the journey in this wave were homemade, often nothing more than wooden planks and the inner tubes of tractor tires bound together with rope. Death rates were high, with some estimates suggesting a death rate of 75 percent (Ackerman 1996).

More recently, between 2015 and '21, an estimated 12,000 people died trying to cross the Mediterranean during the European migrant crisis (Missing Migrants Project 2021). Destination countries are faced with the dilemma of whether to rescue people arriving illegally, with the possibility that this may encourage others to

FIGURE 11.10 Banner in Hamburg (Germany) in support of rescuing migrants at sea and against the criminalization of those who rescue people

attempt the crossing. In 2014, for instance, over 800 migrants died at sea close to the Italian island of Lampedusa. As a response, the Italian government carried out Operation Mare Nostrum and saved thousands of people attempting a Mediterranean crossing. However, the operation was quickly stopped through fears that such rescue efforts would encourage more people to attempt the trip, presenting Italy with even more undocumented migrants (Reisch 2019). Non-governmental organizations such as SOS Méditerranée then stepped in to try to save lives, but their ships were often denied entry to Italian ports. Meanwhile, the captains of the rescue ships were labeled as smugglers by the Italian government and faced trials (Reisch 2019; Kossert 2021). Even the crews of civilian merchant ships, who had saved several thousand lives over the last few years, were often met with hostilities when they attempted to bring rescued people to port. Rescuers and human rights advocates argue that rescuing people at sea is a critical humanitarian action, but the practice remains controversial (Reisch 2019; figure 11.10).

As an island, Australia has also had to deal with migrants arriving by boat and is notable for treating undocumented migrants particularly harshly. Anybody who reaches Australian soil without a valid visa or who violates the terms of a visa is interned. According to the so-called Pacific Solution, the Australian coast guard collects undocumented migrants arriving by boat and brings them to internment camps, where they may be locked up for years. Many of these camps are on Pacific Islands such as Nauru and Papua New Guinea so that the migrants never set foot on Australian soil and thus cannot apply for asylum there. The system has been widely criticized for not hearing cases on an individual basis to see whether migrants meet the refugee criteria of the Geneva Convention (Kossert 2021). The Australian government has launched an "Operation Sovereign Borders" campaign with the tag lines "No Way" and "Zero Chance" to deter potential migrants from even attempting to reach Australia illegally (Australian Government, n.d.).

Other countries are investing in massive border fortifications. Examples include the fence around the Spanish exclave of Ceuta to prevent migration from Morocco; the fortifications built between Greece and Turkey in 2012, Malaysia and Brunei in 2015, and Oman and the United Arab Emirates in 2018; as well as the barrier between India and Bangladesh currently under construction (Jones 2012). The most well-known example, however, is the barrier between the United States and Mexico, begun during the Clinton administration in the 1990s. By 2020 there were about 2,000 miles of walls and fences along the US–Mexico border. After the terrorist attacks of September 11, 2001, the US National Guard was deployed to the border, the number of border patrol agents increased, and more elaborate technology was installed. An unfortunate side effect of the border wall has been to force migrants to cross in more remote and dangerous areas, leading to an increase in the number of migrant deaths.

To navigate the challenges of these increasingly difficult crossings, migrants may turn to smugglers to assist them across the border, which can lead to a suite of additional human rights abuses. **Smuggling** occurs when a person helps another person (or product) to cross a border illegally. Migrants are often subject to abuses because the services that smugglers provide are illegal and therefore unregulated. Smugglers may, for example, extort money, demand sexual favors, or abandon people once they have been paid. There have been numerous reports of the smugglers who help people cross the Mediterranean Sea extorting money from the people who hire them. Similarly, the *coyotes* who smuggle people from Mexico to the United States have been reported to rob migrants of their possessions, abandon them in the desert, and sexually assault women (Fernandez 2019).

Human trafficking of individuals represents an even more extreme form of abuse of migrants. Human trafficking involves the use of force, fraud, or coercion to obtain some type of labor or commercial sex act and tends to feed on individuals' desire to seek an escape from poverty or achieve a better life. Human trafficking is a global phenomenon, with specific geographical patterns (figure 11.11). In some countries, especially in sub-Saharan Africa, trafficking primarily occurs in conflict zones; for example, children being kidnapped and forced to become child soldiers (box 11.5). In sub-Saharan Africa, the Middle East, and parts of Asia, there is also trafficking for hard physical labor in mining and agriculture; for example, children trafficked from Mali to Côte d'Ivoire to harvest cocoa. In most other places, the most common form of trafficking is for sexual exploitation. Trafficking of individuals is often considered to be yet another form of modern slavery.

There are several main ways in which people become victims of sex trafficking. In some cases, young girls are literally kidnapped and moved to another country, such as in parts of Southeast Asia, but there are also cases of destitute families selling girls (and occasionally boys) to traffickers to pay off debts. In such cases the children are often transported clandestinely across international borders, making it harder for them to run away once they are put to work. In other parts of the world, most notably Ukraine, Russia, and other Eastern European countries, girls and women are more often lured under false pretenses. Women are encouraged to sign up for jobs as nannies, waitresses, or dancers; false boyfriends may promise women a better future abroad; or bogus marriage agencies promise them a husband and a better

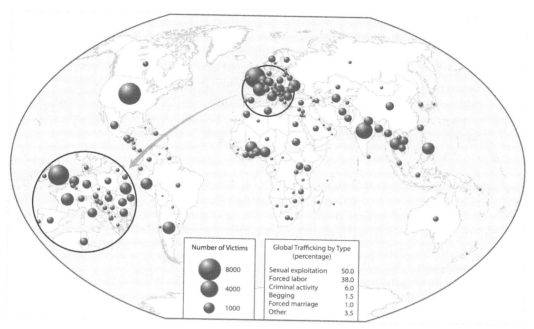

FIGURE 11.11 Detected human trafficking in 2018
Data source: CTDC (2022).

BOX 11.5 CHILD SOLDIERS

Children have participated in wars for a long time, but since about the 1980s child soldiers have become more numerous and can now be found in wars across the world from Peru to Palestine, from Sri Lanka to Sierra Leone. The scale of the problem is unprecedented, in terms of both the sheer number of children involved and the number of conflicts in which child soldiers are used. One of the reasons for this is simply that weapons have become lighter—even children can handle AK-47 assault rifles. As a result, children's roles in war have expanded from being drummers or messengers to being fighters. Child soldiers have also become younger (Honwana 2006). Girls are now also child soldiers. In many cases they work in support functions such as cooking, washing, and cleaning for the fighters and carrying supplies, but in some armed groups such as in Colombia and Sri Lanka girls also fight. In some cases, girls may become sex slaves for male commanders or soldiers.

In most cases, children are compelled to become soldiers because of poverty or coercion. In some places, children have even been kidnapped to join armed forces, as, for example, by the Lord's Resistance Army (LRA) in Uganda, which has abducted at least 20,000 children since the late 1980s. To avoid abduction, numerous children walked from villages to the relative safety of cities at night and slept in churches, public buildings, or on front porches—a phenomenon known as the "night commuters" (Médicins Sans Frontières 2004).

Children are often made pliable by extreme violence. For example, reports about the LRA revealed that children often had to witness the murder of family members or were even forced to kill their own parents. Many are threatened with being killed themselves for falling behind on a march or not fighting effectively. Children are also routinely given alcohol or drugs to give them more stamina and reduce inhibitions. Training is minimal, the discipline of a regular army is absent, and children do not have psychological support. As a result, child soldiers can become brutal looters and killers (Honwana 2006).

life. Because the women sign up willingly, they are usually transported across borders legally, but once they arrive in the destination country their passports are taken away and they often find themselves working as prostitutes. The trafficked women are frequently subjected to abuses but are bound to the trafficker by the precariousness of their legal status, as well as often threats of violence against their families. Rich countries are the main destination for human trafficking for sexual exploitation and are beginning to take a role in fighting human trafficking, although much more still needs to be done.

Conclusion

Migrants undertake a wide range of migrations under a variety of conditions. These migrations form a clear spectrum from being entirely voluntary to outright coerced, with a huge range of circumstances in between characterized by different pressures and constraints operating on the migrants. From a social justice perspective, it is important to note that though all migrants are constrained in their migration decisions by the patchwork of immigration laws found across the world, these policies constrain some migrants more than others, with poor, low-skilled, and otherwise marginalized groups typically having the least agency over their migration decisions. Furthermore, these disempowered groups are often those who are experiencing some of the greatest pressures to migrate, owing to challenging political, economic, social, and environmental conditions in their home countries, setting up a situation that leaves many migrants open to human rights abuses.

Discussion questions

1   Slavery is now illegal everywhere in the world, yet many millions of people remain enslaved. What could be done to end modern slavery?

2   Some European countries generously admitted refugees during the 2015 refugee crisis, whereas others accepted only small numbers. Should the European Union attempt to distribute refugees more equally among its member countries?

3   Are countries responsible for rescuing migrants who attempt dangerous land or sea crossings? Should private rescue organizations be allowed to carry out rescue missions if governments are unwilling to do so?

4   If you could construct a citizenship test for immigration to your country, what questions would you include? Would you incorporate historical or cultural questions? What degree of language proficiency would you require? Explain the rationale behind your decisions.

Suggested readings

Davies, L. 2022. "'Huge Spike' in Global Conflict Caused Record Number of Displacements in 2021. *The Guardian*, May 19, 2022. https://www.theguardian.com/global-development/2022/may/19/huge-spike-in-global-conflict-caused-record-number-of-displacements-in-2021

Hodal, K. 2019. "One in 200 People Is a Slave: Why?" *The Guardian*, February 25, 2019. https://www.theguardian.com/news/2019/feb/25/modern-slavery-trafficking-persons-one-in

Jones, R. 2012. "Why Build a Border Wall?" North American Congress on Latin America, November 8, 2012. https://nacla.org/article/why-build-border-wall

Pattison, P., and N. McIntyre. 2021. "Revealed: 6,500 Migrant Workers Have Died since World Cup Awarded." *The Guardian*, February 23, 2021. https://www.theguardian.com/global-development/2021/feb/23/revealed-migrant-worker-deaths-qatar-fifa-world-cup-2022

Glossary

agency: the ability of an individual to make their own choices; used in contrast to structural constraints

asylum: protection granted by a country to a person fleeing from hardships

asylum seeker: a person requesting the protection of a country; when an asylum application is successful, the person is granted official refugee status

chattel slavery: form of modern slavery similar to traditional forms where enslaved people are owned by others

contract slavery: form of modern slavery where people have contracts but are forced to work under conditions very different from those specified in the contract

debt bondage: most common form of modern slavery, where an individual is bound to another by a debt

economic migrant: a migrant whose migration is mostly motivated by economic concerns such as finding a better job

ethnic cleansing: mass expulsion or killing of an ethnic group, often with the goal of making a population ethnically homogeneous

forced migration: people migrating against their will

guestworker: a person invited to work in a country without the expectation of permanent immigration

human trafficking: moving people across international borders using force, fraud, or coercion to obtain some type of labor or sexual exploitation

kafala system: a migrant labor system in the Gulf countries that binds workers to specific employers

modern slavery: present-day forms of slavery that avoid outright ownership of others but binds people to others through force or debt; the most common form is debt bondage

refugee: a person who has been forced to leave their country owing to a well-founded fear of persecution

repatriation (of refugees):   returning refugees to their home countries

resettlement (of refugees): relocating refugees from the country where they first found protection to another (usually richer) country
smuggling (of people): the illegal movement of a person across a border
structuralist perspective: theoretical approach that views people's lives as shaped by broader societal structures
structures (in contrast to agency): political or economic contexts that constrain people's agency
voluntary migration: migrations where people move of their own free will

Works cited

Ackerman, H. 1996. "The Balsero Phenomenon, 1991–1994." *Cuban Studies* 26: 169–200. http://www.jstor.org/stable/24487714

Androff, D. 2011. "The Problem of Contemporary Slavery: An International Human Rights Challenge for Social Work." *International Social Work* 54 (2): 209–22.

Australian Government. n.d. "Australia's Borders Are Closed to Illegal Migration." Accessed March 16, 2022. https://osb.homeaffairs.gov.au/

Bähr, J. 2010. *Bevölkerungsgeographie*. Stuttgart: Verlag Eugen Ulmer.

Bales, K. 1999. *Disposable People: New Slavery in the Global Economy*. Berkeley: University of California Press.

Berliner Zeitung. 2014. "Flucht und Vertreibung-Übersicht." November 19, 2014. http://www.heimatkreis-stargard.de/Allgemein/Flucht-und-Vertreibung.htm

Bhabha, J. 2014. "Staying Home: The Elusive Benefits of Child Citizenship." In *Child Migration and Human Rights in a Global Age*, 60–95. Princeton: Princeton University Press.

Brian, T., and F. Laczko. 2014. *Fatal Journeys. Tracking Lives Lost During Migration*. International Organization for Migration. https://publications.iom.int/system/files/pdf/fataljourneys_countingtheuncounted.pdf

Bundesminister für Vertriebene. 1953. *Vertriebene, Flüchtlinge, Kriegsgefangene, heimatlose Ausländer: 1949–1952*. Bonn: Bundesminister für Vertriebene.

Chin, R. 2007. *The Guest Worker Question in Postwar Germany*. Cambridge: Cambridge University Press.

Costa, D., and P. Martin. 2018. "Temporary Labor Migration Programs." Economic Policy Institute, August 1, 2018. https://www.epi.org/publication/temporary-labor-migration-programs-governance-migrant-worker-rights-and-recommendations-for-the-u-n-global-compact-for-migration/

CTDC [Counter Trafficking Data Collaborative]. 2022. "Global Data Hub on Human Trafficking." https://www.ctdatacollaborative.org/global-dataset-0

de Haas, H., S. Castles, and M. Miller. 2020. *The Age of Migration: International Population Movements in the Modern World*. New York: The Guilford Press.

EUROSTAT. 2022. "Asylum Applicants by Type of Applicant, Citizenship, Age and Sex." https://ec.europa.eu/eurostat/databrowser/product/page/MIGR_ASYAPPCTZA__custom_1446860

Fernandez, M. 2019. "'You Have to Pay with Your Body': The Hidden Nightmare of Sexual Violence on the Border." *The New York Times*, March 3, 2019. https://www.nytimes.com/2019/03/03/us/border-rapes-migrant-women.html

FRONTEX [European Border and Coast Guard Agency]. 2021. "Risk Analysis (2015–2020)." https://frontex.europa.eu/assets/Publications/Risk_Analysis/Risk_Analysis/Risk_Analysis_2021.pdf

Gates, H. 2014. "Slavery, by the Numbers." *The Root*, February 10, 2014. https://www.theroot.com/slavery-by-the-numbers-1790874492

Gottschlich, J., and S. am Orde. 2011. *Europa macht dicht. Wer zahlt den Preis für unseren Wohlstand?* Frankfurt am Main: Westend Verlag.

Government of Canada. 2022. "Hire a Temporary Worker through the Seasonal Agricultural Worker Program: Overview." February 8, 2022. https://www.canada.ca/en/employment-social-development/services/foreign-workers/agricultural/seasonal-agricultural.html

History in Charts. 2020. "Trail of Tears Indian Removal Statistics." December 4, 2020. https://historyincharts.com/trail-of-tears-indian-removal-statistics

Hodal, K. 2019. "One in 200 People Is a Slave: Why?" *The Guardian*, February 25, 2019. https://www.theguardian.com/news/2019/feb/25/modern-slavery-trafficking-persons-one-in

Honwana, A. 2006. *Child Soldiers in Africa*. Philadelphia: University of Pennsylvania Press.

Jones, R. 2012. *Border Walls: Security and the War on Terror in the United States, India, and Israel*. London: Zed Books.

Kara, S. 2012. *Bonded Labor: Tackling the System of Slavery in South Asia*. New York: Columbia University Press.

Kaya, B. 2002. *The Changing Face of Europe—Population Flows in the 20th Century*. Strasbourg: Council of Europe Publishing.

Kossert, A. 2008. *Kalte Heimat. Die Geschichte der deutschen Vertriebenen nach 1945*. München: Siedler Verlag.

———. 2021. *Flucht. Eine Menschheitsgeschichte*. Bonn: Bundeszentrale für politische Bildung.

Krogstad, J. 2019. "Key Facts about Refugees to the U.S." Pew Research Center, October 7, 2019. https://www.pewresearch.org/fact-tank/2019/10/07/key-facts-about-refugees-to-the-u-s/

Mahdavi, P. 2011. *Gridlock: Labor, Migration, and Human Trafficking in Dubai*. Stanford: Stanford University Press.

Masud-Piloto, F. 1996. *From Welcomed Exiles to Illegal Immigrants: Cuban Migration to the U.S.* Lanham: Rowman and Littlefield.

McLeman, R. 2014. *Climate and Human Migration: Past Experiences, Future Challenges*. New York: Cambridge University Press.

Médicins Sans Frontières. 2004. "Seeking Shelter for the Night: Uganda's 'Night Commuter' Children." MSF, March 22, 2004. https://www.msf.org/seeking-shelter-night-ugandas-night-commuterchildren

Meier-Braun, K.-H. 2015. *Einwanderung und Asyl: Wichtige Fragen*. Bonn: Bundeszentrale für politische Bildung.

Missing Migrants Project. 2021. "Tracking Deaths along Migratory Routes." https://missingmigrants.iom.int/region/Mediterranean

Murphy, D. 2018. "Hidden Chains: Rights Abuses and Forced Labor in Thailand's Fishing Industry." Human Rights Watch, January 23, 2018. https://www.hrw.org/report/2018/01/23/hidden-chains/rights-abuses-and-forced-labor-thailands-fishing-industry

NPS [US National Park Service]. 2020. "Trail of Tears." NPS, July 10, 2020. https://www.nps.gov/trte/learn/historyculture/index.htm

———. 2022. "Trail of Tears National Historic Trail 100K." NPS, March 31, 2022. https://www.arcgis.com/home/item.html?id=3730e8a68d4e40e09b3dcb4d950f2b28

Orchowski, M. 2015. *The Law That Changed the Face of America: The Immigration and Nationality Act of 1965*. Lanham, MD: Rowman and Littlefield.

Ottmer, J. 2016. *Globale Migration. Geschichte und Gegenwart*. Bonn: Bundeszentrale für politische Bildung.

Pauls, E. n.d. "Trail of Tears." Britannica. Accessed January 23, 2022. https://www.britannica.com/event/Trail-of-Tears

Reisch, C.-P. 2019. *Das Meer der Tränen*. München: Riva Verlag.

Robinson, K. 2021. "What Is the Kafala System?" Council on Foreign Relations, March 23, 2021. https://www.cfr.org/backgrounder/what-kafala-system

Rückert, U. 2020. "Vor 190 Jahren: 'Indian Removal Act' wird unterzeichnet." *Deutschlandfunk*. https://www.deutschlandfunk.de/vor-190-jahren-indian-removal-act-wird-unterzeichnet-100.html

Saleh, F. 2016. "Jetzt auch noch die Syrer." In *Planet der Flüchtlinge. Warum es kein Zurück mehr gibt*, edited by Candid Foundation, 60–1. Bonn: Bundeszentrale für politische Bildung.

Schwartz, M. 2021. "Die dunkle Seite der Moderne." *Tagesspiegel*, June 19, 2021.

Singer, A., and J. Wilson. 2011. "From 'There' to 'Here': Refugee Resettlement in Metropolitan America." In *Race, Ethnicity and Place in a Changing America*, edited by J. Frazier, E. Tettey-Fio, and N. Henry, 363–88. Albany: State University of New York Press.

SlaveVoyages. n.d. "Trans-Atlantic Slave Trade—Database." Accessed November 25, 2021. https://www.slavevoyages.org/voyage/database

UNHCR [UN High Commissioner for Refugees]. 2021. "Resettlement." https://www.unhcr.org/resettlement.html

———. 2022. "Refugee Data Finder." https://www.unhcr.org/refugee-statistics/download/?url=DGV2ml

US Department of Homeland Security. 2020. "Yearbook of Immigration Statistics 2020." https://www.dhs.gov/immigration-statistics/yearbook/2020#*

Walk Free. 2018a. "Global Findings." Global Slavery Index. https://www.globalslaveryindex.org/2018/findings/global-findings/

———. 2018b. "Global Slavery Index 2018." https://www.globalslaveryindex.org/resources/downloads/

12

Sustainability and social justice in population issues

After reading this chapter, a student should be able to:

1 use geographic approaches to analyze major global challenges such as urbanization and climate change as population issues;

2 consider major global challenges through different lenses of analysis such as economic, ecological, and social equity;

3 identify sustainability and social justice issues related to major global challenges.

Throughout this book, we have argued that considering population issues through different lenses of analysis—ecological, economic, and social equity—can help develop a sophisticated understanding of population issues, as well as assist with navigating controversies. In the introduction, we used these lenses to explore why some people see population growth as a threat to ecological systems, whereas others see it as an economic opportunity. At the same time, we noted that population topics are rife with social justice challenges, such as whether we should attempt to influence people's fertility decisions to achieve a particular population size. Here, we return to these three lenses of analysis to consider two concluding case studies in the light of the methods and approaches outlined in the preceding chapters. First, we examine cities, given that urbanization has been a defining feature of the modern world; we then turn our attention to global climate change, which will have significant influences on populations into the future.

PART I: CITIES AND POPULATION

Despite a long history of urban development dating back to antiquity, until relatively recently, the vast majority of people lived in rural areas, following an agricultural lifestyle. As recently as 1800, just 5 percent of the global population was urban. The Industrial Revolution brought dramatic changes that caused this balance to shift

DOI: 10.4324/9781003143253-12

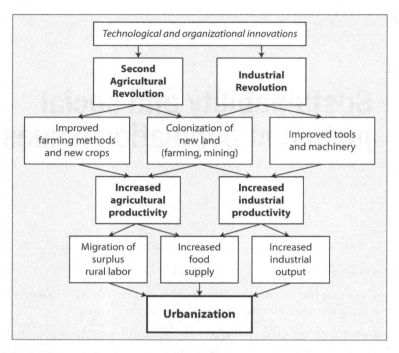

FIGURE 12.1 Agricultural and industrial innovations leading to urban growth

dramatically, however, resulting in a period of intense **urbanization**, initially in Europe and North America. The introduction of new agricultural techniques, including high-yielding seed varieties and mechanization, led to an increase in the efficiency of agricultural production, even as the proportion of workers in the fields declined. Meanwhile, in early industrial cities, more laborers were needed to work in the growing number of factories, generating a flood of rural-to-urban migrants (figure 12.1). Rapid population growth further fueled this flow. The growth of cities began slowly in the 19th century, with perhaps 13 percent of the global population living in urban areas in 1900. During the 20th century, the number and size of cities grew rapidly as urbanization spread also to the Global South, with more than half of the global population living in cities by 2010 (figure 12.2). By 2022, more than 500 cities had populations of at least 1 million (World Population Review 2022). This rural-to-urban migration represents one of the most significant movements of people in human history.

This movement has involved profound cultural changes in how people live and has had impacts on population factors such as fertility, health, migration, and resource use. Here we provide a short introduction to some of these points of intersection between population and urban issues, focusing on three main aspects: (1) ecological impacts of the city, (2) economic and social justice perspectives on the city, and (3) the impact of urbanization on population health.

Ecological impacts of the city

Urbanization redistributed people across space with significant ecological impacts. Where once most people had lived at low population densities across large areas in

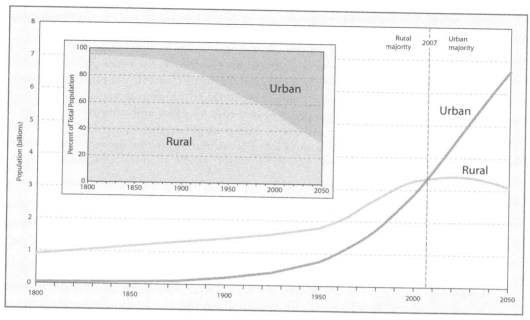

FIGURE 12.2 Urban population growth over time
Data sources: UNDESA (2018) and Grauman (1977).

dispersed villages, increasingly people lived in dense congregations. In one respect, cities can be seen as one of the least natural ways in which people can live, with urban areas offering little space for non–human species and grossly distorting natural ecosystems and nutrient flows. Furthermore, concentrating people in a city puts huge demands on a large area surrounding it as resources are drawn in to supply the city. High densities of people also create challenges with disposal of wastes, as waterways become insufficient to dilute and transport away the biological matter and chemical toxins that can make people sick. Similarly, air quality is often poor owing to high concentrations of industrial facilities, household combustion of fuel, and traffic exhaust fumes. Many cities now have such poor air quality that advisories are periodically issued urging people to stay indoors. In some of China's booming industrial cities, people keep windows shut as a matter of course and may even use air filtration systems to improve the quality of their indoor air. In short, the **ecological footprint** of many of today's cities is immense.

Cities are so influential ecologically that they can even generate their own microclimate, via an **urban heat island effect**. The concentration of dark brick and asphalt in urban areas absorbs heat during the day and reradiates it at night, keeping cities warmer than surrounding rural areas. The situation is made more extreme by the scarcity of vegetation in urban areas, given that vegetation helps moderate climate. Many city dwellers already face periodic heatwaves and unpleasant summer heat, and this situation is only likely to worsen as climate change proceeds.

Though cities undoubtedly face significant ecological challenges, researchers have also noted that cities represent an extremely efficient way to house large populations, leading some commentators to suggest that cities could be the most environmentally sound way of housing people on a planet of 8 billion. Concentrating people in cities makes the provision of infrastructure and public transit very efficient and reduces the overall need for transportation—a major contributor of greenhouse gases. Settling people in dense concentrations also allows for other places to be left open for wild plants and animals. Today, an urban sustainability movement is pressing for conditions in cities to be made more livable through innovations such as bike lanes to reduce vehicle traffic, green spaces to cool cities and improve recreational opportunities, and investment in public transport. Though these strategies are bearing fruit in more affluent regions, a lot of work remains to be done to improve the livability and sustainability of cities in low- and middle-income countries.

Economic and social justice perspectives on the city

The rise of cities represents a fundamental shift not only in *where* people live but also in *how* they live, with economic and social justice implications. With urbanization, employment changes from primarily agricultural to increasingly industrial and post-industrial activities, which are often better paid and less physically strenuous. Access to goods is easier in the city, contributing to a rise in consumer culture. Cities also act as service centers, with urban populations typically having better access to education and healthcare than their rural counterparts. A stark rural–urban divide can therefore develop, reflected in different levels of education, health status, and even voting patterns between urban and rural populations.

Though cities offer huge opportunities and amenities, it is critical to note that these benefits may not be available to all urban dwellers. In poorer countries, recent rural-to-urban migrants often cannot find work in formal jobs and may end up struggling to make ends meet with informal employment such as petty trading or street-side services like shoeshining. For these marginalized groups, many services of the city may be unaffordable. Furthermore, the rapidity of rural-to-urban migration has historically outpaced the development of urban infrastructure, leading to problems with availability of housing and quality of **sanitation**. This was as true in London, Paris, and New York in the 1800s and early 1900s as it is for many cities of the Global South today. The term **megacity** has been coined to refer to our biggest cities, with more than 30 cities housing over 10 million people in 2022 and a handful over 20 million (World Population Review 2022; table 12.1). Many of these megacities suffer acutely from challenges associated with their huge populations, including inadequate infrastructure, high rates of unemployment, shortage of affordable housing, and heavy traffic and pollution.

For poorer populations in the cities of the Global South life is particularly challenging, with over a billion people living in overcrowded urban slums and squatter settlements (UN 2022; figure 12.3). **Slum** households are defined as lacking at least one of the following: durable housing and sufficient living space, easy access to affordable safe water, and adequate sanitation (UN-HABITAT 2007). **Squatter settlements**, or shanty towns, present the additional challenge that people are living in improvised settlements without the permission of the landowner and are frequently located in undesirable or dangerous settings such as in floodplains or along major highways or industrial areas. Because the settlements are built by the residents themselves with scrap materials, the structures may be unsafe and typically

TABLE 12.1 World's most populous cities, 2022

City	Population
Tokyo, Japan	37,435,191
Delhi, India	29,399,141
Shanghai, China	26,317,104
Sao Paulo, Brazil	21,846,507
Mexico City, Mexico	21,671,908
Cairo, Egypt	20,484,965
Dhaka, Bangladesh	20,283,552
Mumbai, India	20,185,064
Beijing, China	20,035,455
Osaka, Japan	19,222,665

Note the predominance of very large cities in the Global South, particularly in Asia. Bear in mind also that how we define the boundaries of a city can significantly influence this list, with some metropolitan areas extending far beyond the limits of what may be defined as the city in official sources.

Data source: World Population Review (2022).

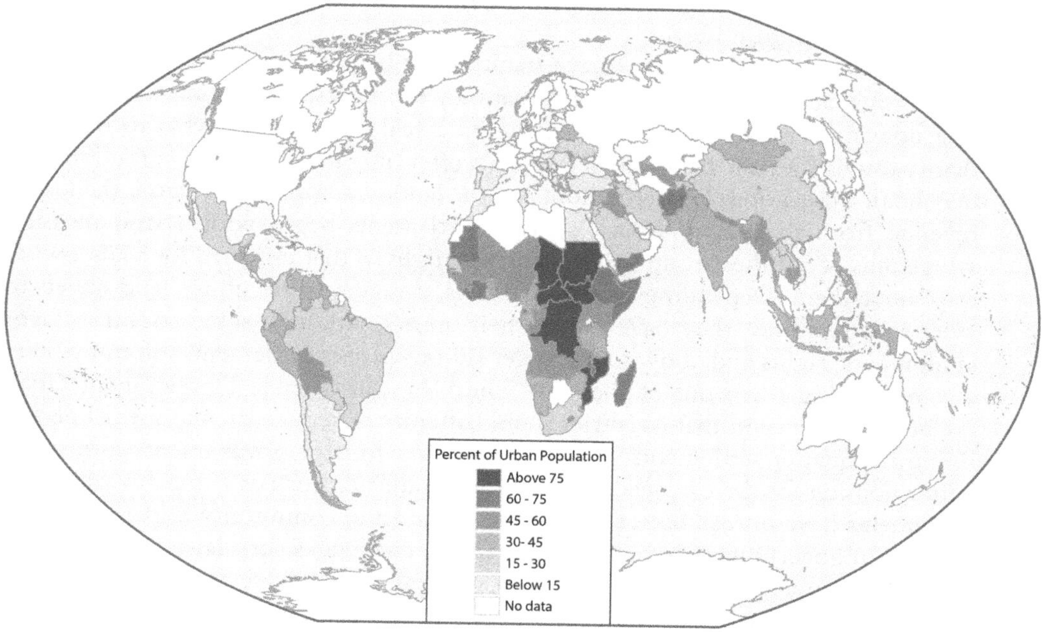

FIGURE 12.3 Proportion of households living in slums by country, 2018
Data source: Ritchie and Roser (2018).

FIGURE 12.4 Squatter settlement in Cape Town, South Africa

lack public services such as electricity, piped water, and sewage removal. Residents often find temporary solutions to these issues, such as illicitly hooking up to a nearby power grid or buying water from a water truck, but these alternatives are often more dangerous, more labor intensive, or more expensive than using formal infrastructure (figure 12.4).

Though squatter settlements have traditionally been considered an urban blight, increasingly activists are pointing out that they can provide a solution to the affordable housing crisis in poor cities, setting up a real tension over what to do about shanty towns. On one hand, the settlements offer only meager living conditions and may be in a precarious legal situation, and the landowner may suffer from the loss of revenue from their land. Many squatter settlements begin with a land invasion where tens of families set up shacks over one night so that it is hard for a landowner to remove them. Depending on national laws, it may be difficult for landowners to evict squatters once they have set up their homes, although some countries favor landowners' rights and allow them to get rid of squatter settlements, often by literally bulldozing them out of the way. The UN's Sustainable Development Goals include the provision that countries should attempt to reduce the number of people living in squatter settlements. The intention was to urge governments to provide better housing for their people, but in several countries, such as Zimbabwe, squatter settlements have instead been razed to the ground and community leaders incarcerated when governments have responded to the challenges they raise (Zárate 2016). Though removing shanty towns is one potential approach, research has suggested that giving squatters title to the land where they are living can allow squatter settlements to become reasonable working-class neighborhoods over time, with the community itself working to improve housing and bring in infrastructure once residents have confidence that their settlement will not be destroyed (Neuwirth 2005).

Squatter settlements can in this way be the first step toward providing affordable housing for low-income communities, at little cost to government.

The issue of affordable housing and informal settlement in poorer countries is clearly a significant social justice question. A wide variety of other facets of urban life similarly raise equity issues, in rich and poor cities alike. Many challenges, such as air pollution, heat, and natural hazards affect all urban dwellers, but the better-off can often afford measures to dampen their effects. This is a foundational premise of the notion of **environmental injustice**, which argues that poor and otherwise marginalized communities bear the brunt of environmental challenges. For example, rich people can more freely decide where within a city to live and so can avoid areas close to polluting industries or other potentially hazardous places like waste disposal sites. The better-off can also afford to build structures strong enough to better withstand hazards such as earthquakes. They often also have the political power to lobby to protect their neighborhoods from environmentally damaging impacts. The poor, by contrast, often have little money or political power to prepare for or address environmental problems. Traffic highlights another huge disparity. Although the rich may be doing much of the driving in many cities, it is often poorer communities that have the highest exposure to the pollution it causes, as well as high rates of road accidents, owing to their greater likelihood of living near major roads. Poorer communities may also suffer more from road building projects, which may displace them or cut through their neighborhoods—the United States' highway building projects of the 1950s and '60s, for instance, involved the displacement of many low-income communities and populations of color as highways were brought right into the center of busy cities.

Many urban areas in the Global North experience further problems associated with decaying urban and industrial infrastructure. In many countries this has led to the movement of more affluent individuals out of struggling neighborhoods, leaving high proportions of older and poorer people, as well as disproportionate numbers of recent immigrants in declining and polluted parts of the city. These communities frequently have greater need of services, even as local services may be being cut owing to a declining population and tax base. In the United States, the situation is exacerbated by the fact that taxes are allocated at a local level, so that rich neighborhoods have tax money to provide extensive services to their residents, whereas poor neighborhoods in the inner city often suffer from substandard services such as poor school systems. This can set a downward spiral into motion through triggering people to leave the poorer neighborhood, depriving it of even more of its tax base, thereby encouraging even more people to leave.

Inequalities associated with these issues are amplified by **segregation**—the physical separation of people according to particular—usually ethnic or racial—characteristics. Ethnic neighborhoods develop in many countries with immigrant populations, associated with the desire for ethnic communities to live together as well as shield themselves from discrimination in the broader population. In this respect, ethnic neighborhoods can be a positive phenomenon, offering cultural support and familiar foods, language, and customs. However, segregation can also be imposed on communities if dominant populations perpetuate the marginalization of ethnic minorities from mainstream society, limiting their access to services and opportunities for social advancement. In the United States, the problem is especially acute owing to institutionalized systems of forced segregation such as redlining, which precluded African Americans from buying houses in more desirable neighborhoods for decades during the mid-20th century. As a result, African American families were hindered in their efforts to buy their own homes, reducing the wealth they were able to pass on to future generations and constraining their opportunities for economic advancement for decades. High levels of segregation persist in many US cities to this day, with minority communities still often found in the most under-resourced parts of the city.

Health and the city

Urbanization is part and parcel of processes of industrialization, modernization, and globalization, with numerous interconnections among all of these processes. It is therefore often hard to disentangle the impact of urbanization specifically on population health, but we know that urbanization is correlated with declining fertility as well as significant changes in patterns of disease and mortality.

In terms of fertility, city dwellers generally have lower fertility than their rural counterparts, as well as lower rates of adolescent marriage and higher use of modern contraception (PRB 2015). The change from an agricultural to an industrial and increasingly post-industrial economy shifts children from being an economic benefit to an economic burden for far longer, owing to the need for long years of education. Urban employment is often less easy to combine with parenthood, encouraging many families to restrict their family size to enable both parents to remain in the workforce. Higher costs of living in urban areas may also necessitate that both parents work outside the home, leaving many families dependent on childcare, which can be very expensive. High land values also discourage large family size because living space is at a premium in urban areas. At the same time, access to healthcare is often better in urban areas, potentially providing more reliable access to contraception. Even the cultural environment of the city is probably important in that later marriage and small family size are now the norm in cities, providing a cultural template for having fewer children. Overall, this combination of factors has led to smaller average family size in urban compared with rural populations across the world.

The influence of city living on health and mortality is more variable. As we have already outlined, rapid urbanization and inadequate infrastructure often lead to higher rates of disease owing particularly to poor sanitation but also poorer air quality, overcrowding, and exposure to industrial pollutants. This is sometimes referred to as the **urban penalty**. Although some cities in the ancient world had quite sophisticated sanitation systems, the cities that developed in Europe in the Middle Ages were largely devoid of formal sanitation systems, with rivers used as open sewers for centuries (De Feo et al. 2014). Urban areas therefore have a long history of waterborne diseases. The concentration of large numbers of people in one place means that fecal contamination of water supplies is more likely and that tainted water is then potentially able to infect more people because larger numbers of people are dependent on that same water supply. The sheer volume of human waste produced by dense urban populations also presents a huge logistical problem. Early solutions such as the building of cesspits and paying people to haul away "nightsoil" (human waste) quickly became inadequate in cities such as London, which grew so rapidly that it became impossible to transport waste to the city

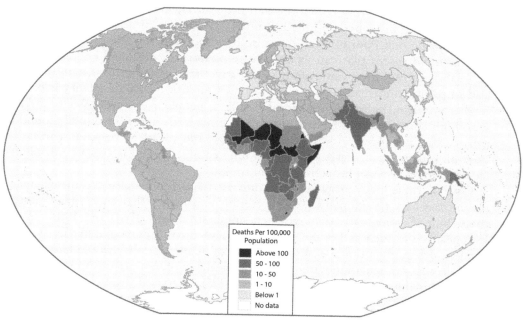

FIGURE 12.5 Death rate from diarrheal diseases, 2019
Data source: Dadonaite, Ritchie, and Roser (2019).

limits. London's cholera epidemics of the 1850s provide clear evidence of fecal contamination of the water supply. It was not until the mid-1800s that progress began to be made to improve sanitation in Europe's cities, with the building of the world's first large-scale sewer systems in places like London, Copenhagen, and Hamburg (De Feo et al. 2014). Today, sanitation measures have largely solved the problem of fecal contamination of water in affluent countries, but diarrheal disease remains a significant problem in the Global South, particularly in sub-Saharan Africa and South and Southeast Asia (figure 12.5). The WHO estimates that 1.7 billion people still do not have access to basic sanitation and only just over half of all people globally have safely managed sanitation (WHO 2022b). Though urban populations are generally better served than rural populations, in 2017, 15 percent of urban residents still lacked access to any sanitation infrastructure and almost 50 percent lacked access to a safely managed system (UNICEF 2020).

The buildup of refuse in urban areas is another significant source of disease because it can act as a breeding ground for pests. The famous outbreaks of bubonic plague in the cities of the Middle Ages were probably enabled by huge populations of rats feeding on urban waste, supporting plague-carrying fleas. Today, a variety of diseases continue to be associated with urban pests, including Chagas disease—spread by triatomine ("kissing") bugs that live in cracks in buildings—and dengue fever, associated with *Aedes aegypti* mosquitos that thrive in urban areas by breeding in small pools of water caught in refuse.

Poor-quality housing and overcrowded conditions also present risk factors for airborne diseases such as tuberculosis. Poor ventilation facilitates the spread of many airborne diseases, as demonstrated most recently during the COVID-19 pandemic.

Additionally, dark, damp conditions can help pathogens to persist, because many disease-causing microbes are killed through exposure to ultraviolet light from the sun or drying out. Poor housing may also be responsible for other airborne threats such as mold spores that can exacerbate respiratory conditions like asthma. The very fabric of buildings can even have negative health implications, ranging from lead in paints to formaldehyde from plastics. Outdoor urban air quality is often a further problem, killing an estimated 7 million people every year (WHO 2022a), with high concentrations of contaminants such as lead and sulfur dioxide found in urban areas, associated with industrial pollution and vehicle exhausts in particular. Poor air quality exacerbates a variety of lung conditions and is also a contributary factor to some types of heart disease and stroke.

Overall, these many problems that comprise the urban penalty explain why people in cities have traditionally had a shorter life expectancy than their rural counterparts. Indeed, some early demographers emphasized how 17th- and 18th-century European cities such as London could not have thrived demographically without rural-to-urban migration, because high rates of infant mortality would have diminished the urban population without a constant influx of new arrivals from the countryside to maintain the population (Galley 1995). Diarrheal disease associated with contaminated water had an especially profound impact on urban life expectancies, proving deadly for many children given their immature immune systems. Once sanitation systems became established, however, urban health improved rapidly.

Today, in the affluent world at least, cities offer some of the highest standards of health and wellbeing in many countries. Perhaps most significant, cities offer people the opportunity to make high wages and provide access to education, offering people the tools needed to raise their socio-economic status, which is strongly associated with longer life expectancy and healthier lives. Access to other services, particularly healthcare, is also often better in cities than in rural areas and, particularly in post-industrial contexts, urban residents tend to have jobs that are not as taxing on the body as hard agricultural labor.

Recently, however, researchers have begun to collect information on a variety of factors that indicate that modern cities are compromising our health in new ways. A large literature documents how the physical layout of many cities is associated with low levels of physical activity via a lack of green space, wide streets with abundant traffic and few pedestrian crossings, and even the perception that crime makes it unsafe to go outside. Today, lack of exercise and its associated conditions such as obesity are more of a threat to urban populations than injuries from strenuous jobs. The feeling that it is unpleasant to be outside in the city has further impacts, with another body of work documenting how disconnection from nature can have negative psychological and physiological impacts on people (Kuo 2015). Notably, the amount of green space and safety of cities tends to vary significantly by neighborhood, with more affluent neighborhoods usually providing more pleasant, safer, greener conditions within the city. This generates social justice concerns as poorer communities experience more of the negative aspects of city living, whereas richer residents can shield themselves by living in neighborhoods that offer the benefits of the greenery of rural areas but with access to the amenities of the city. The psycho-social environment of the city may even damage health. In particular, weaker community ties in cities compared with more tight-knit villages may be associated with higher rates of isolation and loneliness. Loneliness has numerous health implications, ranging from higher rates of depression and dementia to the exacerbation of chronic conditions like type II diabetes and heart disease (Holt-Lunstad, Smith, and Layton 2010). The anonymity of the city compared with village life has even been associated with higher rates of risk-taking behaviors such as alcohol use and risky sexual activities (Stimson 2013).

Overall, therefore, the city is a mixed bag with respect to population health. Though the devastating impact of city living associated with poor sanitation has been alleviated by the development of modern sewage treatment and piped water, the fundamentally different ways in which we live in cities have numerous physical and psychological impacts on health. Nonetheless, today, cities in the affluent world at least are often now relatively healthy places to live, particularly for affluent populations that can shield themselves from the problems of city life. Expanding these benefits to poorer urban populations is essential as the world becomes more urbanized.

PART II: CLIMATE CHANGE

Climate change is one of the most pressing environmental issues of our time, with huge changes predicted for Earth's ecosystems and the livability of the planet for human populations. Climate change is closely tied to population in numerous ways, but we focus here on three main issues: (1) the role of population size and consumption in generating climate change, (2) the impacts of climate change on population health, and (3) the role of climate change in triggering migrations.

Carbon emissions, population size, and consumption patterns

Global climate change provides a prime example of the tensions between economic and environmental goals. From an ecological perspective, it is clear that human activities have destabilized natural atmospheric cycles through adding large quantities of greenhouse gases (particularly carbon dioxide and methane) to the atmosphere associated with the Industrial Revolution (figure 12.6). In addition, landscape clearance and drainage for agriculture has removed many of the ecosystems that have traditionally trapped and stored carbon (especially forests and wetlands), exacerbating this rise in atmospheric carbon dioxide and the warming associated with it. From an ecological perspective, the source of the problem is clear—growing populations and increasing consumer aspirations drive rising consumption, leading to higher carbon emissions.

The relative contribution of population size versus consumption remains controversial, however, illustrating a key theme from the population and resources literature. The huge volume of the global atmosphere means that it is only with billions of people each contributing greenhouse gases that we can see an impact at the global scale. In this respect, climate change can be viewed as a Tragedy of the Commons. We all have an incentive to slightly increase our greenhouse gas emissions to make our lives just a little more comfortable (Why walk when you can drive? Why not turn the heating up?). However, collectively, the emissions from these activities are adding up to have a significant impact on the shared "commons" of the atmosphere through rising temperatures and associated climate change.

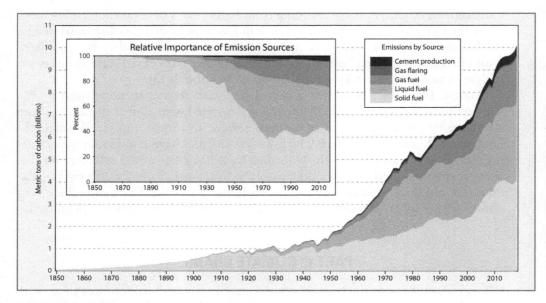

FIGURE 12.6 Global carbon dioxide emissions, 1850–2018
Data source: Gilfillan et al. (2021).

On the other hand, consumption patterns are also pivotal to generating global climate change. Every additional product that we buy has its own carbon footprint through the energy required for its manufacture and transportation; every journey we make has an impact, particularly air travel. Additionally, changes in diet, especially a shift toward eating more meat and dairy, have generated rising carbon emissions. In this respect, it is the collective impact of our rising aspirations and consumer culture that has driven our global carbon footprint to dangerous levels, just as much as population growth per se. This raises significant social equity issues because affluent populations are contributing more significantly to the problem per capita than poorer ones (figure 12.7).

Through an economic lens, it is clear that our expanding industrial economy has generated rising greenhouse gas emissions, but the solution is less clear. In an economic framework, the potential impacts of any diminishment of the global economy are given great weight, and we must balance the ecological need to control greenhouse gas emissions with the economic requirement of maintaining a vibrant economy and high living standards. Additionally, we could add the social justice concern that continuing to raise people out of poverty is critical. Sustainability proponents often counter that to have a vibrant economy in the long term we must pay attention to ecological concerns to avoid undermining our resource base, but in the short term the temptation to over-exploit our environment for economic gain has all too often won out. In economic analyses, the role of technology is often center stage, with many efforts focused around relying on technology to achieve the win–win of reducing greenhouse emissions while maintaining economic productivity. This has generated the notion of a "green economy" that promotes renewable energy sources and greater efficiencies of production. This is also a politically popular approach. Many politicians remain guarded about the idea of reducing consumption,

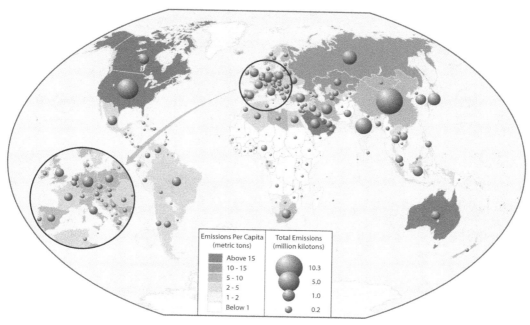

FIGURE 12.7 Global emissions of carbon dioxide, 2020
Data source: World Bank (n.d.).

considering it to be unpopular with voters, and so promoting a "green economy" that protects the environment *and* generates jobs is an appealing solution.

At this point, it is important to note that many climate scientists have argued forcefully that the scale of the problem is so large that it will require significant technological changes *as well as* decreased consumption. The Intergovernmental Panel on Climate Change (IPCC)—an international consortium of climate scientists—has suggested a target of keeping global warming to no more than 1.5°C above pre-industrial levels, although many commentators acknowledge that this will be a huge task (Tollefson 2018). A recent IPCC report noted that it will require a combination of ramping up renewable energy systems, pumping carbon out of the atmosphere and storing it underground, as well as lifestyle changes such as eating less meat and flying less. The report also explores social equity issues, acknowledging that raising people out of poverty will have to be achieved alongside climate targets (IPCC 2018).

Population impacts of global climate change

Though climate change is best known for raising average global temperatures, this warming is predicted to have multiple impacts beyond simple temperature change, including rising sea levels, more extreme weather (including stronger storms and more floods and droughts), ocean acidification, and increased wildfire risk (figure 12.8). Numerous communities are going to have to make significant cultural shifts to adapt their lifestyles to these changes. Indeed, the sheer rapidity of change predicted is likely to be one of the most challenging aspects of climate change. Our

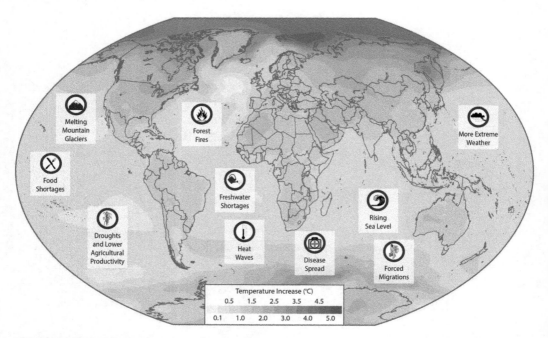

FIGURE 12.8 Global climate change: predicted temperature increase and selected consequences by 2090 (based on RCP 2.6 Model Simulation 2006–2100)
Data source: NCAR (n.d.).

complex human systems are all adapted to current conditions and so climate change will require significant and costly adjustments to be made in everything from agriculture to water distribution, from sanitation systems to how we construct our homes. Human civilization has never experienced climate change of this speed and certainly not in circumstances where we are already stretching many of Earth's natural systems with high-density populations. As such, climate change can be viewed not so much as a threat to the Earth itself—the Earth has experienced high temperatures and dramatically different atmospheric conditions in the geological past—but instead as a threat to the health of human populations. Though we will all have to make adjustments, the impacts of climate change are likely to be greatest for the poor, particularly those most heavily dependent on using resources from marginal environments, generating further social justice critiques. Many Pacific Island communities are already coping with saltwater incursion into fields from rising sea levels, for instance; farmers in the dryland Sahel region of sub-Saharan Africa are experiencing rising variability in rainfall in a region already marginal for agriculture; and indigenous groups in the Arctic are reporting a reduced ability to hunt marine mammals owing to changes in the seasonal extent of sea ice.

The rising likelihood of environmental hazards such as hurricanes, floods, and droughts is also likely to have greater impacts on poorer and marginalized communities. Since the 1970s, the hazards literature has documented numerous ways in which poorer and less empowered communities are more vulnerable to hazards, ranging from living in more vulnerable locations, to having less money to be able to evacuate, to living in structures that are more liable to collapse. We also see racial and gendered inequalities related to hazards. With respect to gender, for instance, disasters have a different impact on men and women, not just because of biological differences but also because of gendered roles and expectations (Neumeyer and Plümper 2007). For instance, expectations of women as caregivers may motivate them to prioritize the protection of others over their own safety, and gender norms in certain communities may mean that women are less likely to have learned to swim or may be unable to evacuate if a male household member is not available to accompany them. Outcomes from hazards can also cleave down racial lines. Hurricane Katrina in 2005 has become the archetypal example. Whereas richer neighborhoods of New Orleans were typically located on higher ground, many poorer neighborhoods, often housing primarily African American communities, were lower lying and thus suffered the most damage from the flooding. These same communities were also less able to evacuate, with New Orleans' Superdome becoming a "shelter of last resort" for those unable to leave, including a disproportionate number of people from the African American community. In short, the significant rise in hazards events that we are likely to see associated with climate change will probably intersect in complex ways with pre-existing vulnerabilities, leading to disproportionate negative impacts on some of the world's most vulnerable communities (box 12.1).

BOX 12.1 CLIMATE CHANGE AND SOCIAL EQUITY IN BANGLADESH

Bangladesh has been described as "ground zero for climate change" (Kara 2012). It is already exposed to both seasonal flood events from the monsoon and Himalayan snowmelt, as well as extreme flood events caused by storm surges from cyclones. It also has a history of tremendous loss of life as well as large-scale population displacement from these events. As a very low-lying country, Bangladesh is also extremely vulnerable to global sea level rise. It has been estimated that a 1m sea level rise would submerge about one-fifth of the country, displacing over 160 million people.

When flooding is severe, a third or more of the country may be under water, with floods destroying crops and houses. Floods often also lead to waterborne diseases as sewage contaminates surface waters. Climate change models suggest that seasonal flooding could worsen, as well as be combined with rising sea levels, pushing farmers to seek other economic opportunities. One of the industries that has been growing rapidly as an alternative to planting crops is the shrimp industry. Ironically, this is contributing to further climate change in an unfortunate feedback loop because the shrimp farms are built on coastal land that was traditionally mangrove forest. These mangroves were the largest carbon sink in Southeast Asia, storing four times more carbon than terrestrial forests. The mangrove forests also offered some protection from cyclones, so their loss is worsening the impacts of seasonal storms. From a population perspective, this shift from agriculture to shrimp farming also has humanitarian implications, potentially exacerbating human trafficking and modern slavery (Kara 2012). Labor exploitation is already widespread in the shrimp industry, and shrimp farms are known to be significant users of bonded labor. Bales (2016) argued that the interconnections between the environment and slavery are so strong in such cases that we have to stop slavery if we want to protect the environment, but we also have to protect the environment in order to protect communities vulnerable to enslavement.

Health impacts of climate change

Just as climate change will have numerous environmental impacts, it is also predicted to have multiple health impacts, providing yet another point of intersection with human populations. As already described, an increased frequency of environmental hazards will put human populations at risk, from both the direct trauma associated with the hazard itself and the potential for infectious disease associated with the aftermath of the event. Although the likelihood of infectious diseases spreading after a hazard event is often overestimated, flooding in particular is a significant risk factor for the spread of waterborne disease (Patz et al. 2014). In some countries, such as the UK, more intense rainfall and flash floods are already leading to periodic contamination of watercourses with sewage (Laville 2021). Warmer water temperatures will also increase the rate at which disease-causing microbes proliferate in water supplies. The increasingly erratic rainfall patterns that are predicted to occur by climate change models could lead to disease in other ways, too. If water supplies become less stable, that could have significant agricultural implications, leading to an increased risk of hunger, as well as shortages of freshwater for domestic use, with disease implications for poorer populations if water for cleaning becomes scarcer. Even for affluent communities, municipal sanitation systems are reliant on having a steady supply of water to flush contaminants away. Ironically, this could mean that drought could also be associated with higher rates of waterborne disease if microbial contaminants become more concentrated in watercourses that no longer receive sufficient rain to flush sewage away. Changes in the distribution of water and rising temperatures could also alter the distribution of vectors of disease such as mosquitos, with many researchers predicting both an increased range and higher incidence of a variety of tropical diseases, including malaria and dengue fever.

Air quality is another area that is predicted to decline as climate change proceeds. Heat and drought are likely to lead to more particulate matter in the air from diverse sources including ash from wildfires, dust from droughts, and more pollen from higher rates of photosynthesis associated with warmer temperatures. Air pollution may also worsen in warmer temperatures because heat and ultraviolet light can trigger chemical reactions among pollutants, generating new compounds. Most notable, ground-level ozone is a major pollutant that is created by the action of ultraviolet light on other contaminants such as nitrous oxides in hot conditions (Tibbets 2015). Air pollution is already a significant problem, with 99 percent of the global population exposed to air quality that periodically exceeds WHO guidelines (WHO 2022a)—this situation is only likely to worsen.

Heat itself can also be a significant health risk. Indeed, heatwaves are a particularly under-appreciated hazard when it comes to human health. In the United States, heatwaves cause more fatalities than any other environmental hazard, with more than three times as many people dying from heat as die in hurricanes (National Weather Service 2020). High temperatures stress the body, leading to an elevated risk of death from pre-existing conditions such as heart disease. High temperatures are also associated with increases in depression and suicide, as well as dementia, psychotic events, and substance abuse (Patz et al. 2014).

Some cities at particular risk of extreme heat, such as Ahmedabad, India, now have early warning systems for heatwaves (*The Economist* 2022). People are perhaps even more vulnerable in places that are *not* used to extreme heat, however, because these populations lack the behavioral adaptations that can help people survive in hot conditions (e.g., use of shutters and staying hydrated). The fact that we are starting to see extreme heatwaves in temperate regions is therefore particular cause for concern. A heatwave that struck Western Europe in 2003, for instance, caused over 30,000 deaths (UNEP 2004). Another in 2021 pushed temperatures in normally mild British Columbia, Canada, to an exceptional 49°C (120°F; *The Economist* 2022). City dwellers are particularly affected by extreme heat, owing to the urban heat island effect. Not only are daytime temperatures increased as energy is absorbed and reradiated by buildings but nighttime temperatures stay high as buildings continue to reradiate heat after the sun goes down. This sort of unrelenting heat is particularly damaging to health compared with having cooler nighttime temperatures to provide temporary relief from a heatwave, as is more likely in rural areas where vegetation can help to moderate temperatures and the heat abates somewhat at night.

Although climate change could improve health for a few—for instance, decreasing cold-related fatalities and improving agricultural productivity in the far North—most scholars agree that the net impacts of climate change on health are likely to be negative. Additionally, poorer and more marginalized communities are already being disproportionately affected by climate change, despite it being a problem largely of the making of the affluent world.

Environmental migrations

These stark inequalities in how different communities will be affected by climate change lead us to a final way in which climate change intersects with population issues: through triggering environmental migrations. Environmental migrants are those who move primarily because of environmental changes that generate challenges for people's ways of life (IOM 2020). Some of the processes that lead to environmental migrations are slow, such as sea level rise or land degradation, whereas others are sudden or catastrophic, such as cyclones, storms, and floods (IOM 2020)—both circumstances show clear potential links to climate change. Notably, weather- and climate-related issues are the most important triggers for environmental migrations, accounting for about 86 percent of environmental migrations (Ionesco, Mokhnacheva, and Gemenne 2017).

In the last decade, about 25 million people per year had to leave their homes because of environmental problems. The distribution of environmental migrants is very uneven, with about 80 percent in Asia, especially China, India, Pakistan, Bangladesh, and the Philippines (figure 12.9). By contrast, the Americas account for only about 10 percent of environmental migrants and Europe less than 1 percent (Ionesco, Mokhnacheva, and Gemenne 2017). Although this distribution is partly a result of the high population of Asia, it is once again notable that people in poorer countries who contributed least to climate change are particularly negatively affected by it.

Environmental change is not new, and people all over the world are resourceful in adapting to changing environments, often developing strategies over centuries to deal with environmental challenges such as floods and droughts. This includes both

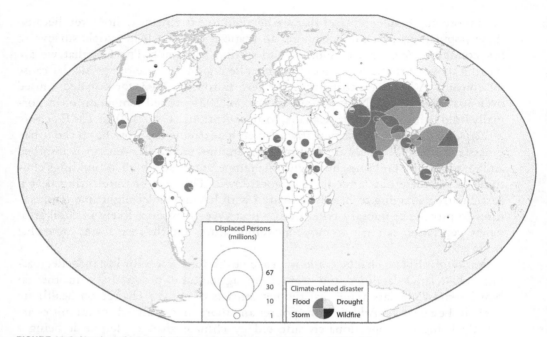

FIGURE 12.9 Number of internally displaced environmental migrants, 2008–20
This map shows the total number of people displaced internally by floods, droughts, storms, and wildfires from 2008 to '20—international migrations are not captured in these data. Note the predominance of floods and storms, particularly in Asia.
Data source: IDMC (2022).

structural adaptations such as building defenses or raising houses to protect them from flooding, as well as cultural responses such as migrating locally in times of drought (Hillmann and Ziegelmayer 2016). In many countries, people use migration as an economic risk diversification strategy such as seeking temporary employment in cities when seasonal droughts or floods make conditions challenging in rural areas and then returning to their villages when conditions improve. However, people's ability to adapt may be overwhelmed if they have to deal with multiple environmental events in quick succession or of increasing magnitude, leaving traditional coping strategies insufficient—the IPCC now talks of an "adaptation gap" when existing adaptations are not enough (*The Economist* 2022).

In public discourse and the media, environmental migrations are often presented as the next major global crisis that will result in "climate refugees" flooding to rich countries. However, no country currently offers asylum to those displaced by climate change (Milko and Watson 2022) and so the notion of climate *refugees* is perhaps a misnomer. At present, environmental migrants are not considered refugees under the Geneva Conventions (Ionesco, Mokhnacheva, and Gemenne 2017), and only a few countries offer a form of temporary protected status to people deemed victims of an environmental disaster (Ottmer 2016). In 2017, New Zealand experimented with offering a limited number of humanitarian visas to Pacific Islanders displaced by climate change, for instance, but soon dropped the program after it became apparent that Pacific Islanders saw emigration as a last resort (Dempster and Ober 2020).

There is therefore no agreement as to how much protection environmental migrants should be entitled to or who is responsible for them (Swiaczny 2015). Furthermore, there seems little willingness on the part of the global community to tackle the root causes of environmental migrations such as climate change and land degradation. Many communities are therefore left to their own devices. In the Maldives, for instance, the government is turning to building artificial islands and dikes to try to hold back the water. In 2014, the Pacific Island nation of Kiribati bought land in Fiji as a place of refuge for residents at risk of flooding (Caramel 2014). These strategies might seem far-fetched at the moment but, unless international law changes, they might become the only options for some countries affected by sea level rise (McLeman 2014). Indeed, because climate change and other environmental issues particularly affect the poorest and most vulnerable populations, who often do not have the resources to migrate, they can become "trapped populations" (Black and Collyer 2014). In this sense, concerns about rich countries being inundated with environmental migrants takes the public's attention away from the poorer regions that are already struggling with localized environmental migrations.

To date, movements within countries are by far the dominant form of environmental migrations. Many coastal cities are already facing the reality of rising sea levels leading to migrations to higher ground (figure 12.10). Indonesia, for example, recently announced that it is relocating its capital city, Jakarta. One-third of the city is predicted to be inundated by sea level rise, driving the relocation of 1.5 million civil servants to a new location on a different island (Milko and Watson 2022). Other cities are building or planning physical defenses against the rising waters such as seawalls and flood gates in the hopes of keeping their populations in place but will likely still have to contend with localized population movements as the city

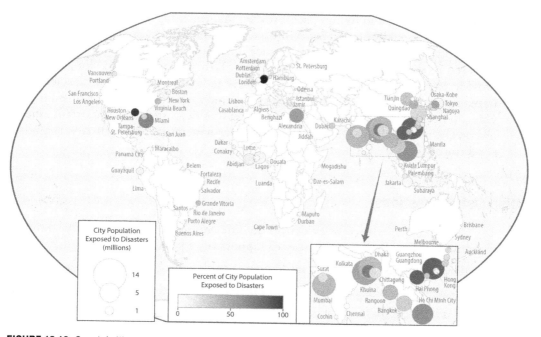

FIGURE 12.10 Coastal cities most likely to be affected by rising sea levels by 2070
Data source: Abadie et al. (2020).

adapts. New York, for instance, is embarking on a massive "climate resilience project" for Lower Manhattan, including permanent seawalls and gates. Though this project claims to be attentive to social justice issues by including diverse populations in the process, critics are concerned that the reshaping of Manhattan's East Side may result in **gentrification** (the displacement of less well-off residents from neighborhoods as they are renovated and become attractive to more affluent residents). This is already occurring in Miami, Florida, where the term **climate gentrification** is used to reflect the fact that affluent residents and commercial developers are beginning to buy properties at higher elevations, displacing communities of color and lower socio-economic status, many of whom have traditionally lived inland where land values were more affordable (Ariza 2020).

In summary, looking at climate change through our three lenses—economic, ecological, and social equity—clearly shows how closely the three are connected. It is economic activities that have triggered the climate change that is threatening the livelihoods of people, forcing them to adapt, including sometimes in ways that further damage the environment. In many cases this results in significant social justice concerns, because it is often the poorest and most vulnerable who suffer the most from the consequences of climate change, whereas it is the affluent who have contributed most to generating the problem.

Conclusion

Population issues are of huge significance to global patterns and processes. As global population continues to grow, the impacts of human populations are only likely to increase in terms of their ecological footprint, as well as through increasing competition for resources and living space. At the same time, rapidly changing economic, social, and ecological landscapes will have profound implications for human populations, often with significant social justice implications—with the poor and disempowered likely to face worsening conditions. Urban processes and climate change provide prime examples of how some of these factors might play out. Focusing on human populations as a central organizing theme for considering global issues provides an important way to focus our attention on one of the most important aspects of current global change—the people who are generating and will face changing conditions.

Discussion questions

1   Consider the balance of evidence: do cities represent an ecologically sound or environmentally damaging way to support large populations?

2   To what degree are squatter settlements an urban problem vs an urban solution?

3   Do affluent nations have a responsibility to accept climate migrants, given that emissions from the Global North have contributed most significantly to generating climate change? Why or why not?

4   What social justice issues do you recognize associated with global climate change?

Suggested readings

Ariza, M. A. 2020. "As Miami Keeps Building, Rising Seas Deepen Its Social Divide." *Yale Environment 360*, September 29, 2020. https://e360.yale.edu/features/as-miami-keeps-building-rising-seas-deepen-its-social-divide

Butler, S., and J. Grabinsky. 2020. "Tackling the Legacy of Persistent Urban Inequality and Concentrated Poverty." Brookings, November 16, 2020. https://www.brookings.edu/blog/up-front/2020/11/16/tackling-the-legacy-of-persistent-urban-inequality-and-concentrated-poverty/

Jarvie, J., and R. Friend. 2016. "Urbanization, Inclusion, and Social Justice." In *State of the World: Can a City Be Sustainable?*, edited by Worldwatch Institute, 343–53. Washington, DC: Island Press. https://doi.org/10.5822/978-1-61091-756-8_29

Lustgarten, A. 2020. "Where Will Everyone Go?" ProPublica, July 23, 2020. https://features.propublica.org/climate-migration/model-how-climate-refugees-move-across-continents/

Patz, J., H. Frumkin, T. Holloway, D. Vimont, et al. 2014. "Climate Change: Challenges and Opportunities for Global Health." *JAMA* 312 (15): 1565–80.

Glossary

climate gentrification: the displacement of poorer or less powerful communities from land becoming newly valuable given climate change; for example, land at higher elevation less prone to sea level rise; *see also* **gentrification**

ecological footprint: the impact of one's activities on the environment; in its original manifestation, the ecological footprint estimated the land area needed to provide all required resources, as well as dispose of wastes, associated with an individual, community, or product

environmental injustice: the idea that communities with less power have greater exposure to unhealthy environments

gentrification: the renovation of a poor urban area, attracting wealthier residents and potentially leading to the displacement of poorer or less powerful communities

megacity: a very large city, currently usually defined as having at least 10 million inhabitants

sanitation: provision of clean drinking water and sewage disposal

segregation: the physical separation of people of different ethnic or racial backgrounds or socio-economic characteristics

slums: poor-quality housing deficient in aspects such as durable construction, sufficient living space, access to affordable safe water, and adequate sanitation

squatter settlement: an informal settlement built on illegally occupied land characterized initially by poor living conditions and lack of public services

urban heat island effect: warmer temperatures in cities associated with dark asphalt, brick, and concrete surfaces absorbing heat during the day and reradiating it at night

urbanization: increasing urban population through rural–urban migration and natural growth

urban penalty: compromised health associated with living in an urban area associated with factors such as overcrowding and contaminated water

Works cited

Abadie, L., L. Jackson, E. Sainz de Murieta, S. Jevrejeva, and I. Galarraga. 2020. "Comparing Urban Coastal Flood Risk in 136 Cities under Two Alternative Sea-Level Projections: RCP 8.5 and an Expert Opinion-Based High-End Scenario." *Ocean & Coastal Management* 193: 105249. https://doi.org/10.1016/j.ocecoaman.2020.105249

Ariza, M. A. 2020. "As Miami Keeps Building, Rising Seas Deepen Its Social Divide." *Yale Environment 360*, September 29, 2020. https://e360.yale.edu/features/as-miami-keeps-building-rising-seas-deepen-its-social-divide

Bales, K. 2016. *Blood and Earth. Modern Slavery, Ecocide, and the Secret to Saving the World.* New York: Spiegel & Grau.

Black, R., and M. Collyer. 2014. "Populations 'Trapped' at Times of Crisis." *Forced Migration Review* 45: 52–6. http://www.fmreview.org/crisis/black-collyer

Caramel, L. 2014. "Besieged by the Rising Tides of Climate Change, Kiribati Buys Land in Fiji." *The Guardian*, June 30, 2014. https://www.theguardian.com/environment/2014/jul/01/kiribati-climate-change-fiji-vanua-levu

Dadonaite, B., H. Ritchie, and M. Roser. 2019. "Diarrheal Diseases." OurWorldinData. https://ourworldindata.org/diarrheal-diseases

De Feo, G., G. Antonious, H. Fardin, F. El-Gohary, et al. 2014. "The Historical Development of Sewers Worldwide." *Sustainability* 6: 3936–74. doi:10.3390/su6063936

Dempster, H., and K. Ober. 2020. "New Zealand's 'Climate Refugee' Visas: Lessons for the Rest of the World." ReliefWeb, UN OCHA, January 10, 2020. https://reliefweb.int/report/world/new-zealands-climate-refugee-visas-lessons-rest-world

The Economist. 2022. "The Latest UN Climate Report Is Gloomy with Some Sunny Patches." March 5, 2022. https://www.economist.com/science-and-technology/the-latest-un-climate-report-is-gloomy-with-some-sunny-patches/21807952

Galley, C. 1995. "A Model of Early Modern Urban Demography." *The Economic History Review* 48 (3): 448–69.

Gilfillan, D., G. Marland, T. Boden, and R. Andres. 2021. "Global, Regional, and National Fossil-Fuel CO_2 Emissions: 1751–2018." CDIAC-FF, Research Institute for Environment, Energy, and Economics, Appalachian State University. https://energy.appstate.edu/cdiac-appstate/data-products

Grauman, J. 1977. "Orders of Magnitude of the World's Urban Population in History." *Population Bulletin of the United Nations* 8: 16–33. https://www.un.org/development/desa/pd/content/population-bulletin-united-nations-special-issue-nos-8

Hillmann, F., and U. Ziegelmayer 2016. "Environmental Change and Migration in Coastal Regions: Examples from Ghana and Indonesia." *Die Erde* 147 (2): 119–38.

Holt-Lunstad, J., T. Smith, and J. Layton. 2010. "Social Relationships and Mortality Risk: A Meta-analytic Review." *PLOS Medicine* 7 (7): e1000316. https://doi.org/10.1371/journal.pmed.1000316

IDMC [Internal Displacement Monitoring Centre]. 2022. "Global Internal Displacement Database." https://www.internal-displacement.org/database/displacement-data

IOM [International Organization for Migration]. 2020. Environmental Migration Portal—Environmental Migration. https://environmentalmigration.iom.int

Ionesco, D., D. Mokhnacheva, and F. Gemenne. 2017. *The Atlas of Environmental Migration.* New York: Routledge. https://www.routledge.com/The-Atlas-of-Environmental-Migration/Ionesco-Mokhnacheva-Gemenne/p/book/9781138022065

IPCC [Intergovernmental Panel on Climate Change]. 2018. "Summary for Policymakers of IPCC Special Report on Global Warming of 1.5°C Approved by Governments." https://www.ipcc.ch/2018/10/08/summary-for-policymakers-of-ipcc-special-report-on-global-warming-of-1-5c-approved-by-governments/

Kara, S. 2012. *Bonded Labor: Tackling the System of Slavery in South Asia.* New York: Columbia University Press.

Kuo, M. 2015. "How Might Contact with Nature Promote Human Health? Promising Mechanisms and a Possible Central Pathway." *Frontiers in Psychology* 6: 1093.

Laville, S. 2021. "Climate Crisis Could Trigger Sewage Surge in English Rivers, MPs Told." *The Guardian*, May 26, 2021. https://www.theguardian.com/environment/2021/may/26/climate-crisis-sewage-discharge-english-rivers-environment-investment

McLeman, R. 2014. *Climate and Human Migration: Past Experiences, Future Challenges*. New York: Cambridge University Press.

Milko, V., and J. Watson. 2022. "UN: Climate Change to Uproot Millions, Especially in Asia." AP News, March 2, 2022. https://apnews.com/article/climate-science-asia-indonesia-united-nations-fe0ae4d5e210a3f390ad2a7a3b798d02

National Weather Service. 2020. "Weather Related Fatality and Injury Statistics." NOAA. https://www.weather.gov/hazstat/

NCAR [National Center for Atmospheric Research]. n.d. "Climate Change Scenarios." Geographic Information Systems Program. Accessed April 24, 2022. https://gis.ucar.edu/gis-climatedata.

Neumayer, E., and T. Plümper. 2007. "The Gendered Nature of Natural Disasters." *Annals of the Association of American Geographers* 97 (3): 551–66.

Neuwirth, R. 2005. *Shadow Cities: A Billion Squatters, A New Urban World*. New York and London: Routledge.

Ottmer, J. 2016. *Globale Migration: Geschichte und Gegenwart*. Bonn: Bundeszentrale für politische Bildung.

Patz, J., H. Frumkin, T. Holloway, D. Vimont, et al. 2014. "Climate Change: Challenges and Opportunities for Global Health." *JAMA* 312 (15): 1565–80.

PRB [Population Reference Bureau]. 2015. "The Urban–Rural Divide in Health and Development." https://www.prb.org/wp-content/uploads/2015/05/urban-rural-datasheet.pdf

Ritchie, H., and M. Roser. 2018. "Urbanization." https://ourworldindata.org/urbanization

Stimson, G. 2013. "The Future of Global Health Is Urban Health." *The Lancet* 382 (9903): 1475. https://doi.org/10.1016/S0140-6736(13)62241-2

Swiaczny, F. 2015. "Migration und Umwelt." *Geographische Rundschau* 4: 46–51.

Tibbets, J. 2015. "Air Quality and Climate Change." *Environmental Health Perspectives* 123 (6): A148–3. doi:10.1289/ehp.123-A148

Tollefson, J. 2018. "IPCC Says Limiting Global Warming to 1.5°C Will Require Drastic Action." *Nature* 562 (7726): 172. https://doi.org/10.1038/d41586-018-06876-2

UN [United Nations]. 2022. "Make Cities and Human Settlements Inclusive, Safe, Resilient and Sustainable." https://unstats.un.org/sdgs/report/2019/goal-11/

UNDESA [United Nations Department of Economic and Social Affairs]. 2018. "World Urbanization Prospects." https://population.un.org/wup/

UNEP [United Nations Environment Programme]. 2004. "Impacts of Summer 2003 Heat Wave in Europe." https://www.unisdr.org/files/1145_ewheatwave.en.pdf

UN-HABITAT. 2007. "Slums: Some Definitions." State of the World's Cities 2006/7. http://mirror.unhabitat.org/documents/media_centre/sowcr2006/SOWCR%205.pdf

UNICEF [United Nations Children's Fund]. 2020. *Snapshot of Global and Regional Urban Water, Sanitation and Hygiene Inequalities*. New York: UNICEF. https://www.unicef.org/media/91561/file/Snapshot-of-global-and-regional-urban-water-sanitation-and-hygiene-inequalities.pdf

WHO [World Health Organization]. 2022a. "Air Pollution." https://www.who.int/health-topics/air-pollution#tab=tab_1

———. 2022b. "Sanitation." WHO, 21 March, 2022. https://www.who.int/news-room/fact-sheets/detail/sanitation

World Bank. n.d. "CO_2 Emissions (kt)." Accessed December 3, 2021. https://data.worldbank.org/indicator/EN.ATM.CO2E.KT

World Population Review. 2022. "World City Populations 2022." https://worldpopulationreview.com/world-cities

Zárate, L. 2016. "They Are Not 'Informal Settlements'—They Are Habitats Made by People." The Nature of Cities, April 26, 2016. https://www.thenatureofcities.com/2016/04/26/they-are-not-informal-settlements-they-are-habitats-made-by-people/

Index

Note: Page locators in **bold** refer to tables and page locators in *italics* refer to figures.

Aborigines 79–80
abortion 73–4, 170–3, 176–7, 182–3, 186–8, 191–5, 208, 210; acceptance of 157; policy 170, 176, 186, 195; restrictions by country *177*; unsafe **74**
adolescents 171–2, 175, 178, 187, 212
adults 65–6, 86, 105–9, 223, 240, 242, 246, 248
Aedes aegypti 227, 311
affirmative action 82, 84–5
Afghanistan 106, 210–11, 222–4, 244, 259, 267
Africa 58, 73–7, 97–9, 128–30, 150, 210–12, **227**, 229–30, 260–1, 278
age 64–72, 83–5, 108–9, 112–13, 146–7, 149–51, 158–60, 165–8, 178–80, 199–205; cohorts 65, 67, 108–9, 150; and sex structure 65, 85, 149; structure 55, 65, 68, 85, 87, 108–9, 149–50, 202–5
agricultural extensification 133, 139
agricultural intensification 133, 139, 141
agricultural revolution 90, 94–5, 112, 231
air pollution 127, 142, 309, 318, 325
Amazon region 63
Americas 52, **227**, 229, 247–8, 257, 276–8; pre–Columbian 229
Angola 77, 97, 119, 216, 223
apartheid 81

Appalachian region 204
aquaculture 132, 139, 141–2
Arab Spring 256, 261, 267, 285
Arctic 79, 316
arithmetic density *see* population density
Asia 17–18, 58, 68–9, 73–4, 129–30, 162–3, 190, **227**, 257–61, 319–20
Australia 17, 39–40, 58–9, 78–80, 87, 119–20, 254, 259, 289–90, 294–6
Avian flu **233**
Azerbaijan 73, 283

baby boom 3, 67, 148, 194
Bangladesh 33, 58–9, 73–4, 126, 273, 317, 319
Belgium 36, 252, 262, 291
Bills of Mortality 35
biocapacity 119–20, 139
biodiversity 95, 124, 126, 131, 144
birth control 103, 117, 155, 173–4, 183, 194–5
birth rates 11, 18–19, 68–72, 96–7, 102, 105–8, 153–4, 158–61, 164, 180
Black Death *93*, 113, 115
Boserup, E. 123, 141
Bracero Program 291
brain drain 250, 261, 272
Brazil 61, 63, 82, 107, 109, 175, 193, 256
Brundtland Report 124, 139
bubonic plague 311
burden of disease 115, 219, 221

Canada 24, 33, 40–1, 53, 78–80, 119–20, 254, 258, 289–91
Cancer Alley 81
carbon dioxide emissions 128, 314–15
carbon footprint 135, 143, 314
carrying capacity 13–14, 16, 20, 96, 103–4, 112, 117
Catholicism 35, 155, 173, 184, 186, 289–90
causes of death 205, 213, 219, 221, 236
CDC see US Centers for Disease Control and Prevention
census 20, 24–7, 29–38, 50–4, 87–8, 189, 191, 193; Australia 30, 33, 78; Canada 31, 33, 36, 53, 78; changing topics covered in 31; characteristics 26–7, 30, 34, 52; errors 31, 33, 51; France 34; Germany 26; Lebanon 25, 52; Myanmar 36–8, 52–3; Nigeria 36; questions pertaining to race and ethnic origin 32; rolling 34, 52–3; Soviet Union 26, 36, 38, 68; topics 25, 27, 29, 31, 36, 53; United Kingdom 24, 53; United States 26, 29–32, 35–6, 50, 54, 61, 290
Central African Republic 49, 52, 202
Chad 202, 205, 254
Chagas disease 311
child mortality 98, 205, 221; see also infant mortality
child soldiers 296–8, 301
childbearing 65–7, 98, 146–50, 153–4, 158–60, 165–6, 168, 178
childcare 12, 65, 85, 158, 160–1, 164–5
childhood 66, 72, 90, 200
childrearing 6, 12, 65–6, 157–60, 175, 217
children 10–12, 65–9, 100–3, 105–7, 145–7, 149–59, 161–6, 181–4, 188–9, 296–8
China 1–2, 33, 47, 52, 72–3, 86, 100, 187–96, 231–3, 307; one–child policy 188, 193
Chinese Exclusion Act 290
chronic diseases see diseases
cities 59, 61–2, 64, 85–6, 226–7, 264, 288–9, 303–13, 320–3, 325; coastal 321, 323; most populous 307
civil registration systems 34
civil war 25, 36, 213, 216, 256, 261
classic immigration countries 254, 276, 289

climate change 59–60, 127, 130, 233, 235, 303, 305, 313–25; health impacts of 318; population impacts 315; and social equity 19, 303, 317, 322
climate refugees 320
cohabitation 157
colorism 81, 84
Columbian Exchange 229
Commoner, B. 121, 141
communities 4–8, 74–82, 91–2, 126–7, 228–30, 238, 242–5, 309–10, 315–19, 321–3; affluent 127, 130, 245, 309, 312, 318–19, 322; of color 6, 15–16, 81–2, 84, 95, 127; ethnic 30, 37, 81–2, 84, 176, 309; global 20, 95, 124, 130, 228, 230, 236, 315–16; low income 6, 176; marginalized 16, 238, 244, 309, 316, 319; poor 138, 181, 187, 199, 206, 213, 309
completed fertility 150, 165
conflicts 99, 113, 259–60, 286, 288, 297
congenital anomalies 206, 221
content error see census, errors
contraception 10, 146, 153–7, 170–3, 175–80, 182–4, 187, 191–2, 194, 209–10; access to 3, 171, 173, 175–6, 178–9, 182–3, 310
contraceptive methods 172, **174**
contraceptives 95, 171–3, 175, 179–80, 182, 184, 188, 194–5
coronavirus see COVID–19
Costa Rica 107, 207, 209, 212, 222, 224
counterurbanization 264, 272
coverage error see census, errors
COVID–19 1–3, 6, 19–22, 43, 64, 226, 228–9, 231, 233–5, 247–9, 274; in India 43
crude birth rates see fertility rates
crude death rates see mortality rates
Cuba 48, 107, 206–7, 285, 288

death rates 104–8, 112, 160, 163, 165, 201–3, 212–13; age–adjusted 203; crude 165, 202–3, 221
deaths 205–6, 210–12, 214, 216–18, 220–2, 226–7, 229–30, 232–3, 235–6, 245–6; of despair 220, 242, 245; infant 153, 183, 200, 205–6, 208, 222
debt bondage see slavery, modern
deforestation 62–3, 122, 124, 127, 131, 133
degenerative diseases see diseases

demographic data 23–54; cartographic display of 45; graphical display of 40
demographic dividend 70, 84, 107, 112, 180, 192
demographic momentum 102, 107, 112
demographic transition model 104–8, 111–12, 152–3, 156, 213, 220
demography 4, 20, 24, 52, 86, 105
Dengue fever **227**
dependency ratios 39, 69, 251
derived data *see* population data, types of
digital nomads 64
discrimination 73, 76–82, 84–5, 87, 200, 202, 250–1, 293; gender–based 73, 77, 293; racial 80–2, 251, 278, 293, 309
diseases 204–6, 209, 213–22, 225–9, 231, 233–6, 245–8, 310–11, 317–18, 324; of affluence 235–6, 245; airborne 217, 226, 311; chronic 80, 202, 209, 217, 219, 225, 235–6; degenerative 217, 221, 225, 235; diarrheal 209, *311*, 324; double burden of 219, 221; infectious 213, 216–21, 225–6, 228, 231, **233**, 235, 245–7, 318; lifestyle 219, 222, 235–6, 245; non–infectious 217, 219, 221, 225, 235, 246; nutrient deficiency 213–15; tropical 227, 277, 318; vector–borne 226–8, 234, 246, 248; waterborne 217, 226, 246, 310, 317–18
divorce 35, 157, 161
doubling time 107, 112

Earth Overshoot Day 120, 142
East Asia 17–18, 66, 68–9, 74, 156, 162–3
Eastern Europe 68, 73, 152, 159–60, 282, 292–3
Ebola 229–31, **233**, 246–7
ecological footprint 119–20, 140, 142, 305, 322–3
ecological overshoot 119, 140
ecological pyramid *137*
ecological reserve and deficit *121*
economic lens 4, 13, 314
economic migrants 266, 276, 285, 288–9, 291, 294
ecumene 57, 84
Egypt 7, 58, 61–2, 85–6, 88, 98
Ehrlich, P. 103, 113, 117, 121, 141

elderly populations 69, 202, 232
emigrants 19, 259, 261
emigration 68, 256, 258–9, 271–2, 291, 320
encephalitis 248
England and Wales 41–2
environmental justice/injustice 126–7, 134, 140, 143, 219, 221, 309, 323
environmental Kuznets curve 122, 140
environmental migrants 319–21
environmental problems 96, 119, 122–3, 125, 309, 319
environmental racism 81, 84, 127, 140
environmental scarcity 142
epidemics 92–3, 212, 216, 220, 227–9, 230–1, 244–7, 311
epidemiological transition 197, 213, 217–21, 223
Equatorial Guinea 74, 207–8
equity 12–13, 15, 19, 95–6, 138, 158–9, 303, 314–15
ethnic cleansing 279, 282, 299
ethnic groups 37–8, 79–80, 82–3, 181, 239
ethnic identity 37, 251
ethnicity 28–30, 36–9, 50–1, 53, 55, 78–9, 82, 84
eugenics 101, 112, 180–1, 192
Eurasia 92, **227**, 229
Europe 15–18, 68–9, 72–5, 93–5, 156–62, 226–30, 251–2, 256–62, 281–7, 289–94; diseases 229, 277; migrant crisis 256, 282, 285, 294; migrants 254, 271, 285, 287, 289–90, 293–4, 298
eutrophication 135, 140

family-friendly policies 145–6, 158–9, 164–5, 191
family planning 3, 22, 96, 101, 171–3, 175–6, 178–80, 183–8, 191–3, 195–6; campaigns 96, 103, 172–3, 175, 179–80
famine 103–4, 114, 117, 213–14, 223, 291
fecundity 145–7, 156, 165
fertility 3–4, 10–12, 21–2, 36, 44, 65, 87, 96–103, 105–15, 145–96; below replacement 98, 101, 107, **109**, 155, 157, 160; choices 12, 96, 175, 187, 192; decisions 6, 99, 164, 170, 187, 191; ethical issues in 170–196;

influences on 145–8; limitation 101,
188, 190–1; lowest **162**; measures 36,
110, 145–69, 188; and mortality 28,
50, 160, 167, 183, 188; patterns 4, 6,
65, 70, 145–69, 185, 191; policies 12,
96–7, 145–6, 158–9, 165, 167–8, 170,
178–84, 186–93, 195; replacement
level 97–8, 101, 107, 109–12, 145,
155–61; transition 101, 105–7, 111–12,
152–7, 164–5, 167–9; treatments 147,
150–2, 189
fertility rates 97–9, 107–8, 110–11,
113–14, 148–52, 154–6, 159, 161–2,
165–8, 178–9; age–specific 41, 150–1
Finland 35, 75, 162, 244
fish catch *132*, 142
food 16, 90–2, 94–5, 103–4, 116–19,
132–3, 135–9, 141–3, 213–17, 221–3;
chain 136–7, 139; production 92,
94–5, 103, 117–18, 130, 133, 135–6,
138–9; security 138–9, 141–2, 153,
214–16, 223
forced migration 253, 276, 279, *281*,
283, 299, 324
foreign-born population 40–1, 257; in
Europe 257
France 34, 82, 84, 86, 252, 254, 256, 291
freshwater 129–31, 140, 142–3, 318;
resources 129–30, *132*, 140, 142–3;
withdrawals 129–31, 143

Gapminder 24, 44, 52
gender 15, 33, 35–6, 52–3, 70–1, 73–9,
84–8, 158–9, 185, 268–9; identity 20,
33, 53, 76–8; inequality 15, 74–5, 210
gender-affirming healthcare 77
gender-based violence 73, 75, 171, 266
gender gap index 75, 88
general fertility rate 149, 165
genocide 38, 182, 192, 195, 266, 287–8
geographic scale *9*, 21
German Basic Law 282
Germans 254, 256, 264, 281–3, 289–90,
292
Germany 180–1, 251–2, 254, 256, 262,
264, 281–3, 291–3
GIS 49–51
Global Footprint Network 119–21, 142
Global North 17, 95–6, 98, 155, 219,
235–6, 238–9, 241–2
Global South 16–18, 95–6, 106–7, 124,
171, 184–6, 217, 236–8, 306–7

globalization 112, 117, 228, 241, 244–5,
270, 272
graphs 23, 39–45, 50, 65, 89; area 42;
bar 40–2, 65; bubble 44–5; line 41–2;
pie charts 42; stacked bar 40–2
Graunt, J. 200
Great Depression 148, 181, 256, 258
Green Revolution 95, 104, 112, 114
greenhouse gas emissions 135–8, 306,
313–14
groundwater 57, 95, 130–1, 134;
depletion 131, 140–1
guestworker programs 291, 293
Gulf countries 254, 259, 261, 271, 293,
299

Hardin, G. 118, 140–2
health 21–2, 126–8, 137–9, 166–9, 171–2,
182–9, 194–249, 312–13, 318–19,
323–5; indicators 197–8, 205–6, 209,
212, 220, 239, 243; inequalities 75,
202, 212, 220, 223–4, 235, 238–40,
325; reproductive 168, 171–2, 176,
178, 180, 183–5, 187, 189; social
determinants of 210, 222–3
health and wellbeing 4, 243, 312
healthcare 18, 65–6, 69–70, 77, 174–6,
205–12, 221–2, 239, 243, 287–8;
access to 175, 205, 207–10, 212, 243,
306, 310, 312; Costa Rica 207, 209,
212, 222; primary 207, 209, 212, 222;
systems 18, 65, 69, 175; universal
206–7, 209, 212, 222, 239, 243
heart disease 199, 202–5, 217, 219–21,
235–6, 238–9, 242, 312–13
hearths of agriculture *91*, 112
high fertility 65, 145–6, 150, 152–3,
176, 291
HIV/AIDS 25, 106, 155–6, 186, 205,
212
Holocaust 38
honor killings 74–5, 77
households 27, 32–4, 66, 70, 306–7;
non–traditional 66
human development index *110*, 152–3
human rights 12–13, 22, 75, 87, 142–3,
184–5, 195, 252–3, 292–6, 300–1
human trafficking 280, 296–301, 317

Iceland 25, 53, 74–5, 164, 206, 243
immigrants 30, 40–2, 159, 250–2, 254–6,
262, 274, 289–94

immigration 19, 53–4, 68, 251, 262–3, 274, 289–90, 292, 298–9, 301–2
Immigration and Nationality Act 290, 301
indentured laborers 257–8
India 43, 73–4, 86–7, 100–2, 107, 111, 113–14, 133–4, 152, 283
Indian Removal Act 279, 302
indigenous peoples 28, 62, 79–80, 84–5, 88, 127, 277, 316
Indonesia 58, 61–2, 86–8, 133, 173, 321, 324; transmigration 62, 86
Industrial Revolution 94, 102, 105, 217, 303, 313
infant mortality 43–5, 98, 101, 107, 205–9, 220, 222–3, 239–40
infant mortality rates 43–5, 101, 107, 205–9, 217, 220, 222, 239–40
infectious diseases see diseases, infectious
infertility 146–7, 151, 164–5, 180; treatments 147, 151, 180
influenza 3, 228, 231–2, 247–8; spread of 232
injuries 126, 213–14, 216–17, 223, 312
internally displaced people 267
International Conference on Population and Development 96, 185; see also United Nations world population conferences
international migrations 27, 254, 262, 272, 320
intersectionality 6, 20
in vitro fertilization 151
Iran 11, 70, 75–6, 179–80, 192–5
Ireland 24, 173, 262, 291
Islamophobia 251–2, 272
Israel 11, 151, 243, 261–2, 274, 284
Italy 2, 22, 81, 269, 292, 295

Japan 18, 21, 68, 70, 109–10, 162, 202–3, 307
Japanese Americans 38
Japanese encephalitis 227
Jews 38, 181, 262
Jordan 284, 286, 293

kafala system 293, 299, 301

labor migrants 259, 261, 263, 274, 293; Gulf countries 261, 293
labor migrations 258–9, 263, 271–2, 291, 300

Latin America 17, 98, 150, 171, 173, 258
Latino populations 3, 239
Lebanon 25, 52, 254, 284, 287, 293
lenses of analysis 1, 4, 12–13, 15–16, 18–19, 303; ecological 1, 4, 12–13, 303; economic 1, 4, 12–13, 15, 18–19, 303; social equity 1, 12, 15, 303
LGBTQ+ 71, 76–8, 85, 242, 264
life cycle analysis 135, 140
life expectancy 44–5, 69, 90, 107–8, 163–4, 197–200, 208–9, 212–13, 220, 239–41
life tables 200, 201, 222
lifeboat ethics 118, 140, 142–3
lifestyle diseases see diseases
long-acting reversible contraceptives (LARCs) 184
Lyme disease 227

malaria 186, 226–7, 229, 318
malnutrition 104–5, 206, 213–15, 217, 221–2, 224
Malthus, T. 16, 94–5, 102–4, 111–12, 114, 116–17, 140, 143
Malthusianism see neo–Malthusianism
maps 21, 23–4, 38–9, 45–7, 49–50, 87; choropleth 45–7, 49; dot density 46–8; flow 49; proportional symbol 47–9; thematic 45, 50
marriage 72–4, 78, 100–1, 157–8, 161–3, 171, 179, 181, 188, 310
Marx, K. 89, 102, 104, 111–12, 140
Marxist 104, 119, 185
maternal mortality 74, 199, 208, 210–12, 222, 224; rates 74, 199, 208, 211
median age 68, 85
MERS (Middle East respiratory syndrome) 231, 233–4, 248
Mexico 7–8, 131, 134, 185–6, 289, 291, 296, 307
Mexico City Policy 186, 194
Middle East 75–6, 129–30, 227, 231, 233–4, 261
migrant crisis 256, 282, 285–6, 294; in Europe 256, 282, 285–6
migrants 6–8, 80–1, 250–6, 258–71, 273–6, 285, 287–96, 298, 319–22; distribution 80, 255, 319; Filipino 259–60; international 253–5, 258, 263, 267, 283, 289; undocumented 260, 262, 265–6, 269, 294–5

migration 3–4, 6–10, 18–19, 24–5, 27, 65, 70–1, 250–302, 304, 323–5; agency in 259, 270, 275–7, 298, 300; chain 270, 272; economic 65, 70, 258–60, 266, 268, 270–5, 289, 291–2, 298–300; environmental 268, 272, 275, 298, 324–5; family 27, 70, 250, 266, 272–3, 282; forced 272–3, 275–6, 279, 281, 289, 291, 293–4, 299; illegal 261, 265–6, 268, 272–3, 294, 300–1; international 24, 27, 253–4, 258, 262–4, 271–3, 299–300; labor 250, 256, 258–9, 261, 263, 270–4, 291, 300–1; lifestyle 264, 272, 275; rural–to–urban 59, 85, 304, 306, 312; settler 291; types of 250, 253, 262–3, 268, 272, 285; voluntary 275, 287, 289, 291, 293–4, 300

migration rates 261

migration theories 250, 253, 268; globalization theory 270, 272; human capital theory 270, 272; neoclassical economic theory 270, 272; network theory 270, 272; new economics of labor theory 270, 273; push–pull model: structuralist theories 270, 273; transnationalism 270, 273; world systems theory 270, 273

mobility 1, 3, 9, 228, 232, 262

Mongolia 39, 59, 120

morbidity 161, 171, 197–225, 246

mortality 2–3, 43–5, 74, 98, 107, 112–13, 197–225, 239–40, 246–8, 310; neo–natal 206, 222; peri–natal 205–6, 221–2

mortality rates 43–5, 197, 201–9, 211, 213, 216–17, 220–1, 239–40; age–adjusted 203–4

mosquitos 226–9, 234, 311, 318

Muslim countries 74, 251–2

Myanmar 36–8, 52–3, 199–200, 259, 273, 288; census 36–8, 52–3

National Health Service (UK) 175, 194, 256, 259

Native Americans 30, 79–80, 229, 239, 279

natural increase 25, 43–4, 51, 97–8, 112, 153

Nazi Germany 11, 181, 281

neo-Malthusianism 103, 117–18, 121, 139, 185

Netherlands 62, 78, 167, 252, 256, 291

New World 229, 256–7, 276–7, 289–91

New Zealand 17, 24, 78–9, 289, 320, 324

Niger 74, 107, 145, 149, 153–4, 167–8

Nigeria 68, 73–4, 202, 206, 211, 246

North Africa 129–30, 219, 278

North America 64, 68, **227**, 230, 254, 258–9, 261

obesity 8–9, 21–2, 214–16, 219–20, 222, 235–9, 243, 245–8; and affluence 238; patterns 8; rates 8–9, 80, 219–20, 236–7, 239, 243

old age 65, 69–70, 72, 108, 153, 155, 199–200

Operation Sovereign Borders 295

outbreaks 58, 92–3, 216, 227–9, 231, 233–4

overpopulation 13–15, 20, 112, 117–18, 140, 143

overshoot see ecological overshoot

Pacific Islanders 320

Pakistan 33, 73–5, 78, 119, 126, 131, 283–4

Palestinians 261, 284

pandemic 1–3, 6, 19, 21–2, 64, 212, 228, 231–2, 245–8

pathogens 220, 226–9, 231, 234, 246, 312

patriarchal societies 72–3, 150, 175

Philippines 58, 190, 259–60, 271, 274, 319

physiological density see population density

pill 95, 171, 174–5, 177, 179; see also contraceptives

plague 92–3, 113–15, 246, 311

Poland 46, 53, 71, 162, 173, 283

pollution 95–6, 122, 124–5, 127–8, 135–6, 140, 309, 318, 325

population: centroid 59–61, 63, 85; composition 19–20, 39–42, 55–88, 271, 281, 283, 292–3

population aging 99, 111, 149, 180, 213

population and environment 116, 127–8

Population Bomb 113, 117, 141

population data 19, 23–5, 36, 38–40, 50, 53, 58; challenges of collecting 36; key sources of 23–4; types of 23, 39–40, 45, 51; visualization of 39

population decline 16, 18, 21–2, 85, 99–100, 113–14, 161, 166–8
population density 46, 50, 59, 85, 226, 228; arithmetic 59–60; physiological *60*, 85
population distribution 19–20, 48–50, 55–88; Brazil 61, 63; Central African Republic 49; Egypt 61–2, 85, 88; by elevation above sea level *57*; by hemisphere 56; Indonesia 58, 61–2, 87; measures of 59; world 54, 56–9, 61–2, 65, 68, 75, 82
population explosion 16, 94, 117, 140, 184
population growth 4, 11, 16, 18–19, 89–115, 117–19, 121, 123, 184–5, 303–5; history of 89–90; India 43, 89, 98, 100–2, 107, 113–14; theories of 89, 102; urban *305*
population health 201, 203, 205, 213, 238, 240, 243–4, 313
population issues 4, 6, 8, 10–13, 19, 184–5, 191–2, 303–25
population policies 96–7, 179, 186–7, 189–91, 195; China 187, 189–90, 195; Iran 179
population projections 97, 99–100, 111–12, 114
population pyramids 65, 67, 107–8, 114; Russia 67–8; selected US counties *67*, *72*; United Arab Emirates *71*
population redistribution 59, 64
population registers 35
population stabilization *98*
population structure 55, *109*, 150, 155; global 109, 155
populations of color 95, 117, 159, 181, 184, 239
potential support ratio 69, 85
poverty 14–16, 21–2, 79–81, 96–7, 102–4, 121, 124, 154–5, 210, 238–9
pregnancy 146, 170–4, 176–7, 192, 195–6, 208, 212
primary healthcare 207, 209, 212, 222
pro-natal policies 11–12, 20, 170, 180, 192; in Romania 182
puberty 147, 167–8
public health systems 216, 228
Puerto Rico 162, 175

race 2–3, 19–20, 29–30, 32–3, 36–7, 78–9, 81–2, 84–5, 200, 238–9

race and ethnicity 30, 55, 78–9, 239
racial groups 29, 80–2, 85
racism 81–2, 84–6, 127, 240, 250, 252; environmental 81, 84, 127, 140; structural 81, 85
rates 18–19, 43–6, 68–74, 95–102, 104–8, 145–68, 201–9, 211–14, 219–21, 234–44
ratios 39, 47, 51, 69, 71–3, 86
raw data *see* population data, types of
refugees 24–5, 71, 256, 260–1, 265–7, 269, 273–4, 282–9, 298–302, 320; climate 289, 320, 323; distribution of 47, *284*; resettled 287–8
repatriation 287, 299
replacement level fertility 97–8, 107, 109–10, 112, 156–7, 159
reproductive freedom 95–6, 171–2, 185, 191–2
reproductive health 168, 171–2, 176, 178, 180, 183–5, 187, 189
reproductive technologies 151–2
resettlement 287–8, 300, 302
resources 12–16, 20–1, 59, 61–2, 104, 115–44, 183–5, 247–8, 277–8, 321–3; finite 13–14, 16, 20, 116–18, 123–4, 127–8, 139–40; renewable 14, 21, 128–31
rising sea levels 59, 315–17, 321
Rohingya 37–8, 259, 273
Romania 182–4, 193–5
rural areas 188–9, 210, 264, 303, 305, 312, 319–20
Russia 11, 67–8, 86, 159–62, 166–8, 199, 241, 254, 266; fertility and mortality rates *160*
Rwanda 38, 74, 266, 287

Sahel 153–4, 316
same-sex sexual activities *see* LGBQT+
sample surveys 23, 25, 36
sanitation 205–7, 216–17, 226, 228, 306, 310–13, 323, 325
SARS (severe acute respiratory syndrome) 228, 231, **233**, 247
scatterplots *9*, 43–4
second demographic transition 157, 164–5, 167–9
segregation 81, 85, 309–10, 323
settlements 81, 92, 306, 308–9, 322, 325
sex and gender 55, 70–1

sex ratios 39, 47, 71–3, 86
sex structure 65, 85, 149
sexual orientation *see* LGBQT+
ship-breaking industry 126
Singapore 82–3, 86–7, 190, 193, 254, 259
slave trade 276–8, 302
slavery 80, 276–80, 293, 296, 298–302, 317, 324; chattel 280, 299; contract 280, 299; modern 279–80, 293, 296, 298–9, 301, 317, 324
slums 306–7, 323, 325; *see also* squatter settlements
social determinants of health 210, 222
social environment 4, 6, 235, 238, 240, 243–4, 246
social equity 12–13, 15, 95–6, 124, 127, 303, 314–15, 317
social infertility 151, 164–5
social justice 4, 15–17, 96, 103–4, 118–19, 123–6, 138–40, 170–2, 220–1, 275–325; perspective 16, 19, 96, 103, 111, 298, 300
socio–economic status 2–3, 6, 197, 200, 202, 238–9, 241–2, 245
South Africa 24, 77, 81, 212, 261, 289
South America 77, 94, **227**, 256–7
South Asia 58, 73, 86, 126, 211–12, 254
South Korea 110, 145, 149, 162–3, 165–8, 190
Southeast Asia 152, 233, 293, 296, 311, 317
Soviet Union 17, 36, 38, 159–60, 240–1, 283–4, 288
Spain 22, 109, 162, 243, 264, 292
squatter settlements 269, 306, 308–9, 322–3; *see also* slums
Sri Lanka 252, 297
sterilization 96, 101, 172–5, 181–2, 193–4, 217
sub-Saharan Africa 97, 106, 205, 208, 210–12, 217, 220, 296
suicide 11, 80, 238–9, 242, 245, 247
sustainability 4, 14–15, 21–2, 96–7, 116, 124–5, 137–40, 142–3, 244, 303–25; three pillars of *125*
sustainable development 14–15, 21–2, 132, 135, 138–9, 141
sustainable intensification 135, 141, 143; *see also* agricultural intensification
Sweden 35, 101, 166, 206, 243

Switzerland 48, 53, 74, 162, 243, 252
Syria 199–200, 256, 261, 267, 269, 273
Syrian refugees 284

Taliban 106, 211, 223, 267
Thailand 107, 109, 153–5, 162, 164, 167–9
total fertility rates *146*, **150**, 154, 178
Tragedy of the Commons 118, 141–2, 313
Trail of Tears 279–80, 301
Trans-Atlantic Slave Trade 276–8, 302
transgender 76–7, 84, 271; *see also* LGBTQ+
transmigration 62, 86

Ukraine 71, 161–2, 166, 266, 288, 296
undernourished population *134*
undernutrition 134, 214–15, 222
United Arab Emirates (UAE) 71, 162, 207, 234, 254, 296
United Kingdom (UK) 2, 175–6, 241, 254, 256, 258–60, 274, 291
United Nations (UN) 14–15, 22, 24–5, 138, 184–5, 187, 195, 253–5, 260–1, 274, 306, 324–5
United Nations world population conferences **185**
United States (US) 29–32, 77–82, 148–51, 181, 186, 206–7, 238–43, 262–4, 284–5, 287–91; age-adjusted mortality rates *204*; age-specific fertility rates *151*; black/white racial disparities in infant mortality *240*; fertility fluctuations *149*; health inequalities 238–40; immigration to *290*
unsustainable resource extraction 127–8
urban heat island effect 305, 319, 323
urban populations 4, 179, 226, 306, 310–13
urbanization 43–4, 88, 105–6, 228, 303–4, 310, 323, 325
US Centers for Disease Control and Prevention (CDC) 149, 151, 220, 222, 224, 232–3, 239–40, 246–7

virus 156, 179, 226–9, 231–4, 246, 248
vital events 25, 34–5, 52
vital registration systems 23, 25
vital statistics 34–6, 38, 53, 166, 222, 224

war 25, 67, 216, 232, 256, 258, 268–70,
  281–5, 288, 291–2
welfare states 18
West Africa 58, 128, 132, 230, 233, 247
Western countries 33, 147, 162, 178, 219,
  242–3
Western world 16, 77, 137, 242
White Australia Policy 290
white privilege 82, 85
women 6, 70–6, 83–8, 145–7, 149–54,
  156–65, 170–9, 181–94, 198–9,

206–12; empowerment of 101, 106,
  187; rights violations 74; single 151–2,
  162; white 6, 20, 159, 206–7, 219
World War I 216, 232, 256, 281
World War II 38, 256, 258, 262, 281–2,
  284, 291

yellow fever 227, 248–9

Zika virus 228
Zimbabwe 308